DATE DUE

BRODART, CO.   Cat. No. 23-221-003

Practical Linear Algebra

Practical Linear Algebra
A Geometry Toolbox

Gerald Farin
Dianne Hansford

A K Peters
Wellesley, Massachusetts

Editorial, Sales, and Customer Service Office

A K Peters, Ltd.
888 Worcester Street, Suite 230
Wellesley, MA 02482
www.akpeters.com

Library of Congress Cataloging-in-Publication Data

Farin, Gerald E.
    Practical linear algebra : a geometry toolbox / Gerald Farin, Dianne Hansford.
        p. cm.
    Includes bibliographical references and index.
    ISBN 1-56881-234-5
        1. Algebras, Linear—Study and teaching. 2. Geometry, Analytic—Study and
teaching. 3. Linear operations. I. Hansford, Dianne. II. Title.

QA184.2.F37   2004
512'.5—dc22   2004053353

Printed in the United States of America
09 08 07 06 05   10 9 8 7 6 5 4 3 2 1

To our advisors
R.E. Barnhill and W. Boehm

Contents

Preface

We assume just about everyone has watched animated movies, such as *Toy Story* or *Shrek*, or is familiar with the latest three-dimensional computer games. Enjoying 3D entertainment sounds like more fun than studying a Linear Algebra book, right? But it is because of Linear Algebra that those movies and games can be brought to a TV or computer screen. When you see a character move on the screen, it's animated using some equation straight out of this book. In this sense, Linear Algebra is a driving force of our new digital world: it is powering the software behind modern visual entertainment and communication.

But this is not a book on entertainment. We start with the fundamentals of Linear Algebra and proceed to various applications. So it doesn't become too dry, we replaced mathematical proofs with motivations, examples, or graphics. For a beginning student, this will result in a deeper level of understanding than standard theorem-proof approaches. The book covers all of undergraduate-level linear algebra in the classical sense—except it is not delivered in a classical way. Since it relies heavily on examples and pointers to applications, we chose the title *Practical Linear Algebra*, or *PLA* for short.

The subtitle of this book is *A Geometry Toolbox*; this is meant to emphasize that we approach linear algebra in a geometric and algorithmic way. This book grew out of a previous one, namely *The Geometry Toolbox for Graphics and Modeling*. Our goal was to bring the material of that book to a broader audience, motivated in a large part by our observations of how little engineers and scientists (non-math majors) retain from classical linear algebra classes. Thus, we set out to fill a void in the linear algebra textbook market. We feel

that we have achieved this, as well as maintaining the spirit of our
first effort: present the material in an intuitive, geometric manner
that will lend itself to retention of the ideas and methods.

Review of Contents

As stated previously, one clear motivation we had for writing PLA
was to present the material so that the reader would retain the information. In our experience, approaching the material first in two and
then in three dimensions lends itself to visualizing and then to understanding. Incorporating many illustrations, Chapters 1–7 introduce
the fundamentals of linear algebra in a 2D setting. These same concepts are revisited in Chapters 10–13 in a 3D setting. The 3D world
lends itself to concepts that do not exist in 2D, and these are explored
there too.

Higher dimensions, necessary for many real-life applications and the
development of abstract thought, are visited in Chapters 14–16. The
focus of these three chapters includes linear system solvers (Gauss
elimination, LU decomposition, Householder's method, and iterative
methods), determinants, inverse matrices, revisiting "eigen things,"
linear spaces, inner products, and the Gram-Schmidt process.

Conics, discussed in Chapter 9, are such a fundamental geometric
entity, and since their development provides a wonderful application
for affine maps, "eigen things," and symmetric matrices, they really
shouldn't be missed. Triangles in Chapter 8 and polygons in Chapter
17 are discussed because they are fundamental geometric entities and
are important in generating computer images. The basics of generating curves are presented in Chapter 18; this, too, is a nice example of
how linear algebra may be applied.

The illustrations in the book come in two forms: figures and
sketches. The figures are computer generated and tend to be complex.
The sketches are hand-drawn and illustrate the core of a concept.
Both are great teaching and learning tools! We made all of them available on the book's website (http://vidya.prism.asu.edu/~farin/pla).
Many of the figures were generated using PostScript, an easy-to-use
geometric language. We have provided a tutorial to this language
in Appendix A. This brief tutorial gives enough information for the
reader to modify the figures in this book, as well as create their own.
However, this book can certainly be used without getting involved
with PostScript.

At the end of each chapter, we have included a list of topics, *What
You Should Know* (*WYSK*), marked by the icon on the left. This list

is intended to encapsulate the main points of each chapter. It is not uncommon for a topic to appear in more than one chapter. We have made an effort to revisit some key ideas more than once. Repetition is useful for retention!

Exercises are listed at the end of each chapter. Solutions to selected exercises are given in Appendix B. More solutions may be found on the book's website. PostScript figures are also used as a basis for some exercises—PostScript offers an easy way to get hands-on experience with many geometry concepts. Often these exercises are labeled *PS* to indicate that they involve PostScript.

Classroom Use

PLA is meant to be used at the freshman/sophomore undergraduate level. It serves as an introduction to Linear Algebra for engineers or computer scientists, as well as a general introduction to geometry. It is also an ideal preparation for Computer Graphics and Geometric Modeling.

As a one-semester course, we recommend choosing a subset of the material that meets the needs of the students. In the table below, LA refers to an introductory Linear Algebra course and CG refers to a course tailored to those planning to work in Computer Graphics or Geometric Modeling. We have created shortened chapter titles to make the table more readable.

	Chapter	**LA**	**CG**
1	2D Coordinate Systems	•	•
2	2D Points & Vectors	•	•
3	2D Lines		•
4	2D Linear Maps	•	•
5	2×2 Linear Systems	•	•
6	2D Affine Maps	•	•
7	2D Eigen Things	•	
8	Triangles		•
9	Conics	•	
10	3D Geometry	•	•
11	3D Interactions		•
12	3D Linear Maps	•	•

	Chapter	LA	CG
13	3D Affine Maps	•	•
14	Linear Systems	•	•
15	Linear Spaces	•	
16	Numerical Methods	•	
17	Polygons		•
18	Curves		•

Website

Practical Linear Algebra, A Geometry Toolbox has a website:
http://vidya.prism.asu.edu/~farin/pla.

This website provides:

- teaching materials,

- additional solutions to exercises,

- the PostScript files illustrated in the book,

- data files referred to in the text,

- errata,

- and more!

A K Peters, Ltd. also maintains a website:
http://www.akpeters.com.

Acknowledgements

We certainly appreciate the positive feedback that we received on *The Geometry Toolbox for Graphics and Modeling*. Some of that feedback may be found on 'The Toolbox's' website.

A few 3D images were generated using GEOMVIEW from The Geometry Center at the University of Minnesota, http://www.geom.uiuc.edu/. (It is now closed.)

Lastly, thanks to the great team at A K Peters! They are a pleasure to work with.

Gerald Farin   August 2004
Dianne Hansford   Arizona State University

1

Descartes' Discovery

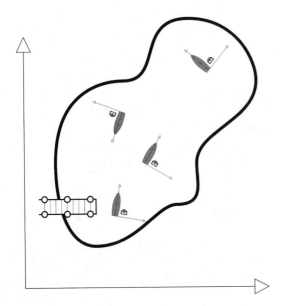

Figure 1.1.
Local and global coordinate systems: the treasure's local coordinates do not change
as the boat moves, however the treasure's global coordinates, defined relative to the
lake, do change.

There is a collection of old German tales that take place sometime
in the 17th century, and they are about an alleged town called Schilda,

1

whose inhabitants were not known for their intelligence. Here is one story [15]:

> An army was approaching Schilda and would likely conquer it. The town council, in charge of the town treasure, had to hide it from the invaders. What better way than to sink it in the nearby town lake? So the town council members board the town boat, head for the middle of the lake, and sink the treasure. The town treasurer gets out his pocket knife and cuts a deep notch in the boat's rim, right where the treasure went down. Why would he do that, the other council members wonder? "So that we will remember where we sunk the treasure, otherwise we'll never find it later!" replies the treasurer. Everyone is duly impressed at such planning genius!
>
> Eventually, the war is over and the town council embarks on the town boat again, this time to reclaim the treasure from the lake. Once out on the lake, the treasurer's plan suddenly does not seem so smart anymore. No matter where they went, the notch in the boat's rim told them they had found the treasure!

The French philosopher René Descartes (1596-1650) would have known better: he invented the theory of *coordinate systems*. The treasurer recorded the sinking of the treasure accurately by marking it on the boat. That is, he recorded the treasure's position relative to a *local* coordinate system. But by neglecting the boat's position relative to the lake, the *global* coordinate system, he lost it all! (See Figure 1.1.) The remainder of this chapter is about the interplay of local and global coordinate systems.

1.1 Local and Global Coordinates: 2D

This book is written using the LaTeX typesetting system (see [10] or [16]) which converts every page to be output to a page description language called PostScript (see [13]). It tells a laser printer where to position all the characters and symbols that go on a particular page. For the first page of this chapter, there is a PostScript command that positions the letter **D** in the chapter heading.

In order to do this, one needs a two-dimensional, or 2D, coordinate system. Its origin is simply the lower left corner of the page, and the x- and y-axes are formed by the horizontal and vertical paper edges meeting there. Once we are given this coordinate system, we can position objects in it, such as our letter **D**.

The **D**, on the other hand, was designed by font designers who obviously did not know about its position on this page or of its actual size. They used their own coordinate system, and in it, the letter **D** is described by a set of points, each having coordinates relative to **D**'s coordinate system, as shown in Sketch 1.1.

We call this system a *local coordinate system,* as opposed to the *global coordinate system* which is used for the whole page. Positioning letters on a page thus requires mastering the interplay of the global and local systems.

Following Sketch 1.2, let's make things more formal: Let (x_1, x_2) be coordinates[1] in a global coordinate system, called the $[\mathbf{e}_1, \mathbf{e}_2]$-system.[2] Let (u_1, u_2) be coordinates in a local system called the $[\mathbf{d}_1, \mathbf{d}_2]$-system. Let an object in the local system be enclosed by a box with lower left corner $(0, 0)$ and upper right corner $(1, 1)$. This means that the object "lives" in the *unit square* of the local system, i.e., a square of edge length one, and with its lower left corner at the origin.[3]

We wish to position our object into the global system so that it fits into a box with lower left corner $(\min_1, \min_2)$ and upper right corner $(\max_1, \max_2)$ called the *target box* (drawn with heavy lines in Sketch 1.2). This is accomplished by assigning to coordinates (u_1, u_2) in the local system the corresponding target coordinates (x_1, x_2) in the global system. This correspondence is characterized by preserving each coordinate value with respect to their extents. The local coordinates are also known as *parameters.* In terms of formulas, these parameters[4] are written as quotients,

$$\frac{u_1 - 0}{1 - 0} = \frac{x_1 - \min_1}{\max_1 - \min_1}$$
$$\frac{u_2 - 0}{1 - 0} = \frac{x_2 - \min_2}{\max_2 - \min_2}.$$

Thus, the corresponding formulas for x_1 and x_2 are quite simple:

$$x_1 = (1 - u_1)\min_1 + u_1\max_1, \tag{1.1}$$
$$x_2 = (1 - u_2)\min_2 + u_2\max_2. \tag{1.2}$$

We say that the coordinates (u_1, u_2) are *mapped* to the coordinates (x_1, x_2). Sketch 1.3 illustrates how the letter **D** is mapped.

---

[1] You may be used to calling coordinates (x, y), however the (x_1, x_2) notation will streamline the material in this book; it also makes writing programs easier.

[2] The boldface notation will be explained in the next chapter.

[3] Restricting ourselves to the unit square for the local system makes this first chapter easy—we will later relax this restriction.

[4] This concept of a parameter is reintroduced in Section 2.5.

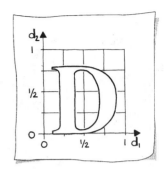

Sketch 1.1.
A local coordinate system.

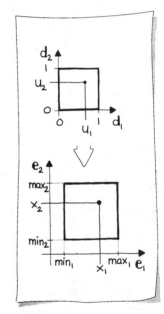

Sketch 1.2.
Global and local systems.

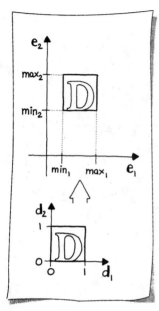

Let's check that this actually works: the coordinates $(u_1, u_2) = (0, 0)$ in the local system must go to the coordinates $(x_1, x_2) = (\min_1, \min_2)$ in the global system. We obtain

$$x_1 = (1 - 0) \cdot \min_1 + 0 \cdot \max_1 = \min_1,$$
$$x_2 = (1 - 0) \cdot \min_2 + 0 \cdot \max_2 = \min_2.$$

Similarly, the coordinates $(u_1, u_2) = (1, 0)$ in the local system must go to the coordinates $(x_1, x_2) = (\max_1, \min_2)$ in the global system. We obtain

$$x_1 = (1 - 1) \cdot \min_1 + 1 \cdot \max_1 = \max_1,$$
$$x_2 = (1 - 0) \cdot \min_2 + 0 \cdot \max_2 = \min_2.$$

Example 1.1

Let the target box be given by

$$(\min_1, \min_2) = (1, 3) \quad \text{and} \quad (\max_1, \max_2) = (3, 5),$$

see Sketch 1.4. The coordinates $(1/2, 1/2)$ can be thought of as the "midpoint" of the local unit square. Let's look at the result of the mapping:

$$x_1 = (1 - \frac{1}{2}) \cdot 1 + \frac{1}{2} \cdot 3 = 2,$$
$$x_2 = (1 - \frac{1}{2}) \cdot 3 + \frac{1}{2} \cdot 5 = 4.$$

This is the "midpoint" of the target box. You see here how the geometry in the unit square is replicated in the target box.

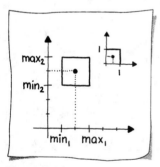

A different way of writing (1.1) and (1.2) is as follows: Define $\Delta_1 = \max_1 - \min_1$ and $\Delta_2 = \max_2 - \min_2$. Now we have

$$x_1 = \min_1 + u_1 \Delta_1, \qquad (1.3)$$
$$x_2 = \min_2 + u_2 \Delta_2. \qquad (1.4)$$

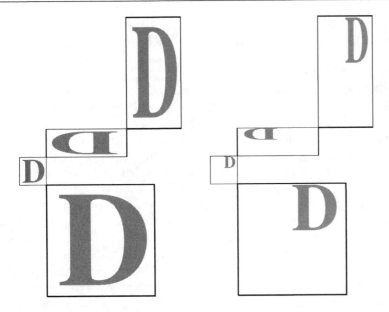

Figure 1.2.
Target boxes: the letter **D** is mapped several times. Left: centered in the unit square.
Right: not centered.

A note of caution: if the target box is not a square, then the object from the local system will be distorted. We see this in the following example, illustrated by Sketch 1.5. The target box is given by

$$(\text{min}_1, \text{min}_2) = (-1, 1) \quad \text{and} \quad (\text{max}_1, \text{max}_2) = (2, 2).$$

You can see how the local object is stretched in the $\mathbf{e}_1$-direction by being put into the global system. Check for yourself that the corners of the unit square (local) still get mapped to the corners of the target box (global)!

In general, if $\Delta_1 > 1$, then the object will be stretched in the $\mathbf{e}_1$-direction, and it will be shrunk if $0 < \Delta_1 < 1$. The case of max_1 smaller than min_1 is not often encountered: it would result in a reversal of the object in the $\mathbf{e}_1$-direction. The same applies, of course, to the $\mathbf{e}_2$-direction. An example of several boxes containing the letter **D** is shown in Figure 1.2. Just for fun, we have included one target box with max_1 smaller than min_1!

Another characterization of the change of shape of the object may be made by looking at the change in *aspect ratio*, which is the ratio of the width to the height, or Δ_1/Δ_2 for the target box. The aspect ratio in the local system is one. Revisiting Example 1.1, the aspect

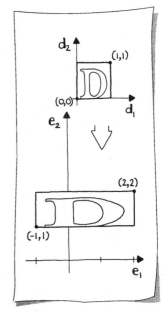

Sketch 1.5.
A distortion.

ratio of the target box is one, therefore there is no distortion of the letter **D**, although it is stretched uniformly in both coordinates. In Sketch 1.5, a target box is given which has aspect ratio 3, therefore the letter **D** is distorted.

This method, by the way, acts strictly on a "don't need to know" basis: we do not need to know the relationship between the local and global systems. In many cases (as in the typesetting example), there actually isn't a known correspondence at the time the object in the local system is created. Of course, one must know where the actual object is located in the local unit square. If it is not nicely centered, we might have the situation shown in Figure 1.2 (right).

You experience our "unit square to target box" mapping whenever you use a computer. When you open a window, you might want to view a particular image in it. The image is stored in a local coordinate system; if it is stored with extents $(0,0)$ and $(1,1)$, then it utilizes *normalized coordinates*. The target box is now given by the extents of your window, which are given in terms of *screen coordinates* and the image is mapped to it using (1.1) and (1.2). Screen coordinates are typically given in terms of *pixels*;[5] a typical computer screen would have about 700×1000 pixels.

1.2   Going from Global to Local

When discussing global and local systems in 2D, we used a target box to position (and possibly distort) the unit square in a local $[\mathbf{d}_1, \mathbf{d}_2]$-system. For given coordinates (u_1, u_2), we could find coordinates (x_1, x_2) in the global system using (1.1) and (1.2), or (1.3) and (1.4).

How about the inverse problem: given coordinates (x_1, x_2) in the global system, what are its local (u_1, u_2) coordinates? The answer is relatively easy: compute u_1 from (1.3), and u_2 from (1.4), resulting in

$$u_1 = \frac{x_1 - \min_1}{\Delta_1}, \qquad (1.5)$$

$$u_2 = \frac{x_2 - \min_2}{\Delta_2}. \qquad (1.6)$$

Applications for this process arise any time you use a mouse to communicate with a computer. Suppose several icons are displayed in a window. When you click on one of them, how does your computer actually know which one? The answer: it uses equations (1.5) and (1.6) to determine its position.

---

[5]The term is short for "picture element."

Figure 1.3.
Selecting an icon: global to local coordinates.

Example 1.2

Let a window on a computer screen have screen coordinates

$$(\min_1, \min_2) = (120, 300) \quad \text{and} \quad (\max_1, \max_2) = (600, 820).$$

The window is filled with 21 icons, arranged in a 7×3 pattern (see Figure 1.3). A mouse click returns screen coordinates $(200, 709)$. Which icon was clicked? The computations that take place are as follows:

$$u_1 = \frac{200 - 120}{480} \approx 0.17,$$
$$u_2 = \frac{709 - 300}{520} \approx 0.79,$$

according to (1.5) and (1.6).

The u_1-partition of normalized coordinates is

$$0, \quad 0.33, \quad 0.67, \quad 1.$$

The value 0.17 for u_1 is between 0.0 and 0.33, so an icon in the first column was picked. The u_2-partition of normalized coordinates is

$$0, \quad 0.14, \quad 0.29, \quad 0.43, \quad 0.57, \quad 0.71, \quad 0.86, \quad 1.$$

The value 0.79 for u_2 is between 0.71 and 0.86, so the "Display" icon in the second row of the first column was picked.

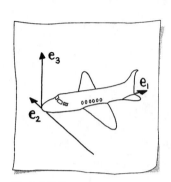

Sketch 1.6.

Airplane coordinates.

1.3   Local and Global Coordinates: 3D

These days, almost all engineering objects are designed using a *Computer Aided Design (CAD)* system. Every object is defined in a coordinate system, and usually many individual objects need to be integrated into one large system. Take designing a large commercial airplane, for example. It is defined in a three-dimensional (or 3D) coordinate system with its origin at the frontmost part of the plane, the $\mathbf{e}_1$-axis pointing toward the rear, the $\mathbf{e}_2$-axis pointing to the right (that is, if you're sitting in the plane), and the $\mathbf{e}_3$-axis is pointing upward. See Sketch 1.6.

Before the plane is built, it undergoes intense computer simulation in order to find its optimal shape. As an example, consider the engines: these may vary in size, and their exact locations under the wings need to be specified. An engine is defined in a local coordinate system, and it is then moved to its proper location. This process will have to be repeated for all engines. Another example would be the seats in the plane: the manufacturer would design just one—then multiple copies of it are put at the right locations in the plane's design.

Following Sketch 1.7, and making things more formal again, we are given a local 3D coordinate system, called the $[\mathbf{d}_1, \mathbf{d}_2, \mathbf{d}_3]$-system, with coordinates (u_1, u_2, u_3). We assume that the object under consideration is located inside the *unit cube*, i.e., all of its defining points satisfy

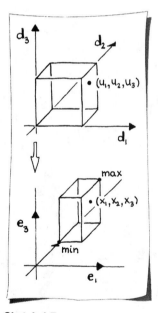

Sketch 1.7.

Global and local 3D systems.

$$0 \leq u_1, u_2, u_3 \leq 1.$$

This cube is to be mapped onto a *3D target box* in the global $[\mathbf{e}_1, \mathbf{e}_2, \mathbf{e}_3]$-system. Let the target box be given by its lower corner $(\min_1, \min_2, \min_3)$ and its upper corner $(\max_1, \max_2, \max_3)$. How do we map coordinates (u_1, u_2, u_3) from the local unit cube into the corresponding target coordinates (x_1, x_2, x_3) in the target box? Exactly as in the 2D case, with just one more equation:

$$x_1 = (1 - u_1)\text{min}_1 + u_1\text{max}_1, \tag{1.7}$$
$$x_2 = (1 - u_2)\text{min}_2 + u_2\text{max}_2, \tag{1.8}$$
$$x_3 = (1 - u_3)\text{min}_3 + u_3\text{max}_3. \tag{1.9}$$

As an easy exercise, check that the corners of the unit cube are mapped to the corners of the target box!

The analog to (1.3) and (1.4) is given by the rather obvious

$$x_1 = \text{min}_1 + u_1\Delta_1, \tag{1.10}$$
$$x_2 = \text{min}_2 + u_2\Delta_2, \tag{1.11}$$
$$x_3 = \text{min}_3 + u_3\Delta_3. \tag{1.12}$$

As in the 2D case, if the target box is not a cube, object distortions will result—this may be desired or not.

1.4 Stepping Outside the Box

We have restricted all objects to be within the unit square or cube; as a consequence, their images were inside the respective target boxes. This notion helps with an initial understanding, but it is not at all essential. Let's look at a 2D example, with the target box given by

$$(\text{min}_1, \text{min}_2) = (1, 1) \quad \text{and} \quad (\text{max}_1, \text{max}_2) = (2, 3)$$

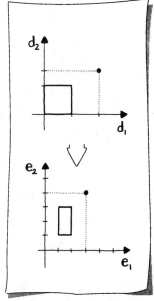

Sketch 1.8.
A 2D coordinate outside a box.

(see Sketch 1.8).

The coordinates $(u_1, u_2) = (2, 3/2)$ are not inside the $[\mathbf{d}_1, \mathbf{d}_2]$-system unit square. Yet we can map it using (1.1) and (1.2):

$$x_1 = -\text{min}_1 + 2\text{max}_1 = 3,$$
$$x_2 = -\frac{1}{2}\text{min}_2 + \frac{3}{2}\text{max}_2 = 4.$$

Since the initial coordinates (u_1, u_2) were not inside the unit square, the mapped coordinates (x_1, x_2) are not inside the target box. The notion of mapping a square to a target box is a useful concept for mentally visualizing what is happening—but it is not actually a restriction to the coordinates that we can map!

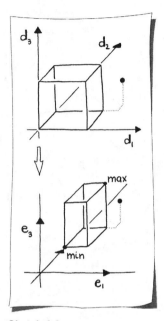

Sketch 1.9.

A point outside a 3D box.

Example 1.3

Without much belaboring, it is clear the same holds for 3D. An example should suffice: the target box is given by

$$(\min_1, \min_2, \min_3) = (1, 0, 1) \quad \text{and}$$
$$(\max_1, \max_2, \max_3) = (2, 1, 2),$$

and we want to map the coordinates

$$(u_1, u_2, u_3) = (-1, -1, -1)$$

The result, illustrated by Sketch 1.9, is computed using (1.7)–(1.9): it is

$$(x_1, x_2, x_3) = (0, -1, 0).$$

1.5   Creating Coordinates

Suppose you have an interesting real object, like a model of a cat. A friend of yours in Hollywood would like to use this cat in her latest hi-tech animated movie. Such movies only use mathematical descriptions of objects—everything must have coordinates! You might recall the movie *Toy Story*. It is a computer-animated movie, meaning that the characters and objects in every scene have a mathematical representation.

So how do you give your cat model coordinates? This is done with a *CMM*, or *coordinate measuring machine*, see Figure 1.4. The CMM is essentially an arm which is able to record the position of its tip by keeping track of the angles of its joints.

Your cat model is placed on a table and somehow fixed so it does not move during digitizing. You let the CMM's arm touch three points on the table; they will be converted to the origin and the $\mathbf{e}_1$- and $\mathbf{e}_2$-coordinate axes of a 3D coordinate system. The $\mathbf{e}_3$-axis (vertical to the table) is computed automatically.[6] Now when you now touch your cat model with the tip of the CMM's arm, it will associate three coordinates with that position and record them. You repeat this for several hundred points, and you have your cat in the box! This process is called *digitizing*. In the end, the cat has been "discretized,"

---

[6] Just how we convert the three points on the table to the three axes is covered in Section 11.8.

Figure 1.4.
Creating coordinates: a cat is turned into math. (Microscribe-3D from Immersion Corporation, http://www.immersion.com.)

or turned into a finite number of coordinate triples. This set of points is called a *point cloud*.

Someone else will now have to build a mathematical model of your cat.[7] The mathematical model will next have to be put into scenes of the movie—but all that's needed for that are 3D coordinate transformations! (See Chapters 12 and 13.)

- local to global coordinates (2D and 3D)
- global to local coordinates (2D and 3D)
- parameter

- unit square
- aspect ratio
- normalized coordinates
- digitizing
- point cloud

---

[7]This type of work is called *Geometric Modeling* or *Computer Aided Geometric Design*, see [7].

1.6 Exercises

1. Let coordinates of triangle vertices in the local $[\mathbf{d}_1, \mathbf{d}_2]$-system unit square be given by

$$(u_1, u_2) = (0.1, 0.1),$$
$$(v_1, v_2) = (0.9, 0.2),$$
$$(w_1, w_2) = (0.4, 0.7).$$

   (a) If the $[\mathbf{d}_1, \mathbf{d}_2]$-system unit square is mapped to the target box with

$$(\min_1, \min_2) = (1, 2) \quad \text{and} \quad (\max_1, \max_2) = (3, 3),$$

   where are the coordinates of the triangle vertices mapped?

   (b) What (u_1, u_2) coordinates correspond to $(x_1, x_2) = (2, 2)$?

2. Let the $[\mathbf{d}_1, \mathbf{d}_2, \mathbf{d}_3]$-system unit cube be mapped to the 3D target box with

$$(\min_1, \min_2, \min_3) = (1, 1, 1) \quad \text{and} \quad (\Delta_1, \Delta_2, \Delta_3) = (1, 2, 4).$$

   Where will the coordinates $(u_1, u_2, u_3) = (0.5, 0, 0.7)$ be mapped?

3. Take the file `Boxes.ps` from the download section of the website[8] and play with the PostScript `translate` commands to move some of the **D**-boxes around. Refer to the PostScript Tutorial (Appendix A) at the end of the book for some basic hints on using PostScript.

4. Given local coordinates $(2, 2)$ and $(-1, -1)$, find the global coordinates with respect to the target box with

$$(\min_1, \min_2) = (1, 1) \quad \text{and} \quad (\max_1, \max_2) = (7, 3).$$

Make a sketch of the local and global systems. Connect the coordinates in each system with a line and compare.

5. In some implementations of the computer graphics viewing pipeline, *normalized device coordinates*, or NDC, are defined as the cube with extents $(-1, -1, -1)$ and $(1, 1, 1)$. The next step in the pipeline maps the (u_1, u_2) coordinates from NDC (u_3 is ignored) to the *viewport*, the area of the screen where the image will appear. Give equations for (x_1, x_2) in the viewport defined by extents $(\min_1, \min_2)$ and $(\max_1, \max_2)$ which correspond to (u_1, u_2) in NDC.

---

[8]The website's URL is listed in the Preface.

2

Here and There: Points and Vectors in 2D

Figure 2.1.
Hurricane Andrew: the hurricane is shown here approaching south Louisiana. (Image courtesy of EarthWatch Communications, Inc., http://www.earthwatch.com.)

In 1992 Hurricane Andrew hit the northwestern Bahamas, south Florida, and south-central Louisiana with such great force that it is rated as the most expensive hurricane to date, causing more than $25 billion of damage.[1] In the hurricane image (Figure 2.1) air is moving rapidly, spiraling in a counterclockwise fashion. What isn't so clear from this image is that the air moves faster as it approaches the eye of the hurricane. This air movement is best described by points and vectors: at any location (point), air moves in a certain direction and with a certain speed (velocity vector).

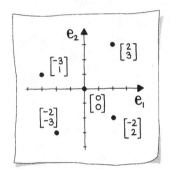

Sketch 2.1.

Points and their coordinates.

This hurricane image is a good example of how helpful 2D geometry can be in a 3D world. Of course a hurricane is a 3D phenomenon; however, by analyzing 2D slices, or cross sections, we can develop a very informative analysis. Many other applications call for 2D geometry only. The purpose of this chapter is to define the two most fundamental tools we need to work in a 2D world: points and vectors.

2.1   Points and Vectors

The most basic geometric entity is the *point*. A point is a reference to a *location*. Sketch 2.1 illustrates examples of points. In the text, boldface lowercase letters represent points, e.g.,

$$\mathbf{p} = \begin{bmatrix} p_1 \\ p_2 \end{bmatrix}. \tag{2.1}$$

The location of $\mathbf{p}$ is p_1-units along the $\mathbf{e}_1$-axis and p_2-units along the $\mathbf{e}_2$-axis. So you see that a point's *coordinates*, p_1 and p_2, are dependent upon the location of the coordinate origin. We use the boldface notation so there is a noticeable difference between a one-dimensional (1D) number, or *scalar* p. To clearly identify $\mathbf{p}$ as a point, the notation $\mathbf{p} \in \mathbb{E}^2$ is used. This means that a 2D point "lives" in 2D Euclidean space $\mathbb{E}^2$.

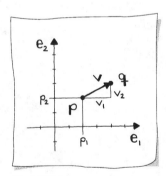

Sketch 2.2.

Two points and a vector.

Now let's move away from our reference point. Following Sketch 2.2, suppose the reference point is $\mathbf{p}$, and when moving along a straight path, our target point is $\mathbf{q}$. The directions from $\mathbf{p}$ would be to follow the *vector* $\mathbf{v}$. Our notation for a vector is the same as for a point: boldface lowercase letters. To get to $\mathbf{q}$ we say,

$$\mathbf{q} = \mathbf{p} + \mathbf{v}. \tag{2.2}$$

---

[1]More on hurricanes can be found at the National Hurricane Center via http://www.nhc.noaa.gov.

To calculate this, add each component separately, that is

$$\begin{bmatrix} q_1 \\ q_2 \end{bmatrix} = \begin{bmatrix} p_1 \\ p_2 \end{bmatrix} + \begin{bmatrix} v_1 \\ v_2 \end{bmatrix} = \begin{bmatrix} p_1 + v_1 \\ p_2 + v_2 \end{bmatrix}.$$

For example, in Sketch 2.2, we have

$$\begin{bmatrix} 4 \\ 3 \end{bmatrix} = \begin{bmatrix} 2 \\ 2 \end{bmatrix} + \begin{bmatrix} 2 \\ 1 \end{bmatrix}.$$

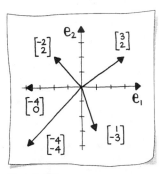

The *components* of $\mathbf{v}$, v_1 and v_2, indicate how many units to move along the $\mathbf{e}_1$- and $\mathbf{e}_2$-axis, respectively. This means that $\mathbf{v}$ can be defined as

$$\mathbf{v} = \mathbf{q} - \mathbf{p}. \qquad (2.3)$$

This defines a vector as a difference of two points which describes a *direction and a distance*, or a *displacement*. Examples of vectors are illustrated in Sketch 2.3.

Sketch 2.3.
Vectors and their components.

How to determine a vector's length is covered in Section 2.4. Above we described this length as a distance. Alternatively, this length can be described as speed: then we have a *velocity vector*.[2] Yet another interpretation is that the length represents acceleration: then we have a *force vector*.

A vector has a *tail* and a *head*. As in Sketch 2.2, the tail is typically displayed positioned at a point, or *bound to a point* in order to indicate the geometric significance of the vector. However, unlike a point, a vector does *not* define a position. Two vectors are equal if they have the same component values, just as points are equal if they have the same coordinate values. Thus, considering a vector as a difference of two points, there are any number of vectors with the same direction and length. See Sketch 2.4 for an illustration.

A special vector worth mentioning is the *zero vector*,

$$\mathbf{0} = \begin{bmatrix} 0 \\ 0 \end{bmatrix}.$$

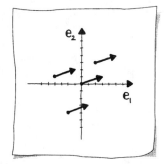

Sketch 2.4.
Instances of one vector.

This vector has no direction or length. Other somewhat special vectors include

$$\mathbf{e}_1 = \begin{bmatrix} 1 \\ 0 \end{bmatrix} \quad \text{and} \quad \mathbf{e}_2 = \begin{bmatrix} 0 \\ 1 \end{bmatrix}.$$

In the sketches, these vectors are not always drawn true to length to prevent them from obscuring the main idea.

To clearly identify $\mathbf{v}$ as a vector, we write $\mathbf{v} \in \mathbb{R}^2$. This means that a 2D vector "lives" in a 2D *linear space* $\mathbb{R}^2$. (Other names for $\mathbb{R}^2$ are *real* or *vector spaces*.)

---

[2]This is what we'll use to continue the Hurricane Andrew example.

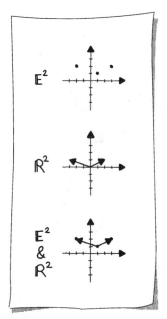

Sketch 2.5.

Euclidean and linear spaces
displayed separately and
together.

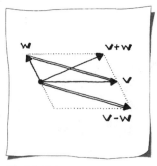

Sketch 2.6.

Parallelogram rule.

2.2   What's the Difference?

When writing a point or a vector we use boldface lowercase letters;
when programming we use the same data structure, e.g., arrays. This
makes it appear that points and vectors can be treated in the same
manner. Not so!

Points and vectors are different geometric entities. This is reiter-
ated by saying they live in different spaces, $\mathbb{E}^2$ and $\mathbb{R}^2$. As shown
in Sketch 2.5, for convenience and clarity elements of Euclidean and
linear spaces are typically displayed together.

The primary reason for differentiating between points and vectors is
to achieve geometric constructions which are *coordinate independent.*
Such constructions are manipulations applied to geometric objects
that produce the same result regardless of the location of the coordinate origin. (Example: the midpoint of two points.) This idea becomes clearer by analyzing some fundamental manipulations of points
and vectors. In what follows, let's use $\mathbf{p}, \mathbf{q} \in \mathbb{E}^2$ and $\mathbf{v}, \mathbf{w} \in \mathbb{R}^2$.

Coordinate Independent Operations:

- Subtracting two points $(\mathbf{p}-\mathbf{q})$ yields a vector as depicted in Sketch
  2.2 and Equation (2.3).

- Adding or subtracting two vectors yields another vector. See
  Sketch 2.6 which illustrates the *parallelogram rule*: the vectors
  $\mathbf{v} - \mathbf{w}$ and $\mathbf{v} + \mathbf{w}$ are the diagonals of the parallelogram defined by
  $\mathbf{v}$ and $\mathbf{w}$. This is a coordinate independent operation since vectors
  are defined as a difference of points.

- Multiplying by a scalar s is called *scaling.* Scaling a vector is a
  well-defined operation. The result $s\mathbf{v}$ adjusts the length by the
  scaling factor. The direction is unchanged if $s > 0$ and reversed
  for $s < 0$. If $s = 0$ then the result is the zero vector. Sketch 2.7
  illustrates some examples of scaling a vector.

- Adding a vector to a point $(\mathbf{p}+\mathbf{v})$ yields another point as in Sketch
  2.2 and Equation (2.2).

Any coordinate independent combination of two or more points
and/or vectors can be grouped to fall into one or more of the items
above. See the Exercises for examples.

Coordinate Dependent Operations:

- Scaling a point (s**p**) is not a well-defined operation because it is not a coordinate independent operation. Sketch 2.8 illustrates that the result of scaling the solid black point by one-half with respect to two different coordinate systems results in two different points.

- Adding two points (**p**+**q**) is not a well-defined operation because it is not a coordinate independent operation. As depicted in Sketch 2.9, the result of adding the two solid black points is dependent on the coordinate origin. (The parallelogram rule is used here to construct the results of the additions.)

However, special combinations of points are allowed, and they are defined in Section 2.5

2.3   Vector Fields

Figure 2.2.
Vector field: simulating hurricane air velocity. Lighter gray indicates greater velocity.

A good way to visualize the interplay between points and vectors is through the example of *vector fields*. In general, we speak of a vector field if every point in a given region is assigned a vector. We have already encountered an example of this in Figure 2.1: Hurricane Andrew! Recall that at each location (point) we could describe the air velocity (vector). Our previous image did not actually tell us anything about the air speed, although we could presume something about the direction. This is where a vector field is helpful. Shown in Figure 2.2 is a vector field simulating Hurricane Andrew. By plotting

Sketch 2.7.
Scaling a vector.

Sketch 2.8.
Scaling of points is ambiguous.

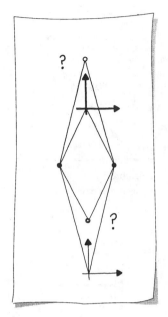

Sketch 2.9.

Addition of points is
ambiguous.

all the vectors the same length and using gray scale to indicate speed, the vector field can be more informative than the photograph.

Other important applications of vector fields arise in the areas of automotive and aerospace design: before a car or an airplane is built, it undergoes extensive aerodynamic simulations. In these simulations, the vectors that characterize the flow around an object are computed from complex differential equations. In Figure 2.3 we have another example of a vector field.

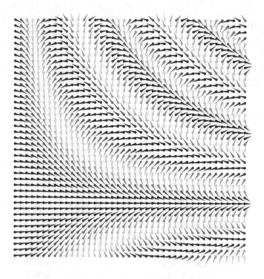

Figure 2.3.

Vector field: every sampled point has an associated vector. Lighter gray indicates greater vector length.

2.4   Length of a Vector

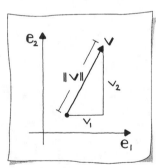

Sketch 2.10.

Length of a vector.

As mentioned in Section 2.1, the length of a vector can represent distance, velocity, or acceleration. We need a method for finding the length of a vector, or the *magnitude*. As illustrated in Sketch 2.10, a vector defines the displacement necessary (with respect to the e_1- and e_2-axis) to get from a point at the tail of the vector to a point at the head of the vector.

In Sketch 2.10 we have formed a right triangle. The square of the length of the hypotenuse of a right triangle is well known from the *Pythagorean theorem*. Denote the *length* of a vector $\mathbf{v}$ as $\|\mathbf{v}\|$. Then

$$\|\mathbf{v}\|^2 = v_1^2 + v_2^2.$$

Therefore, the magnitude of $\mathbf{v}$ is

$$\|\mathbf{v}\| = \sqrt{v_1^2 + v_2^2}. \tag{2.4}$$

Notice that if we scale the vector by an amount k then

$$\|k\mathbf{v}\| = k\|\mathbf{v}\|. \tag{2.5}$$

A *normalized vector* $\mathbf{w}$ has *unit length*, that is

$$\|\mathbf{w}\| = 1.$$

Normalized vectors are also known as *unit vectors*. To *normalize* a vector simply means to scale a vector so that it has unit length. If $\mathbf{w}$ is to be our unit length version of $\mathbf{v}$ then

$$\mathbf{w} = \frac{\mathbf{v}}{\|\mathbf{v}\|}.$$

Each component of $\mathbf{v}$ is divided by the scalar value $\|\mathbf{v}\|$. This scalar value is always *nonnegative*, which means that its value is zero or greater. It can be zero! You must check the value before dividing to be sure it is greater than your *zero divide tolerance*.[3]

We utilized unit vectors in Figures 2.2 and 2.3. The vectors in those figures are drawn as unit vectors. Gray scales are used to indicate their magnitudes: that way the vectors do not overlap, and the figure is easier to understand.

Example 2.1

Start with

$$\mathbf{v} = \begin{bmatrix} 5 \\ 0 \end{bmatrix}.$$

Applying (2.4), $\|\mathbf{v}\| = \sqrt{5^2 + 0^2} = 5$. Then the normalized version of $\mathbf{v}$ is defined as

$$\mathbf{w} = \begin{bmatrix} 5/5 \\ 0/5 \end{bmatrix} = \begin{bmatrix} 1 \\ 0 \end{bmatrix}.$$

Clearly $\|\mathbf{w}\| = 1$, so this is a normalized vector. Since we have only scaled $\mathbf{v}$ by a positive amount, the direction of $\mathbf{w}$ is the same as $\mathbf{v}$.

Figure 2.4.

Unit vectors: they define a circle.

There are infinitely many unit vectors. Imagine drawing them all, emanating from the origin. The figure that you will get is a circle of radius one! See Figure 2.4.

To find the *distance between two points* we simply form a vector defined by the two points, e.g., $\mathbf{v} = \mathbf{q} - \mathbf{p}$, and apply (2.4).

Example 2.2

Let

$$\mathbf{q} = \begin{bmatrix} -1 \\ 2 \end{bmatrix} \quad \text{and} \quad \mathbf{p} = \begin{bmatrix} 1 \\ 0 \end{bmatrix}.$$

Then

$$\mathbf{q} - \mathbf{p} = \begin{bmatrix} -2 \\ 2 \end{bmatrix}$$

and

$$\|\mathbf{q} - \mathbf{p}\| = \sqrt{(-2)^2 + 2^2} = \sqrt{8} \approx 2.83.$$

Sketch 2.11 illustrates this example.

Sketch 2.11.

Distance between two points.

---

[3]The zero divide tolerance is the absolute value of the smallest number by which you can divide confidently. (When we refer to checking that a value is greater than this number, it means to check the absolute value.)

2.5   Combining Points

Seemingly contrary to Section 2.2, there actually is a way to combine two points such that we get a (meaningful) third one. Take the
example of the midpoint **r** of two points **p** and **q**; more specifically,
take

$$\mathbf{p} = \begin{bmatrix} 1 \\ 6 \end{bmatrix}, \quad \mathbf{r} = \begin{bmatrix} 2 \\ 3 \end{bmatrix}, \quad \mathbf{q} = \begin{bmatrix} 3 \\ 0 \end{bmatrix},$$

as shown in Sketch 2.12.

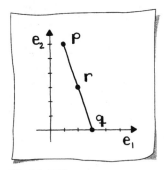

Sketch 2.12.
The midpoint of two points.

Let's start with the known coordinate independent operation of
adding a vector to a point. Define **r** by adding an appropriately
scaled version of the vector $\mathbf{v} = \mathbf{q} - \mathbf{p}$ to the point **p**:

$$\mathbf{r} = \mathbf{p} + \frac{1}{2}\mathbf{v}$$

$$\begin{bmatrix} 2 \\ 3 \end{bmatrix} = \begin{bmatrix} 1 \\ 6 \end{bmatrix} + \frac{1}{2} \begin{bmatrix} 2 \\ -6 \end{bmatrix}.$$

Expanding, this shows that **r** can also be defined as

$$\mathbf{r} = \frac{1}{2}\mathbf{p} + \frac{1}{2}\mathbf{q}$$

$$\begin{bmatrix} 2 \\ 3 \end{bmatrix} = \frac{1}{2} \begin{bmatrix} 1 \\ 6 \end{bmatrix} + \frac{1}{2} \begin{bmatrix} 3 \\ 0 \end{bmatrix}.$$

This is a legal expression for a combination of points!

There is nothing magical about the factor $1/2$, however. Adding
a (scaled) vector to a point is a well-defined, coordinate independent
operation that yields another point. Any point of the form

$$\mathbf{r} = \mathbf{p} + t\mathbf{v} \tag{2.6}$$

is on the line through **p** and **q**. Again, we may rewrite this as

$$\mathbf{r} = \mathbf{p} + t(\mathbf{q} - \mathbf{p})$$

and then

$$\mathbf{r} = (1 - t)\mathbf{p} + t\mathbf{q}. \tag{2.7}$$

Sketch 2.13 gives an example with $t = 1/3$.

The scalar values $(1 - t)$ and t are *coefficients*. A weighted sum
of points where the coefficients sum to one is called a *barycentric
combination*. In this special case, where one point **r** is being expressed
in terms of two others, **p** and **q**, $1 - t$ and t are called the *barycentric
coordinates* of **r**.

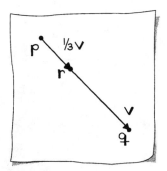

Sketch 2.13.
Barycentric combinations:
$t = 1/3$.

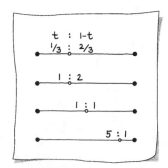

Sketch 2.14.

Examples of ratios.

A barycentric combination allows us to construct $\mathbf{r}$ anywhere on the line defined by $\mathbf{p}$ and $\mathbf{q}$. If we would like to restrict $\mathbf{r}$'s position to the *line segment* between $\mathbf{p}$ and $\mathbf{q}$, then we allow only *convex combinations*: t must satisfy $0 \leq t \leq 1$. To define points outside of the line segment between $\mathbf{p}$ and $\mathbf{q}$, we need values of $t < 0$ or $t > 1$.

The position of $\mathbf{r}$ is said to be in the *ratio* of $t : (1 - t)$ or $t/(1 - t)$. In physics, $\mathbf{r}$ is known as the *center of gravity* of two points $\mathbf{p}$ and $\mathbf{q}$ with weights $1 - t$ and t, respectively. From a constructive approach, the ratio is formed from the quotient

$$\text{ratio} = \frac{\|\mathbf{r} - \mathbf{p}\|}{\|\mathbf{q} - \mathbf{r}\|}.$$

Some examples are illustrated in Sketch 2.14.

Example 2.3

Suppose we have three collinear points, $\mathbf{p}$, $\mathbf{q}$, and $\mathbf{r}$ as illustrated in Sketch 2.15. The points have the following locations.

$$\mathbf{p} = \begin{bmatrix} 2 \\ 4 \end{bmatrix}, \quad \mathbf{r} = \begin{bmatrix} 6.5 \\ 7 \end{bmatrix}, \quad \mathbf{q} = \begin{bmatrix} 8 \\ 8 \end{bmatrix}.$$

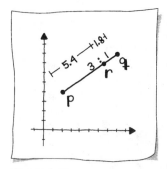

Sketch 2.15.

Barycentric coordinates in relation to lengths.

What are the barycentric coordinates of $\mathbf{r}$ with respect to $\mathbf{p}$ and $\mathbf{q}$?

To answer this, recall the relationship between the ratio and the barycentric coordinates. The barycentric coordinates t and $(1 - t)$ define $\mathbf{r}$ as

$$\begin{bmatrix} 6.5 \\ 7 \end{bmatrix} = (1 - t) \begin{bmatrix} 2 \\ 4 \end{bmatrix} + t \begin{bmatrix} 8 \\ 8 \end{bmatrix}.$$

The ratio indicates the location of $\mathbf{r}$ relative to $\mathbf{p}$ and $\mathbf{q}$ in terms of relative distances. Suppose the ratio is $s_1 : s_2$. If we scale s_1 and s_2 such that they sum to one, then s_1 and s_2 are the barycentric coordinates t and $(1 - t)$, respectively. By calculating the distances between points:

$$l_1 = \|\mathbf{r} - \mathbf{p}\| \approx 5.4,$$
$$l_2 = \|\mathbf{q} - \mathbf{r}\| \approx 1.8,$$
$$l_3 = l_1 + l_2 \approx 7.2,$$

we find that

$$t = l_1/l_3 = 0.75 \quad \text{and}$$
$$(1 - t) = l_2/l_3 = 0.25.$$

These are the barycentric coordinates. Let's verify this:

$$\begin{bmatrix} 6.5 \\ 7 \end{bmatrix} = 0.25 \times \begin{bmatrix} 2 \\ 4 \end{bmatrix} + 0.75 \times \begin{bmatrix} 8 \\ 8 \end{bmatrix}.$$

The barycentric coordinate t is also called a *parameter*. (See Section 3.2 for more details.) This parameter is defined by the quotient

$$t = \frac{\|\mathbf{r} - \mathbf{p}\|}{\|\mathbf{q} - \mathbf{p}\|}.$$

We have seen how useful this quotient can be in Section 1.1 for the construction of a point in the global system that corresponded to a point with parameter t in the local system.

We can create barycentric combinations with *more than two points*. Let's look at three points $\mathbf{p}$, $\mathbf{q}$, and $\mathbf{r}$, which are not collinear. Any point $\mathbf{s}$ can be formed from

$$\mathbf{s} = \mathbf{r} + t_1(\mathbf{p} - \mathbf{r}) + t_2(\mathbf{q} - \mathbf{r}).$$

This is a coordinate independent operation of point + vector + vector. Expanding and regrouping, we can also define $\mathbf{s}$ as

$$\begin{aligned} \mathbf{s} &= t_1\mathbf{p} + t_2\mathbf{q} + (1 - t_1 - t_2)\mathbf{r} \\ &= t_1\mathbf{p} + t_2\mathbf{q} + t_3\mathbf{r}. \end{aligned} \tag{2.8}$$

Thus, the point $\mathbf{s}$ is defined by barycentric combination with coefficients t_1, t_2, and $t_3 = 1 - t_1 - t_2$ with respect to $\mathbf{p}$, $\mathbf{q}$, and $\mathbf{r}$, respectively. This is another special case where the barycentric combination coefficients correspond to *barycentric coordinates*. Sketch 2.16 illustrates. We will encounter barycentric coordinates in more detail in Chapter 8.

We can also *combine points* so that the result is a vector. For this, we need the coefficients to sum to zero. We encountered a simple case of this in (2.3). Suppose you have the equation

$$\mathbf{e} = \mathbf{r} - 2\mathbf{p} + \mathbf{q}, \qquad \mathbf{r}, \mathbf{p}, \mathbf{q} \in \mathbb{E}^2.$$

Does $\mathbf{e}$ have a geometric meaning? Looking at the sum of the coefficients, $1 - 2 + 1 = 0$, we would conclude by the rule above that $\mathbf{e}$ is a vector. How to see this? By rewriting the equation as

$$\mathbf{e} = (\mathbf{r} - \mathbf{p}) + (\mathbf{q} - \mathbf{p}),$$

it is clear that $\mathbf{e}$ is a vector formed from (vector + vector).

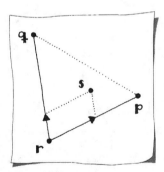

Sketch 2.16.
A barycentric combination of three points.

2.6 Independence

Two vectors $\mathbf{v}$ and $\mathbf{w}$ describe a parallelogram, as shown in Sketch 2.6. It may happen that this parallelogram has zero area; then the two vectors are parallel. In this case, we have a relationship of the form $\mathbf{v} = c\mathbf{w}$. If two vectors are parallel, then we call them *linearly dependent*. Otherwise, we say that they are *linearly independent*.

Two linearly independent vectors may be used to write any other vector $\mathbf{u}$ as a *linear combination*:

$$\mathbf{u} = r\mathbf{v} + s\mathbf{w}.$$

How to find r and s is described in Chapter 5. Two linearly independent vectors in 2D are also called a *basis* for $\mathbb{R}^2$. If $\mathbf{v}$ and $\mathbf{w}$ are linearly dependent, then you cannot write all vectors as a linear combination of them, as the following example shows.

Example 2.4

Let

$$\mathbf{v} = \begin{bmatrix} 1 \\ 2 \end{bmatrix} \quad \text{and} \quad \mathbf{w} = \begin{bmatrix} 2 \\ 4 \end{bmatrix}.$$

If we tried to write the vector

$$\mathbf{u} = \begin{bmatrix} 1 \\ 0 \end{bmatrix}$$

as $\mathbf{u} = r\mathbf{v} + s\mathbf{w}$, then this would lead to

$$1 = r + 2s, \tag{2.9}$$
$$0 = 2r + 4s. \tag{2.10}$$

If we multiply the first equation by a factor of 2, the two right-hand sides will be equal. Equating the new left-hand sides now results in the expression $2 = 0$. This shows that $\mathbf{u}$ cannot be written as a linear combination of $\mathbf{v}$ and $\mathbf{w}$. (See Sketch 2.17.)

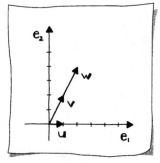

Sketch 2.17.
Dependent vectors.

2.7 Dot Product

Given two vectors $\mathbf{v}$ and $\mathbf{w}$, we might ask:

- Are they the *same* vector?

- Are they *perpendicular* to each other?

- What *angle* do they form?

The *dot product* is the tool to resolve these questions.

To motivate the dot product, let's start with the Pythagorean theorem and Sketch 2.18. We know that two vectors **v** and **w** are perpendicular if and only if

$$\|\mathbf{v} - \mathbf{w}\|^2 = \|\mathbf{v}\|^2 + \|\mathbf{w}\|^2. \tag{2.11}$$

Writing the components in (2.11) explicitly

$$(v_1 - w_1)^2 + (v_2 - w_2)^2 = (v_1^2 + v_2^2) + (w_1^2 + w_2^2),$$

and then expanding, bringing all terms to the left-hand side of the equation yields

$$(v_1^2 - 2v_1w_1 + w_1^2) + (v_2^2 - 2v_2w_2 + w_2^2) - (v_1^2 + v_2^2) - (w_1^2 + w_2^2) = 0,$$

which reduces to

$$v_1w_1 + v_2w_2 = 0. \tag{2.12}$$

We find that perpendicular vectors have the property that the sum of the products of their components is zero. The short-hand vector notation for (2.12) is

$$\mathbf{v} \cdot \mathbf{w} = 0. \tag{2.13}$$

This result has an immediate application: a vector **w** perpendicular to a given vector **v** can be formed as

$$\mathbf{w} = \begin{bmatrix} -v_2 \\ v_1 \end{bmatrix}$$

(switching components and negating the sign of one). Then $\mathbf{v} \cdot \mathbf{w}$ becomes $v_1(-v_2) + v_2v_1 = 0$.

If we take two arbitrary vectors **v** and **w**, then $\mathbf{v} \cdot \mathbf{w}$ will in general not be zero. But we can compute it anyway, and define

$$s = \mathbf{v} \cdot \mathbf{w} = v_1w_1 + v_2w_2 \tag{2.14}$$

to be the *dot product* of **v** and **w**. Notice that the dot product returns a scalar s, which is why it is also called a *scalar product*. (Mathematicians have yet another name for the dot product—an *inner product*. See Section 15.3 for more on these.) From (2.14) it is clear that

$$\mathbf{v} \cdot \mathbf{w} = \mathbf{w} \cdot \mathbf{v}.$$

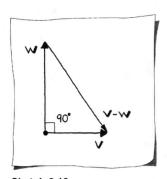

Sketch 2.18.
Perpendicular vectors.

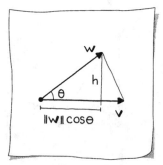

Sketch 2.19.

Geometry of the dot product.

This is called the *symmetry property*. Other properties of the dot product are given in the Exercises.

In order to understand the geometric meaning of the dot product of two vectors, let's construct a triangle from two vectors $\mathbf{v}$ and $\mathbf{w}$ as illustrated in Sketch 2.19.

From trigonometry, we know that the height h of the triangle can be expressed as

$$h = \|\mathbf{w}\| \sin(\theta).$$

Squaring both sides results in

$$h^2 = \|\mathbf{w}\|^2 \sin^2(\theta).$$

Using the identity

$$\sin^2(\theta) + \cos^2(\theta) = 1,$$

we have

$$h^2 = \|\mathbf{w}\|^2(1 - \cos^2(\theta)). \tag{2.15}$$

We can also express the height h with respect to the other right triangle in Sketch 2.19 and by using the Pythagorean theorem:

$$h^2 = \|\mathbf{v} - \mathbf{w}\|^2 - (\|\mathbf{v}\| - \|\mathbf{w}\| \cos\theta)^2. \tag{2.16}$$

Equating (2.15) and (2.16) and simplifying, we have the expression,

$$\|\mathbf{v} - \mathbf{w}\|^2 = \|\mathbf{v}\|^2 + \|\mathbf{w}\|^2 - 2\|\mathbf{v}\|\|\mathbf{w}\| \cos\theta. \tag{2.17}$$

We have just proved the *Law of Cosines*, which generalizes the Pythagorean theorem by correcting it for triangles with an opposing angle different from 90°.

We can formulate another expression for $\|\mathbf{v} - \mathbf{w}\|^2$ by explicitly writing out

$$\begin{aligned}\|\mathbf{v} - \mathbf{w}\|^2 &= (\mathbf{v} - \mathbf{w}) \cdot (\mathbf{v} - \mathbf{w}) \\ &= \|\mathbf{v}\|^2 - 2\mathbf{v} \cdot \mathbf{w} + \|\mathbf{w}\|^2. \end{aligned} \tag{2.18}$$

By equating (2.17) and (2.18) we find that

$$\mathbf{v} \cdot \mathbf{w} = \|\mathbf{v}\|\|\mathbf{w}\| \cos\theta. \tag{2.19}$$

Here is another expression for the *dot product*—it is a very useful one! Rearranging (2.19), the cosine of the angle between the two vectors can be determined as

$$\cos\theta = \frac{\mathbf{v} \cdot \mathbf{w}}{\|\mathbf{v}\|\|\mathbf{w}\|}. \tag{2.20}$$

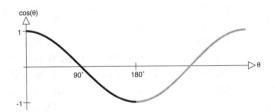

Figure 2.5.
Cosine function: its values at $\theta = 0°$, $\theta = 90°$, and $\theta = 180°$ are important to remember.

By examining a plot of the cosine function in Figure 2.5, some sense can be made of (2.20).

First we consider the special case of perpendicular vectors. Recall the dot product was zero, which makes $\cos(90°) = 0$, just as it should be.

If $\mathbf{v}$ has the same (or opposite) direction as $\mathbf{w}$, that is $\mathbf{v} = k\mathbf{w}$, then (2.20) becomes

$$\cos\theta = \frac{k\mathbf{w} \cdot \mathbf{w}}{\|k\mathbf{w}\|\|\mathbf{w}\|}.$$

Using (2.5), we have

$$\cos\theta = \frac{k\|\mathbf{w}\|^2}{k\|\mathbf{w}\|\|\mathbf{w}\|} = \pm 1.$$

Again, examining Figure 2.5, we see this corresponds to either $\theta = 0°$ or $\theta = 180°$, for vectors of the same or opposite direction, respectively.

The cosine values from (2.20) range between ± 1; this corresponds to angles between $0°$ and $180°$ (or 0 and π radians). Thus, the smaller angle between the two vectors is measured. This is clear from the derivation: the angle θ enclosed by completing the triangle defined by the two vectors must be less than $180°$. Three types of angles can be formed:

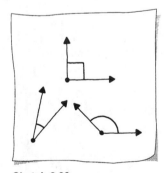

- *right*: $\cos(\theta) = 0 \to \mathbf{v} \cdot \mathbf{w} = 0$;

- *acute*: $\cos(\theta) > 0 \to \mathbf{v} \cdot \mathbf{w} > 0$;

- *obtuse*: $\cos(\theta) < 0 \to \mathbf{v} \cdot \mathbf{w} < 0$.

Sketch 2.20.
Three types of angles.

These are illustrated in counterclockwise order from twelve o'clock in Sketch 2.20.

If the actual angle θ needs to be calculated, then the arccosine function has to be invoked: $\theta = \text{acos}(s)$. One word of warning: in some math libraries, if $s > 1$ or $s < -1$ then an error occurs and a non-usable result (NaN—Not a Number) is returned.

Thus, if s is calculated, it is best to check that its value is within the appropriate range. It is not uncommon that an intended value of $s = 1.0$ is actually something like $s = 1.0000001$ due to *round-off*. Thus, the arccosine function should be used with caution. In many instances, as in comparing angles, the cosine of the angle is all you need!

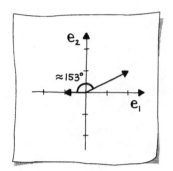

Sketch 2.21.

The angle between two vectors.

Example 2.5

Let's calculate the angle between the two vectors illustrated in Sketch 2.21, forming an obtuse angle:

$$\mathbf{v} = \begin{bmatrix} 2 \\ 1 \end{bmatrix} \quad \text{and} \quad \mathbf{w} = \begin{bmatrix} -1 \\ 0 \end{bmatrix}.$$

Calculate the length of each vector,

$$\|\mathbf{v}\| = \sqrt{2^2 + 1^2} = \sqrt{5}$$
$$\|\mathbf{w}\| = \sqrt{-1^2 + 0^2} = 1.$$

The cosine of the angle between the vectors is calculated using (2.20) as

$$\cos(\theta) = \frac{(2 \times -1) + (1 \times 0)}{\sqrt{5} \times 1} = \frac{-2}{\sqrt{5}} \approx -0.8944.$$

Then

$$\text{arccos}(-0.8944) \approx 153.4°.$$

To convert an angle given in degrees to radians multiply by $\pi/180°$. (Recall that $\pi \approx 3.14159$ radians.) This means that

$$2.677 \text{ radians} \approx 153.4° \times \frac{\pi}{180°}.$$

2.8   Orthogonal Projections

Sketch 2.19 illustrates that the projection of the vector $\mathbf{w}$ onto $\mathbf{v}$ creates a footprint of length $b = \|\mathbf{w}\| \cos(\theta)$. This is derived simply by equating (2.20) and $\cos(\theta) = b/||\mathbf{w}||$ and solving for b. The *orthogonal projection* of $\mathbf{w}$ onto $\mathbf{v}$ is then the vector

$$\mathbf{u} = (||\mathbf{w}|| \cos(\theta)) \frac{\mathbf{v}}{||\mathbf{v}||} = \frac{\mathbf{v} \cdot \mathbf{w}}{||\mathbf{v}||^2} \mathbf{v}. \qquad (2.21)$$

This projection is revisited in Section 3.7 to find the point at the end of the footprint.

Using the orthogonal projection, it is easy to decompose the 2D vector $\mathbf{w}$ into a sum of two perpendicular vectors, namely $\mathbf{u}$ and $\mathbf{u}^\perp$ (a vector perpendicular to $\mathbf{u}$), such that

$$\mathbf{w} = \mathbf{u} + \mathbf{u}^\perp. \qquad (2.22)$$

Already having found the vector $\mathbf{u}$, we now set

$$\mathbf{u}^\perp = \mathbf{w} - \frac{\mathbf{v} \cdot \mathbf{w}}{||\mathbf{v}||^2} \mathbf{v}.$$

(See the Exercises of Chapter 10 for a 3D version of this decomposition.)

2.9 Inequalities

Here are two important inequalities when dealing with vector lengths.

Let's start with the expression from (2.19), i.e.,

$$\mathbf{v} \cdot \mathbf{w} = ||\mathbf{v}|| \, ||\mathbf{w}|| \cos \theta.$$

Squaring both sides gives

$$(\mathbf{v} \cdot \mathbf{w})^2 = ||\mathbf{v}||^2 ||\mathbf{w}||^2 \cos^2 \theta.$$

Noticing that $0 \leq \cos^2 \theta \leq 1$, we conclude that

$$(\mathbf{v} \cdot \mathbf{w})^2 \leq ||\mathbf{v}||^2 ||\mathbf{w}||^2. \qquad (2.23)$$

This is called the *Cauchy-Schwartz Inequality*. This inequality is fundamental in the study of more general vector spaces than those presented here. See a text such as [1] for a more general and thorough derivation of this inequality.

Suppose we would like to find an inequality which describes the relationship between the length of two vectors $\mathbf{v}$ and $\mathbf{w}$ and the length of their sum $\mathbf{v} + \mathbf{w}$. In other words, how does the length of the third

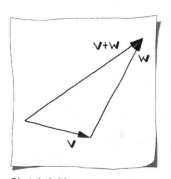

Sketch 2.22.

The triangle inequality.

side of a triangle relate to the lengths of the other two? Let's begin with expanding $\|\mathbf{v} + \mathbf{w}\|^2$:

$$\begin{aligned}
\|\mathbf{v} + \mathbf{w}\|^2 &= (\mathbf{v} + \mathbf{w}) \cdot (\mathbf{v} + \mathbf{w}) \\
&= \mathbf{v} \cdot \mathbf{v} + 2\mathbf{v} \cdot \mathbf{w} + \mathbf{w} \cdot \mathbf{w} \\
&\leq \mathbf{v} \cdot \mathbf{v} + 2|\mathbf{v} \cdot \mathbf{w}| + \mathbf{w} \cdot \mathbf{w} \\
&\leq \mathbf{v} \cdot \mathbf{v} + 2\|\mathbf{v}\|\|\mathbf{w}\| + \mathbf{w} \cdot \mathbf{w} \\
&= \|\mathbf{v}\|^2 + 2\|\mathbf{v}\|\|\mathbf{w}\| + \|\mathbf{w}\|^2 \\
&= (\|\mathbf{v}\| + \|\mathbf{w}\|)^2.
\end{aligned} \tag{2.24}$$

Taking square roots gives

$$\|\mathbf{v} + \mathbf{w}\| \leq \|\mathbf{v}\| + \|\mathbf{w}\|,$$

which is known as the *triangle inequality*. It states the intuitively obvious fact that the sum of any two edge lengths in a triangle is never smaller than the length of the third edge; see Sketch 2.22 for an illustration.

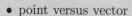

- point versus vector
- coordinate independent
- vector length
- unit vector
- zero divide tolerance
- Pythagorean theorem
- distance between two points
- parallelogram rule
- scaling
- ratio
- barycentric combination

- convex combination
- barycentric coordinates
- linearly dependent vectors
- linear combination
- basis for $\mathbb{R}^2$
- dot product
- Law of Cosines
- perpendicular vectors
- angle between vectors
- orthogonal projection
- Cauchy-Schwartz inequality
- triangle inequality

2.10 Exercises

1. Illustrate the parallelogram rule applied to the vectors

$$\mathbf{v} = \begin{bmatrix} -2 \\ 1 \end{bmatrix} \quad \text{and} \quad \mathbf{w} = \begin{bmatrix} 2 \\ 1 \end{bmatrix}.$$

2. Define your own $\mathbf{p}, \mathbf{q} \in \mathbb{E}^2$ and $\mathbf{v}, \mathbf{w} \in \mathbb{R}^2$. Determine which of the following expressions are geometrically meaningful. Illustrate those that are.

   (a) $\mathbf{p} + \mathbf{q}$ (b) $\frac{1}{2}\mathbf{p} + \frac{1}{2}\mathbf{q}$
   (c) $\mathbf{p} + \mathbf{v}$ (d) $3\mathbf{p} + \mathbf{v}$
   (e) $\mathbf{v} + \mathbf{w}$ (f) $2\mathbf{v} + \frac{1}{2}\mathbf{w}$
   (g) $\mathbf{v} - 2\mathbf{w}$ (h) $\frac{3}{2}\mathbf{p} - \frac{1}{2}\mathbf{q}$

3. Illustrate a point with barycentric coordinates $(1/2, 1/4, 1/4)$ with respect to three other points.

4. Consider three noncollinear points. Form the set of all convex combinations of these points. What is the geometry of this set?

5. What is the length of the vector $\mathbf{v} = \begin{bmatrix} -4 \\ -3 \end{bmatrix}$?

6. Find the distance between the points

$$\mathbf{p} = \begin{bmatrix} 3 \\ 3 \end{bmatrix} \quad \text{and} \quad \mathbf{q} = \begin{bmatrix} -2 \\ -3 \end{bmatrix}.$$

7. Normalize the vector $\mathbf{v} = \begin{bmatrix} -4 \\ -3 \end{bmatrix}$.

8. Show that the dot product has the following properties for vectors $\mathbf{u}, \mathbf{v}, \mathbf{w} \in \mathbb{R}^2$.

$$\mathbf{u} \cdot \mathbf{v} = \mathbf{v} \cdot \mathbf{u} \qquad \text{symmetric}$$

$$\mathbf{v} \cdot (s\mathbf{w}) = s(\mathbf{v} \cdot \mathbf{w}) \qquad \text{homogeneous}$$

$$(\mathbf{v} + \mathbf{w}) \cdot \mathbf{u} = \mathbf{v} \cdot \mathbf{u} + \mathbf{w} \cdot \mathbf{u} \qquad \text{distributive}$$

$$\mathbf{v} \cdot \mathbf{v} > 0 \quad \text{if} \quad \mathbf{v} \neq \mathbf{0} \quad \text{and} \quad \mathbf{v} \cdot \mathbf{v} = 0 \quad \text{if} \quad \mathbf{v} = \mathbf{0}$$

9. Compute the angle (in degrees) formed by the vectors

$$\mathbf{v} = \begin{bmatrix} 5 \\ 5 \end{bmatrix} \quad \text{and} \quad \mathbf{w} = \begin{bmatrix} 3 \\ -3 \end{bmatrix}.$$

10. Given the vectors

$$\mathbf{v} = \begin{bmatrix} 1 \\ -1 \end{bmatrix} \quad \text{and} \quad \mathbf{w} = \begin{bmatrix} 3 \\ 2 \end{bmatrix},$$

   find the orthogonal projection $\mathbf{u}$ of $\mathbf{w}$ onto $\mathbf{v}$. Decompose $\mathbf{w}$ into components $\mathbf{u}$ and $\mathbf{u}^{\perp}$.

3

Lining Up: 2D Lines

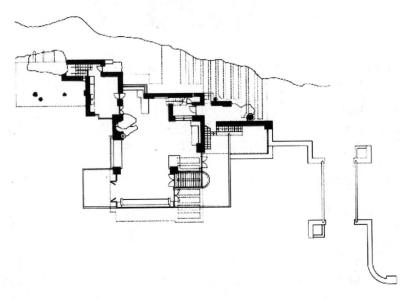

Figure 3.1.
Fallingwater: the floorplan of this building by Frank Lloyd Wright. (Source: http://www
.archinform.de/projekte/974.htm.)

"Real" objects are three-dimensional, or 3D. So why should we
consider 2D objects, such as the 2D lines in this chapter? Consider
Figure 3.1. It shows North America's most celebrated residential
buildling: *Fallingwater*, designed by Frank Lloyd Wright. Clearly,

Figure 3.2.
Fallingwater: a picture of this building by Frank Lloyd Wright. (Source: http://www
.archinform.de/projekte/974.htm.)

the building illustrated in Figure 3.2 is 3D; yet in order to describe it, we need 2D floor plans outlining the building's structure. These floorplans consist almost entirely of 2D lines.

3.1 Defining a Line

Sketch 3.1.
Elements to define a line.

As illustrated in Sketch 3.1, two elements of 2D geometry define a line:

- two points;

- a point and a vector parallel to the line;

- a point and a vector perpendicular to the line.

The unit vector that is perpendicular to a line is referred to as the *normal* to the line. Figure 3.3 shows two *families of lines*: one family of lines shares a common point and the other family of lines shares the same normal. Just as there are different ways to specify a line geometrically, there are different mathematical representations: parametric, implicit, and explicit. Each representation will be examined

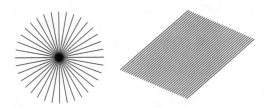

Figure 3.3.

Families of lines: one family shares a common point and the other shares a common normal.

and the advantages of each will be explained. Additionally, we will explore how to convert from one form to another.

3.2 Parametric Equation of a Line

The *parametric equation of a line* $\mathbf{l}(t)$ has the form

$$\mathbf{l}(t) = \mathbf{p} + t\mathbf{v}, \tag{3.1}$$

where $\mathbf{p} \in \mathbb{E}^2$ and $\mathbf{v} \in \mathbb{R}^2$. The scalar value t is the *parameter*. (See Sketch 3.2.) Evaluating (3.1) for a specific parameter $t = \hat{t}$, generates a point on the line.

We encountered (3.1) in Section 2.5 in the context of barycentric coordinates. Interpreting $\mathbf{v}$ as a difference of points, $\mathbf{v} = \mathbf{q} - \mathbf{p}$, this equation was reformulated as

$$\mathbf{l}(t) = (1 - t)\mathbf{p} + t\mathbf{q}. \tag{3.2}$$

A parametric line can be written either in the form of (3.1) or (3.2). The latter is typically referred to as *linear interpolation*.

One way to interpret the parameter t is as *time*; at time $t = 0$ we will be at point $\mathbf{p}$ and at time $t = 1$ we will be at point $\mathbf{q}$. Sketch 3.2 illustrates that as t varies between zero and one, $t \in [0, 1]$, points are generated on the line between $\mathbf{p}$ and $\mathbf{q}$. Recall from Section 2.5 that these values of t constitute a *convex combination*, which is a special case of a *barycentric combination*. If the parameter is a negative number, that is $t < 0$, the direction of $\mathbf{v}$ reverses, generating points on the line "behind" $\mathbf{p}$. The case $t > 1$ is similar: this scales $\mathbf{v}$ so that it is elongated, which generates points "past" $\mathbf{q}$. In the context of linear interpolation, when $t < 0$ or $t > 1$, it is called *extrapolation*.

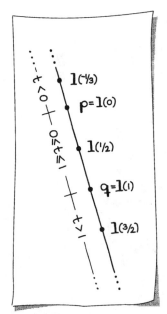

Sketch 3.2.

Parametric form of a line.

The parametric form is very handy for computing points on a line. For example, to compute ten equally spaced points on the line segment between $\mathbf{p}$ and $\mathbf{q}$, simply define ten values of $t \in [0,1]$ as

$$t = i/9, \qquad i = 0 \ldots 9.$$

(Be sure this is a floating point calculation when programming!) Equally spaced parameter values correspond to equally spaced points. See the Exercises for a programming problem.

Example 3.1

Compute five points on the line defined by the points

$$\mathbf{p} = \begin{bmatrix} 1 \\ 2 \end{bmatrix} \quad \text{and} \quad \mathbf{q} = \begin{bmatrix} 6 \\ 4 \end{bmatrix}.$$

Define $\mathbf{v} = \mathbf{q} - \mathbf{p}$, then the line is defined as

$$\mathbf{l}(t) = \begin{bmatrix} 1 \\ 2 \end{bmatrix} + t \begin{bmatrix} 5 \\ 2 \end{bmatrix}.$$

Generate five t-values as

$$t = i/4, \qquad i = 0 \ldots 4.$$

Plug each t-value into the formulation for $\mathbf{l}(t)$:

$$i = 0, \quad t = 0, \qquad \mathbf{l}(0) = \begin{bmatrix} 1 \\ 2 \end{bmatrix};$$

$$i = 1, \quad t = 1/4, \quad \mathbf{l}(1/4) = \begin{bmatrix} 9/4 \\ 5/2 \end{bmatrix};$$

$$i = 2, \quad t = 2/4, \quad \mathbf{l}(2/4) = \begin{bmatrix} 7/2 \\ 3 \end{bmatrix};$$

$$i = 3, \quad t = 3/4, \quad \mathbf{l}(3/4) = \begin{bmatrix} 19/4 \\ 7/2 \end{bmatrix};$$

$$i = 4, \quad t = 1, \qquad \mathbf{l}(1) = \begin{bmatrix} 6 \\ 4 \end{bmatrix}.$$

Plot these values for yourself to verify them.

As you can see, the position of the point **p** and the direction and length of the vector **v** determine which points on the line are generated as we increment through $t \in [0, 1]$. This particular artifact of the parametric equation of a line is called the *parametrization*. The parametrization is related to the speed at which a point traverses the line. We may affect this speed by scaling **v**: the larger the scale factor, the faster the point's motion!

3.3 Implicit Equation of a Line

Another way to represent the same line is to use the *implicit equation of a line*. For this representation, we start with a point **p**, and as illustrated in Sketch 3.3, construct a vector **a** that is perpendicular to the line.

For any point **x** on the line, it holds that

$$\mathbf{a} \cdot (\mathbf{x} - \mathbf{p}) = 0. \tag{3.3}$$

This says that **a** and the vector $(\mathbf{x} - \mathbf{p})$ are perpendicular. If **a** has unit length, it is called the *normal* to the line, and then (3.3) is the *point normal form* of a line. Expanding this equation, we get

$$a_1 x_1 + a_2 x_2 + (-a_1 p_1 - a_2 p_2) = 0.$$

Commonly, this is written as

$$a x_1 + b x_2 + c = 0, \tag{3.4}$$

where

$$a = a_1, \tag{3.5}$$
$$b = a_2, \tag{3.6}$$
$$c = -a_1 p_1 - a_2 p_2. \tag{3.7}$$

Equation (3.4) is called the *implicit equation of the line*.

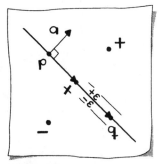

Sketch 3.3.
Implicit form of a line.

Example 3.2

Following Sketch 3.4, suppose we know two points,

$$\mathbf{p} = \begin{bmatrix} 2 \\ 2 \end{bmatrix} \quad \text{and} \quad \mathbf{q} = \begin{bmatrix} 6 \\ 4 \end{bmatrix},$$

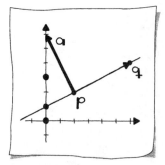

Sketch 3.4.
Implicit construction.

on the line. To construct the coefficients a, b, and c in (3.4), first form the vector

$$\mathbf{v} = \mathbf{q} - \mathbf{p} = \begin{bmatrix} 4 \\ 2 \end{bmatrix}.$$

Now construct a vector $\mathbf{a}$ that is perpendicular to $\mathbf{v}$:

$$\mathbf{a} = \begin{bmatrix} -v_2 \\ v_1 \end{bmatrix} = \begin{bmatrix} -2 \\ 4 \end{bmatrix}. \qquad (3.8)$$

Note, equally as well, we could have chosen $\mathbf{a}$ to be

$$\begin{bmatrix} 2 \\ -4 \end{bmatrix}.$$

The coefficients a and b in (3.5) and (3.6) are now defined as $a = -2$ and $b = 4$. With $\mathbf{p}$ as defined above, solve for c as in (3.7). In this example,

$$c = 2 \times 2 - 4 \times 2 = -4.$$

The implicit equation of the line is complete

$$-2x_1 + 4x_2 - 4 = 0.$$

The implicit form is very useful for deciding if an arbitrary point lies on the line. To test if a point $\mathbf{x}$ is on the line, just plug its coordinates into (3.4). If the value f of the left-hand side of this equation,

$$f = ax_1 + bx_2 + c,$$

is zero then the point is on the line.

A numerical caveat is needed here. Checking equality with floating point numbers should never be done. Instead, a tolerance ϵ around zero must be used. What is a meaningful tolerance in this situation? We'll see in Section 3.7 that

$$d = \frac{f}{\|\mathbf{a}\|} \qquad (3.9)$$

reflects the true distance of $\mathbf{x}$ to the line. Now the tolerance has a physical meaning, which makes it much easier to specify. Sketch 3.3 illustrates the physical relationship of this tolerance to the line.

The sign of d indicates on which side of the line the point lies. This sign is dependent upon the definition of $\mathbf{a}$. (Remember, there were two possible orientations.) Positive d corresponds to the point on the side of the line to which **a** points.

Example 3.3

Let's continue with our example for the line

$$-2x_1 + 4x_2 - 4 = 0,$$

as illustrated in Sketch 3.4. We want to test if the point

$$\mathbf{x} = \begin{bmatrix} 0 \\ 1 \end{bmatrix}$$

lies on the line. First, calculate

$$\|\mathbf{a}\| = \sqrt{-2^2 + 4^2} = \sqrt{20}.$$

The distance is

$$d = (-2 \times 0 + 4 \times 1 - 4)/\sqrt{20} = 0/\sqrt{20} = 0,$$

which indicates the point is on the line.

Test the point

$$\mathbf{x} = \begin{bmatrix} 0 \\ 3 \end{bmatrix}.$$

For this point,

$$d = (-2 \times 0 + 4 \times 3 - 4)/\sqrt{20} = 8/\sqrt{20} \approx 1.79.$$

Checking Sketch 3.3, this is a positive number, indicating that it is on the same side of the line as the direction of **a**. Check for yourself that d does indeed reflect the actual distance of this point to the line.

Test the point

$$\mathbf{x} = \begin{bmatrix} 0 \\ 0 \end{bmatrix}.$$

Calculating the distance for this point, we get

$$d = (-2 \times 0 + 4 \times 0 - 4)/\sqrt{20} = -4/\sqrt{20} \approx -0.894.$$

Checking Sketch 3.3, this is a negative number, indicating it is on the opposite side of the line as the direction of **a**.

Examining (3.4) you might notice that a *horizontal* line takes the form

$$bx_2 + c = 0.$$

This line intersects the $\mathbf{e}_2$-axis at $-c/b$. A *vertical* line takes the form

$$ax_1 + c = 0.$$

This line intersects the $\mathbf{e}_1$-axis at $-c/a$. Using the implicit form, these lines are in no need of special handling.

3.4 Explicit Equation of a Line

The *explicit equation of a line* is the third possible representation. The explicit form is closely related to the implicit form in (3.4). It expresses x_2 as a function of x_1: rearranging the implicit equation we have

$$x_2 = -\frac{a}{b}x_1 - \frac{c}{b}.$$

A more typical way of writing this is

$$x_2 = \hat{a}x_1 + \hat{b}.$$

where $\hat{a} = -a/b$ and $\hat{b} = -c/b$.

The coefficients have geometric meaning: $\hat{a}$ is the *slope* of the line and $\hat{b}$ is the $\mathbf{e}_2$-intercept. Sketch 3.5 illustrates the geometry of the coefficients for the line

$$x_2 = 1/3x_1 + 1.$$

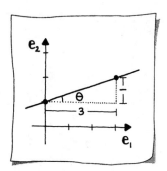

Sketch 3.5.

A line in explicit form.

The slope measures the steepness of the line as a ratio of the change in x_2 to a change in x_1: "rise/run," or more precisely $\tan(\theta)$. The $\mathbf{e}_2$-intercept indicates that the line passes through $(0, \hat{b})$.

Immediately, a drawback of the explicit form is apparent. If the "run" is zero then the (vertical) line has infinite slope. This makes life very difficult when programming! When we study transformations (e.g., changing the orientation of some geometry) in Chapter 6, we will see that infinite slope actually arises often.

The primary popularity of the explicit form comes from the study of calculus. Additionally, in computer graphics, this form is popular when *pixel* calculation is necessary. Examples are Bresenham's line drawing algorithm and scan line polygon fill algorithms (see [11]).

3.5 Converting Between Parametric and Implicit Equations

As we have discussed, there are advantages to both the parametric and implicit representations of a line. Depending on the geometric algorithm, it may be convenient to use one form rather than the other. We'll ignore the explicit form, since as we said, it isn't very useful for general 2D geometry.

3.5.1 Parametric to Implicit

Given: The line l in parametric form,

$$l : l(t) = \mathbf{p} + t\mathbf{v}.$$

Find: The coefficients a, b, c that define the implicit equation of the line

$$l : ax_1 + bx_2 + c = 0.$$

Solution: First form a vector $\mathbf{a}$ that is perpendicular to the vector $\mathbf{v}$. Choose

$$\mathbf{a} = \begin{bmatrix} -v_2 \\ v_1 \end{bmatrix}.$$

This determines the coefficients a and b, as in (3.5) and (3.6), respectively. Simply let $a = a_1$ and $b = a_2$. Finally, solve for the coefficient c as in (3.7). Taking $\mathbf{p}$ from $l(t)$ and $\mathbf{a}$, form

$$c = -(a_1p_1 + a_2p_2).$$

We stepped through a numerical example of this in the derivation of the implicit form in Section 3.3, and it is illustrated in Sketch 3.4. In this example, $l(t)$ is given as

$$l(t) = \begin{bmatrix} 2 \\ 2 \end{bmatrix} + t \begin{bmatrix} 4 \\ 2 \end{bmatrix}.$$

3.5.2 Implicit to Parametric

Given: The line l in implicit form,

$$l : ax_1 + bx_2 + c = 0.$$

Find: The line l in parametric form,

$$l : l(t) = \mathbf{p} + t\mathbf{v}.$$

Solution: Recognize that we need one point on the line and a vector parallel to the line. The vector is easy: simply form a vector perpendicular to $\mathbf{a}$ of the implicit line. For example, we could set

$$\mathbf{v} = \begin{bmatrix} b \\ -a \end{bmatrix}.$$

Next, find a point on the line. Two candidate points are the intersections with the $\mathbf{e}_1$- or $\mathbf{e}_2$-axis,

$$\begin{bmatrix} -c/a \\ 0 \end{bmatrix} \quad \text{or} \quad \begin{bmatrix} 0 \\ -c/b \end{bmatrix},$$

respectively. For numerical stability, let's choose the intersection closest to the origin. Thus, we choose the former if $|a| > |b|$, and the latter otherwise.

Example 3.4

Revisit the numerical example from the implicit form derivation in Section 3.3; it is illustrated in Sketch 3.4. The implicit equation of the line is

$$-2x_1 + 4x_2 - 4 = 0.$$

We want to find a parametric equation of this line,

$$l : l(t) = \mathbf{p} + t\mathbf{v}.$$

First form

$$\mathbf{v} = \begin{bmatrix} 4 \\ 2 \end{bmatrix}.$$

Now determine which is greater in absolute value, a or b. Since $|-2| < |4|$, we choose

$$\mathbf{p} = \begin{bmatrix} 0 \\ 4/4 \end{bmatrix} = \begin{bmatrix} 0 \\ 1 \end{bmatrix}.$$

The parametric equation is

$$l(t) = \begin{bmatrix} 0 \\ 1 \end{bmatrix} + t \begin{bmatrix} 4 \\ 2 \end{bmatrix}.$$

The implicit and parametric forms both allow an infinite number of representations for the same line. In fact, in the example we just finished, the loop

$$\text{parametric} \rightarrow \text{implicit} \rightarrow \text{parametric}$$

produced two different parametric forms. We started with

$$\mathbf{l}(t) = \begin{bmatrix} 2 \\ 2 \end{bmatrix} + t \begin{bmatrix} 4 \\ 2 \end{bmatrix},$$

and ended with

$$\mathbf{l}(t) = \begin{bmatrix} 0 \\ 1 \end{bmatrix} + t \begin{bmatrix} 4 \\ 2 \end{bmatrix}.$$

We could have just as easily generated the line

$$\mathbf{l}(t) = \begin{bmatrix} 0 \\ 1 \end{bmatrix} + t \begin{bmatrix} -4 \\ -2 \end{bmatrix},$$

if $\mathbf{v}$ was formed with the rule

$$\mathbf{v} = \begin{bmatrix} -b \\ a \end{bmatrix}.$$

Sketch 3.6 illustrates the first and third parametric representations of this line.

These three parametric forms represent the same line! However, the manner in which the lines will be *traced* will differ. This is referred to as the *parametrization* of the line. We already encountered this concept in Section 3.2.

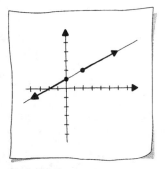

Sketch 3.6.
Two parametric representations for the same line.

3.6 Distance of a Point to a Line

If you are given a point $\mathbf{r}$ and a line $\mathbf{l}$, how far is that point from the line? This problem arises frequently. For example, as in Figure 3.4, a line has been fit to point data. In order to measure how well this line *approximates* the data, it is necessary to check the distance of each point to the line. It should be intuitively clear that the distance $d(\mathbf{r}, \mathbf{l})$ of a point to a line is the *perpendicular distance*.

3.6.1 Starting with an Implicit Line

Suppose our problem is formulated as follows:

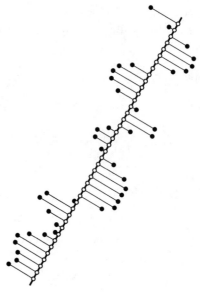

Figure 3.4.
Point to line distance: measuring the distance of each point to the line.

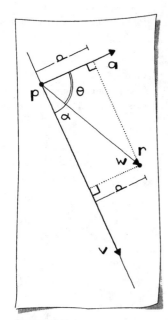

Sketch 3.7.
Distance point to line.

Given: A line l in implicit form, defined by the coefficients $a, b,$ and $c,$ and a point $\mathbf{r}$.

Find: $d(\mathbf{r}, l)$, or d for brevity.

Solution:

$$d = \frac{ar_1 + br_2 + c}{\|\mathbf{a}\|},$$

or in vector notation

$$d = \frac{\mathbf{a} \cdot (\mathbf{r} - \mathbf{p})}{\|\mathbf{a}\|}.$$

Let's investigate why this is so. Recall that the implicit equation of a line was derived through use of the dot product

$$\mathbf{a} \cdot (\mathbf{x} - \mathbf{p}) = 0,$$

as in (3.3); a line is given by a point $\mathbf{p}$ and a vector $\mathbf{a}$ normal to the line. Any point $\mathbf{x}$ on the line will satisfy this equality.

As in Sketch 3.7, we now consider a point $\mathbf{r}$ that is clearly not on the line. As a result, the equality will not be satisfied; however, let's assign a value v to the left-hand side:

$$v = \mathbf{a} \cdot (\mathbf{r} - \mathbf{p}).$$

To simplify, define $\mathbf{w} = \mathbf{r} - \mathbf{p}$, as in Sketch 3.7. Recall the definition of the dot product in (2.19) as

$$\mathbf{v} \cdot \mathbf{w} = \|\mathbf{v}\| \|\mathbf{w}\| \cos \theta.$$

Thus, the expression for v becomes

$$v = \mathbf{a} \cdot \mathbf{w} = \|\mathbf{a}\| \|\mathbf{w}\| \cos(\theta). \qquad (3.10)$$

The right triangle in Sketch 3.7 allows for an expression for $\cos(\theta)$ as

$$\cos(\theta) = \frac{d}{\|\mathbf{w}\|}.$$

Substituting this into (3.10), we have

$$v = \|\mathbf{a}\| d.$$

This indicates that the actual distance of $\mathbf{r}$ to the line is

$$d = \frac{v}{\|\mathbf{a}\|} = \frac{\mathbf{a} \cdot (\mathbf{r} - \mathbf{p})}{\|\mathbf{a}\|} = \frac{ar_1 + br_2 + c}{\|\mathbf{a}\|}. \qquad (3.11)$$

If many points will be checked against a line, it is advantageous to store the line in *point normal form*. This means that $\|\mathbf{a}\| = 1$. This will eliminate the division in (3.11).

Example 3.5

Start with the line and the point

$$\mathbf{l}: 4x_1 + 2x_2 - 8 = 0; \qquad \mathbf{r} = \begin{bmatrix} 5 \\ 3 \end{bmatrix}.$$

Find the distance from $\mathbf{r}$ to the line. (Draw your own sketch for this example, similar to Sketch 3.7.)

First, calculate
$$\|\mathbf{a}\| = \sqrt{4^2 + 2^2} = 2\sqrt{5}.$$

Then the distance is

$$d(\mathbf{r}, \mathbf{l}) = \frac{4 \times 5 + 2 \times 3 - 8}{2\sqrt{5}} = \frac{9}{\sqrt{5}} \approx 4.02.$$

As another exercise, let's rewrite the line in point normal form with coefficients

$$\hat{a} = \frac{4}{2\sqrt{5}} = \frac{2}{\sqrt{5}},$$

$$\hat{b} = \frac{2}{2\sqrt{5}} = \frac{1}{\sqrt{5}},$$

$$\hat{c} = \frac{c}{\|\mathbf{a}\|} = \frac{-8}{2\sqrt{5}} = -\frac{4}{\sqrt{5}},$$

thus making the point normal form of the line

$$\frac{2}{\sqrt{5}}x_1 + \frac{1}{\sqrt{5}}x_2 - \frac{4}{\sqrt{5}} = 0.$$

3.6.2   Starting with a Parametric Line

Alternatively, suppose our problem is formulated as follows:

Given: A line l in parametric form, defined by a point **p** and a vector **v**, and a point **r**.

Find: $d(\mathbf{r}, \mathbf{l})$, or d for brevity. Again, this is illustrated in Sketch 3.7.

Solution: Form the vector $\mathbf{w} = \mathbf{r} - \mathbf{p}$. Use the relationship

$$d = \|\mathbf{w}\| \sin(\alpha).$$

Later in Section 10.2, we will see how to express $\sin(\alpha)$ directly in terms of **v** and **w**; for now, we express it in terms of the cosine:

$$\sin(\alpha) = \sqrt{1 - \cos(\alpha)^2},$$

and as before

$$\cos(\alpha) = \frac{\mathbf{v} \cdot \mathbf{w}}{\|\mathbf{v}\|\|\mathbf{w}\|}.$$

Thus, we have defined the distance d.

Example 3.6

We'll use the same line as in the previous example, but now it will be given in parametric form as

$$\mathbf{l}(t) = \begin{bmatrix} 0 \\ 4 \end{bmatrix} + t \begin{bmatrix} 2 \\ -4 \end{bmatrix}.$$

We'll also use the same point

$$\mathbf{r} = \begin{bmatrix} 5 \\ 3 \end{bmatrix}.$$

Add any new vectors for this example to the sketch you drew for the previous example.

First create the vector

$$\mathbf{w} = \begin{bmatrix} 5 \\ 3 \end{bmatrix} - \begin{bmatrix} 0 \\ 4 \end{bmatrix} = \begin{bmatrix} 5 \\ -1 \end{bmatrix}.$$

Next calculate $\|\mathbf{w}\| = \sqrt{26}$ and $\|\mathbf{v}\| = \sqrt{20}$. Compute

$$\cos(\alpha) = \frac{\begin{bmatrix} 2 \\ -4 \end{bmatrix} \cdot \begin{bmatrix} 5 \\ -1 \end{bmatrix}}{\sqrt{26}\sqrt{20}} \approx 0.614.$$

Thus, the distance to the line becomes

$$d(\mathbf{r}, \mathbf{l}) \approx \sqrt{26}\sqrt{1 - (0.614)^2} \approx 4.02,$$

which rightly produces the same result as the previous example.

3.7 The Foot of a Point

Section 3.6 detailed how to calculate the distance of a point from a line. A new question arises: which point on the line is closest to the point? This point will be called the *foot* of the given point.

If you are given a line in implicit form, it is best to convert it to parametric form for this problem. This illustrates how the implicit form is handy for *testing* if a point is on the line, however it is not as handy for *finding* points on the line.

The problem at hand is thus:

Given: A line l in parametric form, defined by a point $\mathbf{p}$ and a vector $\mathbf{v}$, and another point $\mathbf{r}$.

Find: The point $\mathbf{q}$ on the line which is closest to $\mathbf{r}$. (See Sketch 3.8.)

Solution: The point $\mathbf{q}$ can be defined as

$$\mathbf{q} = \mathbf{p} + t\mathbf{v}, \tag{3.12}$$

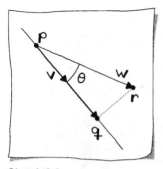

Sketch 3.8.

Closest point **q** on line to point **r**.

so our problem is solved once we have found the scalar factor t. From Sketch 3.8, we see that

$$\cos(\theta) = \frac{\|t\mathbf{v}\|}{\|\mathbf{w}\|}.$$

Using

$$\cos(\theta) = \frac{\mathbf{v} \cdot \mathbf{w}}{\|\mathbf{v}\|\|\mathbf{w}\|},$$

we find

$$t = \frac{\mathbf{v} \cdot \mathbf{w}}{\|\mathbf{v}\|^2}.$$

Example 3.7

Given: The parametric line l defined as

$$\mathbf{l}(t) = \begin{bmatrix} 0 \\ 1 \end{bmatrix} + t \begin{bmatrix} 0 \\ 2 \end{bmatrix},$$

and point

$$\mathbf{r} = \begin{bmatrix} 3 \\ 4 \end{bmatrix}.$$

Find: The point **q** on l that is closest to **r**. This example is easy enough to find the answer by simply drawing a sketch, but let's go through the steps.

Solution: Define the vector

$$\mathbf{w} = \mathbf{r} - \mathbf{p} = \begin{bmatrix} 3 \\ 4 \end{bmatrix} - \begin{bmatrix} 0 \\ 1 \end{bmatrix} = \begin{bmatrix} 3 \\ 3 \end{bmatrix}.$$

Compute $\mathbf{v} \cdot \mathbf{w} = 6$ and $\|\mathbf{v}\| = 2$. Thus, $t = 3/2$ and

$$\mathbf{q} = \begin{bmatrix} 0 \\ 4 \end{bmatrix}.$$

Try this example with $\mathbf{r} = \begin{bmatrix} 2 \\ -1 \end{bmatrix}$.

3.8   A Meeting Place: Computing Intersections

Finding a point in common between two lines is done many times over in a CAD or graphics package. Take for example Figure 3.5: the top part of the figure shows a great number of intersection lines. In order to color some of the areas, as in the bottom part of the figure, it is necessary to know the intersection points. Intersection problems

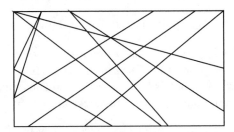

Figure 3.5.
Intersecting lines: the top figure may be drawn without knowing where the shown lines intersect. By finding line/line intersections (bottom), it is possible to color areas— creating an artistic image!

arise in many other applications. The first question to ask is what type of information do you want:

- Do you want to know merely whether the lines intersect?

- Do you want to know the point at which they intersect?

- Do you want a parameter value on one or both lines for the intersection point?

The particular question(s) you want to answer along with the line representation(s) will determine the best method for solving the intersection problem.

3.8.1 Parametric and Implicit

We then want to solve the following:

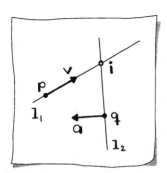

Sketch 3.9.

Parametric and implicit line intersection.

Given: Two lines l_1 and l_2:

$$l_1 : \quad l_1(t) = \mathbf{p} + t\mathbf{v}$$
$$l_2 : \quad ax_1 + bx_2 + c = 0.$$

Find: The intersection point **i**. See Sketch 3.9 for an illustration.[1]

Solution: We will approach the problem by finding the specific parameter $\hat{t}$ with respect to l_1 of the intersection point.

This intersection point, when inserted into the equation of l_2, will cause the left-hand side to evaluate to zero:

$$a[p_1 + \hat{t}v_1] + b[p_2 + \hat{t}v_2] + c = 0.$$

This is one equation and one unknown! Just solve for $\hat{t}$,

$$\hat{t} = \frac{-c - ap_1 - bp_2}{av_1 + bv_2}, \tag{3.13}$$

then $\mathbf{i} = l(\hat{t})$.

But wait—we must check if the denominator of (3.13) is zero before carrying out this calculation. Besides causing havoc numerically, what else does a zero denominator infer? The denominator

$$\text{denom} = av_1 + bv_2$$

can be rewritten as

$$\text{denom} = \mathbf{a} \cdot \mathbf{v}.$$

We know from (2.7) that a zero dot product implies that two vectors are perpendicular. Since **a** is perpendicular to the line l_2 in implicit form, the lines are *parallel* if

$$\mathbf{a} \cdot \mathbf{v} = 0.$$

Of course, we always check for equality within a tolerance! A physically meaningful tolerance is best. Thus, it is better to check the quantity

$$\cos(\theta) = \frac{\mathbf{a} \cdot \mathbf{v}}{\|\mathbf{a}\|\|\mathbf{v}\|}; \tag{3.14}$$

the tolerance will be the cosine of an angle. It usually suffices to use a tolerance between $\cos(0.1°)$ and $\cos(0.5°)$. Angle tolerances are particularly nice to have because they are *dimension independent*.[2]

---

[1]In Section 3.3, we studied the conversion from the geometric elements of a point **q** and perpendicular vector **a** to the implicit line coefficients.

[2]Note that we do not need to use the actual angle, just the cosine of the angle.

If the test in (3.14) indicates the lines are parallel, then we might want to determine if the lines are identical. By simply plugging in the coordinates of $\mathbf{p}$ into the equation of l_2, and computing, we get

$$d = \frac{ap_1 + bp_2 + c}{\|\mathbf{a}\|}.$$

If d is equal to zero (within tolerance), then the lines are identical.

Example 3.8

Given: Two lines l_1 and l_2,

$$l_1 : \quad l_1(t) = \begin{bmatrix} 0 \\ 3 \end{bmatrix} + t \begin{bmatrix} -2 \\ -1 \end{bmatrix}$$
$$l_2 : \quad 2x_1 + x_2 - 8 = 0.$$

Find: The intersection point $\mathbf{i}$. Create your own sketch and try to predict what the answer should be!

Solution: Find the parameter $\hat{t}$ for l_1 as given in (3.13). First check the denominator:

$$\text{denom} = 2 \times (-2) + 1 \times (-1) = -5.$$

This is not zero, so we proceed to find

$$\hat{t} = \frac{8 - 2 \times 0 - 1 \times 3}{-5} = -1.$$

Plug this parameter value into l_1 to find the intersection point:

$$l_1(-1) = \begin{bmatrix} 0 \\ 3 \end{bmatrix} + -1 \begin{bmatrix} -2 \\ -1 \end{bmatrix} = \begin{bmatrix} 2 \\ 4 \end{bmatrix}.$$

3.8.2   Both Parametric

Another method for finding the intersection of two lines arises by using the parametric form for both, illustrated in Sketch 3.10.

Given: Two lines in parametric form,

$$l_1 : \quad l_1(t) = \mathbf{p} + t\mathbf{v}$$
$$l_2 : \quad l_2(s) = \mathbf{q} + s\mathbf{w}.$$

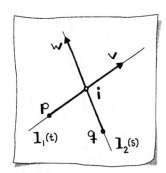

Sketch 3.10.
Intersection of lines in parametric form.

Note that we use two different parameters, t and s, here. This is because the lines are totally independent of each other.

Find: The intersection point $\mathbf{i}$.

Solution: We need two parameter values $\hat{t}$ and $\hat{s}$ such that

$$\mathbf{p} + \hat{t}\mathbf{v} = \mathbf{q} + \hat{s}\mathbf{w}.$$

This may be rewritten as

$$\hat{t}\mathbf{v} - \hat{s}\mathbf{w} = \mathbf{q} - \mathbf{p}. \tag{3.15}$$

We have two equations (one for each coordinate) and two unknowns $\hat{t}$ and $\hat{s}$. To solve for the unknowns, we could formulate an expression for $\hat{t}$ using the first equation, and substitute this expression into the second equation. This then generates a solution for $\hat{s}$. Use this solution in the expression for $\hat{t}$, and solve for $\hat{t}$. (Equations like this are treated systematically in Chapter 5.) Once we have $\hat{t}$ and $\hat{s}$, the intersection point is found by inserting one of these values into its respective parametric line equation.

If the vectors $\mathbf{v}$ and $\mathbf{w}$ are linearly dependent, as discussed in Section 2.6, then it will not be possible to find a unique $\hat{t}$ and $\hat{s}$. The lines are parallel and possibly identical.

Example 3.9

Given: Two lines $\mathbf{l}_1$ and $\mathbf{l}_2$,

$$\mathbf{l}_1: \quad \mathbf{l}_1(t) = \begin{bmatrix} 0 \\ 3 \end{bmatrix} + t \begin{bmatrix} -2 \\ -1 \end{bmatrix}$$

$$\mathbf{l}_2: \quad \mathbf{l}_2(s) = \begin{bmatrix} 4 \\ 0 \end{bmatrix} + s \begin{bmatrix} -1 \\ 2 \end{bmatrix}.$$

Find: The intersection point $\mathbf{i}$. This means that we need to find $\hat{t}$ and $\hat{s}$ such that $\mathbf{l}_1(\hat{t}) = \mathbf{l}_2(\hat{s})$.[3] Again, create your own sketch and try to predict the answer!

Solution: Set up the two equations with two unknowns as in (3.15).

$$\hat{t}\begin{bmatrix} -2 \\ -1 \end{bmatrix} - \hat{s}\begin{bmatrix} -1 \\ 2 \end{bmatrix} = \begin{bmatrix} 4 \\ 0 \end{bmatrix} - \begin{bmatrix} 0 \\ 3 \end{bmatrix}.$$

---

[3]Really we only need $\hat{t}$ or $\hat{s}$ to find the intersection point.

Use the methods in Chapter 5 to solve these equations, resulting in $\hat{t} = -1$ and $\hat{s} = 2$. Plug these values into the line equations to verify the same intersection point is produced for each:

$$l_1(-1) = \begin{bmatrix} 0 \\ 3 \end{bmatrix} + -1 \begin{bmatrix} -2 \\ -1 \end{bmatrix} = \begin{bmatrix} 2 \\ 4 \end{bmatrix}$$

$$l_2(2) = \begin{bmatrix} 4 \\ 0 \end{bmatrix} + 2 \begin{bmatrix} -1 \\ 2 \end{bmatrix} = \begin{bmatrix} 2 \\ 4 \end{bmatrix}.$$

3.8.3 Both Implicit

And yet a third method:

Given: Two lines in implicit form,

$$l_1 : \quad ax_1 + bx_2 + c = 0,$$
$$l_2 : \quad \bar{a}x_1 + \bar{b}x_2 + \bar{c} = 0.$$

As illustrated in Sketch 3.11, each line is geometrically given in terms of a point and a vector perpendicular to the line.

Find: The intersection point

$$\mathbf{i} = \hat{\mathbf{x}} = \begin{bmatrix} \hat{x}_1 \\ \hat{x}_2 \end{bmatrix}$$

that simultaneously satisfies l_1 and l_2.

Solution: We have two equations

$$a\hat{x}_1 + b\hat{x}_2 = -c, \qquad (3.16)$$
$$\bar{a}\hat{x}_1 + \bar{b}\hat{x}_2 = -\bar{c} \qquad (3.17)$$

with two unknowns, $\hat{x}_1$ and $\hat{x}_2$. Equations like this are solved in Chapter 5.

If the lines are parallel then it will not be possible to find $\hat{\mathbf{x}}$. This means that $\mathbf{a}$ and $\bar{\mathbf{a}}$, as in Sketch 3.11, are linearly dependent.

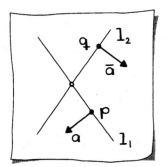

Sketch 3.11.
Intersection of two lines in implicit form.

Example 3.10

Given: Two lines l_1 and l_2,

$$l_1 : \quad x_1 - 2x_2 + 6 = 0$$
$$l_2 : \quad 2x_1 + x_2 - 8 = 0.$$

Find: The intersection point $\hat{\mathbf{x}}$ as above. Create your own sketch and try to predict the answer!

Solution: Reformulate the equations for l_1 and l_2 as in (3.16) and (3.17). Using the techniques in Chapter 5, you will find that

$$\hat{x} = \begin{bmatrix} 2 \\ 4 \end{bmatrix}.$$

Plug this point into the equations for l_1 and l_2 to verify.

- parametric form of a line
- linear interpolation
- point normal form
- implicit form of a line
- explicit form of a line
- equation of a line through two points
- equation of a line defined by a point and a vector parallel to the line
- equation of a line defined by a point and a vector perpendicular to the line
- distance of a point to a line
- line form conversions
- foot of a point
- intersection of lines

3.9 Exercises

1. Find the *computationally* best way to organize the evaluation of n equally spaced points on a line.

2. Find the equation for a line in implicit form which passes through the points

$$\mathbf{p} = \begin{bmatrix} -2 \\ 0 \end{bmatrix} \quad \text{and} \quad \mathbf{q} = \begin{bmatrix} 0 \\ -1 \end{bmatrix}.$$

3. Test if the following points lie on the line defined in Exercise 2:

$$\begin{bmatrix} 0 \\ 0 \end{bmatrix} \quad \begin{bmatrix} -4 \\ 1 \end{bmatrix} \quad \begin{bmatrix} 5 \\ 1 \end{bmatrix} \quad \begin{bmatrix} -3 \\ -1 \end{bmatrix}.$$

4. For the points in Exercise 3, if a point does not lie on the line, calculate the distance from the line.

5. Redefine the line in Exercise 2 using $-\mathbf{a}$ (as per (3.3)). Recompute the distance of each point in Exercise 3 to the line.

6. Given two lines: l_1 defined by points

$$\begin{bmatrix} 1 \\ 0 \end{bmatrix} \quad \text{and} \quad \begin{bmatrix} 0 \\ 3 \end{bmatrix},$$

and l_2 defined by points

$$\begin{bmatrix} -1 \\ 6 \end{bmatrix} \quad \text{and} \quad \begin{bmatrix} -4 \\ 1 \end{bmatrix},$$

find the intersection point using each of the three methods in Section 3.8.

7. Find the intersection of the lines

$$l_1: \quad l_1(t) = \begin{bmatrix} 0 \\ -1 \end{bmatrix} + t \begin{bmatrix} 1 \\ 1 \end{bmatrix} \quad \text{and}$$
$$l_2: \quad -x_1 + x_2 + 1 = 0.$$

8. Find the intersection of the lines

$$l_1: \quad -x_1 + x_2 + 1 = 0 \quad \text{and}$$
$$l_2: \quad l(t) = \begin{bmatrix} 2 \\ 2 \end{bmatrix} + t \begin{bmatrix} 2 \\ 2 \end{bmatrix}.$$

9. Find the closest point on the line

$$l(t) = \begin{bmatrix} 2 \\ 2 \end{bmatrix} + t \begin{bmatrix} 2 \\ 2 \end{bmatrix}$$

to the point

$$\mathbf{r} = \begin{bmatrix} 0 \\ 2 \end{bmatrix}.$$

(Does this problem look familiar? Compare it to Exercise 10 in Chapter 2.)

10. Define a line $l(t)$ defined by points

$$\mathbf{p} = \begin{bmatrix} 0 \\ 1 \end{bmatrix} \quad \text{and} \quad \mathbf{q} = \begin{bmatrix} 4 \\ 2 \end{bmatrix},$$

such that $l(0) = \mathbf{p}$.

11. Again, consider the points

$$\mathbf{p} = \begin{bmatrix} 0 \\ 1 \end{bmatrix} \quad \text{and} \quad \mathbf{q} = \begin{bmatrix} 4 \\ 2 \end{bmatrix}.$$

Now define a line $\mathbf{m}(t)$ which is perpendicular to l and which passes through the midpoint of $\mathbf{p}$ and $\mathbf{q}$.

4

Changing Shapes: Linear Maps in 2D

Figure 4.1.
Linear maps in 2D: an interesting geometric figure constructed by applying 2D linear maps to a square.

Geometry always has two parts to it: one part is the description of the objects that can be generated; the other investigates how these objects can be changed (or transformed). Any object formed by several vectors may be mapped to an arbitrarily bizarre curved or distorted object—here, we are interested in those maps that map vectors to vectors and are "benign" in some well-defined sense. All these maps may be described using the tools of matrix operations. An interesting pattern is generated from a simple square in Figure 4.1 by such "benign" 2D linear maps.

4.1   Skew Target Boxes

In Section 1.1, we saw how to map an object from a unit square to a rectangular target box. We will now look at the part of that mapping which is a linear map, and modify it slightly to make it more general and useful.

First, our unit square will be defined by vectors $\mathbf{e}_1$ and $\mathbf{e}_2$. Thus, a vector $\mathbf{v}$ in this $[\mathbf{e}_1, \mathbf{e}_2]$-system is defined as

$$\mathbf{v} = v_1\mathbf{e}_1 + v_2\mathbf{e}_2. \tag{4.1}$$

If we focus on mapping vectors to vectors, then we will limit the target box to having a lower-left corner at the origin. In Chapter 6 we will re-introduce the idea of an generally positioned target box. Instead of specifying two extreme points for a rectangular target box, we will describe a parallelogram target box by two vectors $\mathbf{a}_1, \mathbf{a}_2$, defining an $[\mathbf{a}_1, \mathbf{a}_2]$-system. A vector $\mathbf{v}$ is now mapped to a vector $\mathbf{v}'$ by

$$\mathbf{v}' = v_1\mathbf{a}_1 + v_2\mathbf{a}_2, \tag{4.2}$$

as illustrated by Sketch 4.1. This simply states that we duplicate the $[\mathbf{e}_1, \mathbf{e}_2]$-geometry in the $[\mathbf{a}_1, \mathbf{a}_2]$-system: $\mathbf{v}'$ has the same components in the new system as $\mathbf{v}$ did in the old one. Reviewing a definition from Section 2.6, we recall that any combination $c\mathbf{u} + d\mathbf{w}$ of two vectors $\mathbf{u}$ and $\mathbf{w}$ is called a *linear combination.*

The components of a subscripted vector will be written with a double subscript as

$$\mathbf{a}_1 = \begin{bmatrix} a_{1,1} \\ a_{2,1} \end{bmatrix}.$$

The vector component index precedes the vector subscript. The next section will clarify the reason for this notation.

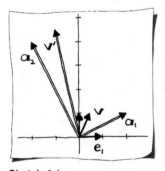

Sketch 4.1.

A skew target box defined by $\mathbf{a}_1$ and $\mathbf{a}_2$.

Example 4.1

Let the origin and

$$\mathbf{a}_1 = \begin{bmatrix} 2 \\ 1 \end{bmatrix}, \qquad \mathbf{a}_2 = \begin{bmatrix} -2 \\ 4 \end{bmatrix}$$

define a new coordinate system, and let

$$\mathbf{v} = \begin{bmatrix} 1/2 \\ 1 \end{bmatrix}$$

be a vector in the $[\mathbf{e}_1, \mathbf{e}_2]$-system. What can we say about the vector $\mathbf{v}'$ in the $[\mathbf{a}_1, \mathbf{a}_2]$-system which has the same components relative to it? We must be able to write $\mathbf{v}'$ as

$$\mathbf{v}' = \frac{1}{2} \times \begin{bmatrix} 2 \\ 1 \end{bmatrix} + 1 \times \begin{bmatrix} -2 \\ 4 \end{bmatrix} = \begin{bmatrix} -1 \\ 9/2 \end{bmatrix}. \qquad (4.3)$$

Thus $\mathbf{v}'$ has components

$$\begin{bmatrix} 1/2 \\ 1 \end{bmatrix}$$

with respect to the $[\mathbf{a}_1, \mathbf{a}_2]$-system; with respect to the $[\mathbf{e}_1, \mathbf{e}_2]$-system, it has coordinates

$$\begin{bmatrix} -1 \\ 9/2 \end{bmatrix}.$$

See Sketch 4.1 for an illustration.

4.2   The Matrix Form

The components for the vector $\mathbf{v}'$ in the $[\mathbf{e}_1, \mathbf{e}_2]$-system from Example 4.1 are expressed as

$$\begin{bmatrix} -1 \\ 9/2 \end{bmatrix} = \frac{1}{2} \times \begin{bmatrix} 2 \\ 1 \end{bmatrix} + 1 \times \begin{bmatrix} -2 \\ 4 \end{bmatrix}. \qquad (4.4)$$

This is strictly an equation between vectors. It invites a more concise notation using *matrix notation*:

$$\begin{bmatrix} -1 \\ 9/2 \end{bmatrix} = \begin{bmatrix} 2 & -2 \\ 1 & 4 \end{bmatrix} \begin{bmatrix} 1/2 \\ 1 \end{bmatrix}. \qquad (4.5)$$

The 2×2 array in this equation is called a *matrix*. It has two columns, corresponding to the vectors $\mathbf{a}_1$ and $\mathbf{a}_2$. It also has two rows, namely the first row with entries $2, -2$ and the second one with $1, 4$.

In general, an equation like this one has the form

$$\mathbf{v}' = \begin{bmatrix} a_{1,1} & a_{1,2} \\ a_{2,1} & a_{2,2} \end{bmatrix} \begin{bmatrix} v_1 \\ v_2 \end{bmatrix}, \tag{4.6}$$

or,

$$\mathbf{v}' = A\mathbf{v}, \tag{4.7}$$

where A is the 2×2 matrix. The vector $\mathbf{v}'$ is called the *image* of $\mathbf{v}$. The *linear map* is described by the matrix A—we may think of A as being the map's coordinates. Just as we do for points and vectors, we will also refer to the linear map itself by A.

The elements $a_{1,1}$ and $a_{2,2}$ form the *diagonal* of the matrix. The product $A\mathbf{v}$ has two components, each of which is obtained as a dot product between the corresponding row of the matrix and $\mathbf{v}$. In full generality, we have

$$\begin{bmatrix} a_{1,1} & a_{1,2} \\ a_{2,1} & a_{2,2} \end{bmatrix} \begin{bmatrix} v_1 \\ v_2 \end{bmatrix} = \begin{bmatrix} v_1 a_{1,1} + v_2 a_{1,2} \\ v_1 a_{2,1} + v_2 a_{2,2} \end{bmatrix}.$$

For example,

$$\begin{bmatrix} 0 & -1 \\ 2 & 4 \end{bmatrix} \begin{bmatrix} -1 \\ 4 \end{bmatrix} = \begin{bmatrix} -4 \\ 14 \end{bmatrix}.$$

Another note on notation. Coordinate systems, such as the $[\mathbf{e}_1, \mathbf{e}_2]$-system, can be interpreted as a matrix with columns $\mathbf{e}_1$ and $\mathbf{e}_2$. Thus,

$$[\mathbf{e}_1, \mathbf{e}_2] \equiv \begin{bmatrix} 1 & 0 \\ 0 & 1 \end{bmatrix}.$$

There is a neat way to write the matrix-times-vector algebra in a way that facilitates manual computation. As explained above, every entry in the resulting vector is a dot product of the input vector and a row of the matrix. Let's arrange this as follows:

$$
\begin{array}{cc|c}
 & & 2 \\
 & & 1/2 \\
\hline
2 & -2 & 3 \\
1 & 4 & 4
\end{array}
$$

Each entry of the resulting vector is now at the intersection of the corresponding matrix row and the input vector, which is written as a column. As you multiply and then add the terms in your dot products, this scheme guides you to the correct position in the result automatically!

4.3   More about Matrices

Matrices were first introduced by H. Grassmann in 1844. They became the basis of *linear algebra*. Most of their properties can be studied by just considering the humble 2×2 case.

We will now demonstrate several matrix properties. For the sake of concreteness, we shall use the example

$$A = \begin{bmatrix} 2 & 1 \\ -1 & 3 \end{bmatrix}, \quad \mathbf{u} = \begin{bmatrix} 1 \\ 2 \end{bmatrix}, \quad \mathbf{v} = \begin{bmatrix} -1 \\ 4 \end{bmatrix}.$$

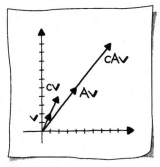

Sketch 4.2.
Matrices preserve scalings.

We have already encountered 2×2 matrices. Sometimes it is convenient if we also think of the vector $\mathbf{v}$ as a matrix. It is a matrix with one column and two rows!

We may multiply all elements of a matrix by one factor; we then say that we have multiplied the matrix by that factor. Using our example, we may multiply the matrix A by a factor, say 2:

$$2 \times \begin{bmatrix} 2 & 1 \\ -1 & 3 \end{bmatrix} = \begin{bmatrix} 4 & 2 \\ -2 & 6 \end{bmatrix}.$$

Matrices are related to *linear* operations, i.e., multiplication by scalar factors or addition of vectors. For example, if we scale a vector by a factor c, then its image will also be scaled by c:

$$A(c\mathbf{v}) = cA\mathbf{v}.$$

Sketch 4.2 shows this for an example with $c = 2$.

Example 4.2

Here are the computations that go along with Sketch 4.2:

$$\begin{bmatrix} 2 & 1 \\ -1 & 3 \end{bmatrix} \times 2 \times \begin{bmatrix} 1 \\ 2 \end{bmatrix} = 2 \times \begin{bmatrix} 2 & 1 \\ -1 & 3 \end{bmatrix} \begin{bmatrix} 1 \\ 2 \end{bmatrix} = \begin{bmatrix} 8 \\ 10 \end{bmatrix}.$$

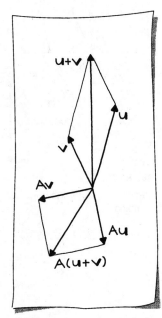

Sketch 4.3.
Matrices preserve sums.

Matrices also preserve summations:

$$A(\mathbf{u} + \mathbf{v}) = A\mathbf{u} + A\mathbf{v}$$

(see Sketch 4.3). This is also called the *distributive law*.

The last two properties taken together imply that matrices preserve *linear combinations*:

$$A(a\mathbf{u} + b\mathbf{v}) = aA\mathbf{u} + bA\mathbf{v}. \qquad (4.8)$$

Example 4.3

The following illustrates that matrices preserve linear combinations

$$\begin{bmatrix} -1 & 1/2 \\ 0 & -1/2 \end{bmatrix} \left(3 \begin{bmatrix} 1 \\ 2 \end{bmatrix} + 2 \begin{bmatrix} -1 \\ 4 \end{bmatrix} \right) =$$

$$3 \begin{bmatrix} -1 & 1/2 \\ 0 & -1/2 \end{bmatrix} \begin{bmatrix} 1 \\ 2 \end{bmatrix} + 2 \begin{bmatrix} -1 & 1/2 \\ 0 & -1/2 \end{bmatrix} \begin{bmatrix} -1 \\ 4 \end{bmatrix} = \begin{bmatrix} 6 \\ -7 \end{bmatrix}.$$

Preservation of linear combinations is a key property of matrices—we will make substantial use of it throughout this book. Sketch 4.4 illustrates Example 4.3.

Another matrix operation is *matrix addition*. Two matrices A and B may be added by adding corresponding elements:

$$\begin{bmatrix} a_{1,1} & a_{1,2} \\ a_{2,1} & a_{2,2} \end{bmatrix} + \begin{bmatrix} b_{1,1} & b_{1,2} \\ b_{2,1} & b_{2,2} \end{bmatrix} = \begin{bmatrix} a_{1,1} + b_{1,1} & a_{1,2} + b_{1,2} \\ a_{2,1} + b_{2,1} & a_{2,2} + b_{2,2} \end{bmatrix}. \qquad (4.9)$$

Notice that the matrices must be of the same dimensions; this is not true for matrix multiplication.

Using matrix addition, we may write

$$A\mathbf{v} + B\mathbf{v} = (A + B)\mathbf{v}.$$

This works because of the very simple definition of matrix addition. This is also called the distributive law.

Yet another matrix operation is forming the *transpose matrix*. It is denoted by A^{T} and is formed by interchanging the rows and columns of A: the first row of A^{T} is A's first column, and the second row of A^{T} is A's second column. For example, if

$$A = \begin{bmatrix} 1 & -2 \\ 3 & 5 \end{bmatrix}, \quad \text{then} \quad A^{\mathrm{T}} = \begin{bmatrix} 1 & 3 \\ -2 & 5 \end{bmatrix}.$$

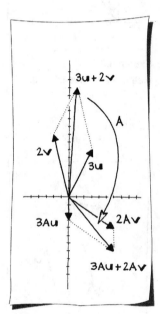

Sketch 4.4.

Matrices preserve linear combinations.

Since we may think of a vector $\mathbf{v}$ as a matrix, we should be able to find $\mathbf{v}$'s transpose. Not very hard: it is a vector with one row and two columns,

$$\mathbf{v} = \begin{bmatrix} -1 \\ 4 \end{bmatrix} \quad \text{and} \quad \mathbf{v}^{\mathrm{T}} = \begin{bmatrix} -1 & 4 \end{bmatrix}.$$

It is not hard to confirm that

$$[A + B]^{\mathrm{T}} = A^{\mathrm{T}} + B^{\mathrm{T}}. \tag{4.10}$$

Two more straightforward identities are

$$A^{\mathrm{T}^{\mathrm{T}}} = A, \quad \text{and} \quad [cA]^{\mathrm{T}} = cA^{\mathrm{T}}. \tag{4.11}$$

A *symmetric matrix* is a special matrix that we will encounter many times. A matrix A is symmetric if $A = A^{\mathrm{T}}$, for example

$$\begin{bmatrix} 5 & 8 \\ 8 & 5 \end{bmatrix}.$$

The columns of a matrix define an $[\mathbf{a}_1, \mathbf{a}_2]$-system. If the vectors $\mathbf{a}_1$ and $\mathbf{a}_2$ are linearly independent then the matrix is said to have *full rank*, or for the 2×2 case, the matrix has *rank* 2. If $\mathbf{a}_1$ and $\mathbf{a}_2$ are linearly dependent then the matrix has rank 1 (see Section 4.8). These two statements may be summarized as: the rank of a matrix equals the number of linearly independent column vectors.

4.4 Scalings

Consider the linear map given by

$$\mathbf{v}' = \begin{bmatrix} 1/2 & 0 \\ 0 & 1/2 \end{bmatrix} \mathbf{v} = \begin{bmatrix} v_1/2 \\ v_2/2 \end{bmatrix}. \tag{4.12}$$

This map will "reduce" $\mathbf{v}$ since $\mathbf{v}' = 1/2\mathbf{v}$. Its effect is illustrated in Figure 4.2. That figure—and more to follow—has two parts. The left part is a Phoenix whose feathers form rays that correspond to a sampling of unit vectors. The right part shows what happens if we map the Phoenix, and in turn the unit vectors, using the matrix from (4.12). In this figure we have drawn the $\mathbf{e}_1$ and $\mathbf{e}_2$ vectors,

$$\mathbf{e}_1 = \begin{bmatrix} 1 \\ 0 \end{bmatrix} \quad \text{and} \quad \mathbf{e}_2 = \begin{bmatrix} 0 \\ 1 \end{bmatrix},$$

Figure 4.2.
Scaling: a uniform scaling.

but in future figures we will not. Notice the positioning of these vectors relative to the Phoenix. Now in the right half, $\mathbf{e}_1$ and $\mathbf{e}_2$ have been mapped to the vectors $\mathbf{a}_1$ and $\mathbf{a}_2$. These are the column vectors of the matrix in (4.12),

$$\mathbf{a}_1 = \begin{bmatrix} 1/2 \\ 0 \end{bmatrix} \quad \text{and} \quad \mathbf{a}_2 = \begin{bmatrix} 0 \\ 1/2 \end{bmatrix}.$$

The Phoenix's shape provides a sense of orientation. In this example, the linear map did not change orientation, but more complicated maps will.

Next, consider

$$\mathbf{v}' = \begin{bmatrix} 2 & 0 \\ 0 & 2 \end{bmatrix} \mathbf{v}.$$

Now, $\mathbf{v}$ will be "enlarged."

In general, a scaling is defined by the operation

$$\mathbf{v}' = \begin{bmatrix} s_{1,1} & 0 \\ 0 & s_{2,2} \end{bmatrix} \mathbf{v}, \tag{4.13}$$

thus allowing for nonuniform scalings in the $\mathbf{e}_1$- and $\mathbf{e}_2$-direction. Figure 4.3 gives an example for $s_{1,1} = 1/2$ and $s_{2,2} = 2$.

A scaling affects the *area* of the object that is scaled. If we scale an object by $s_{1,1}$ in the $\mathbf{e}_1$-direction, then its area will be changed by a factor $s_{1,1}$. Similarly, it will change by a factor of $s_{2,2}$ when we

Figure 4.3.
Scaling: a nonuniform scaling.

apply that scaling to the $\mathbf{e}_2$-direction. The total effect is thus a factor of $s_{1,1}s_{2,2}$.

You can see this from Figure 4.2 by mentally constructing the square spanned by $\mathbf{e}_1$ and $\mathbf{e}_2$ and comparing its area to the rectangle spanned by the image vectors. It is also interesting to note that, in Figure 4.3, the scaling factors result in no change of area, although a distortion did occur.

4.5   Reflections

Consider the scaling

$$\mathbf{v}' = \begin{bmatrix} 1 & 0 \\ 0 & -1 \end{bmatrix} \mathbf{v}. \tag{4.14}$$

We may rewrite this as

$$\begin{bmatrix} v_1' \\ v_2' \end{bmatrix} = \begin{bmatrix} v_1 \\ -v_2 \end{bmatrix}.$$

The effect of this map is apparently a change in sign of the second component of $\mathbf{v}$, as shown in Figure 4.4. Geometrically, this means that the input vector $\mathbf{v}$ is reflected about the $\mathbf{e}_1$-axis, or the line $x_1 = 0$.

Figure 4.4.
Reflections: a reflection about the $\mathbf{e}_1$-axis.

Obviously, reflections like the one above are just a special case of scalings—previously we simply had not given much thought to negative scaling factors. However, a reflection takes a more general form, and it results in the mirror image of the vectors. Mathematically, a reflection maps each vector about a line through the origin.

The most common reflections are those about the coordinate axes, with one such example illustrated in Figure 4.4, and about the lines $x_1 = x_2$ and $x_1 = -x_2$. The reflection about the line $x_1 = x_2$ is achieved by the matrix

$$\mathbf{v}' = \begin{bmatrix} 0 & 1 \\ 1 & 0 \end{bmatrix} \mathbf{v} = \begin{bmatrix} v_2 \\ v_1 \end{bmatrix}.$$

Its effect is shown in Figure 4.5, that is, the components of the input vector are interchanged.

By inspection of the figures in this section, it appears that reflections do not change areas. But be careful—they do change the *sign* of the area due to a change in orientation. If we rotate $\mathbf{e}_1$ into $\mathbf{e}_2$, we move in a counterclockwise direction. Now, rotate

$$\mathbf{a}_1 = \begin{bmatrix} 0 \\ 1 \end{bmatrix} \quad \text{into} \quad \mathbf{a}_2 = \begin{bmatrix} 1 \\ 0 \end{bmatrix},$$

and notice that we move in a clockwise direction. This change in orientation is reflected in the sign of the area. We will examine this in detail in Section 4.10.

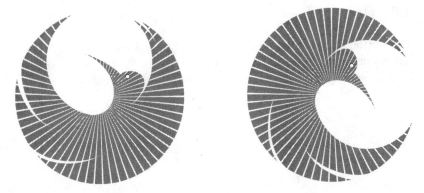

Figure 4.5.
Reflections: a reflection about the line $x_1 = x_2$.

The matrix

$$\mathbf{v}' = \begin{bmatrix} -1 & 0 \\ 0 & -1 \end{bmatrix} \mathbf{v},\qquad (4.15)$$

as seen in Figure 4.6, appears to be a reflection, but it is really a rotation of 180°. (Rotations are covered in Section 4.6.) If we rotate $\mathbf{a}_1$ into $\mathbf{a}_2$ we move in a counterclockwise direction, confirming that this is not a reflection.

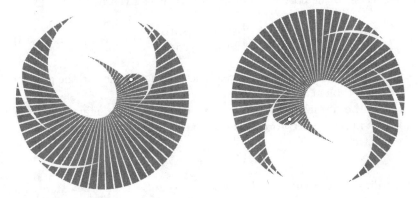

Figure 4.6.
Reflections: what appears to be a reflection about both axes is actually a rotation of 180°.

4.6   Rotations

The notion of rotating a vector around the origin is intuitively clear, but a corresponding matrix takes a few moments to construct. To keep it easy at the beginning, let us rotate the unit vector

$$\mathbf{e}_1 = \begin{bmatrix} 1 \\ 0 \end{bmatrix}$$

by α degrees, resulting in a new (rotated) vector

$$\mathbf{e}_1' = \begin{bmatrix} \cos \alpha \\ \sin \alpha \end{bmatrix}.$$

Notice that $\cos^2 \alpha + \sin^2 \alpha = 1$, thus this is a rotation. Consult Sketch 4.5 to convince yourself of this fact!

Thus, we need to find a matrix R which achieves

$$\begin{bmatrix} \cos \alpha \\ \sin \alpha \end{bmatrix} = \begin{bmatrix} r_{1,1} & r_{1,2} \\ r_{2,1} & r_{2,2} \end{bmatrix} \begin{bmatrix} 1 \\ 0 \end{bmatrix}.$$

Additionally, we know that $\mathbf{e}_2$ will rotate to

$$\mathbf{e}_2' = \begin{bmatrix} \sin \alpha \\ \cos \alpha \end{bmatrix}.$$

This leads to the correct *rotation matrix*; it is given by

$$R = \begin{bmatrix} \cos \alpha & -\sin \alpha \\ \sin \alpha & \cos \alpha \end{bmatrix}. \tag{4.16}$$

But let's verify that we have already found the solution to the general rotation problem.

Let $\mathbf{v}$ be an arbitrary vector. We claim that the matrix R from (4.16) will rotate it by α degrees to a new vector $\mathbf{v}'$. If this is so, then we must have

$$\mathbf{v} \cdot \mathbf{v}' = \|\mathbf{v}\|^2 \cos \alpha$$

according to the rules of dot products (see Section 2.7). Here, we made use of the fact that a rotation does not change the length of a vector, i.e., $\|\mathbf{v}\| = \|\mathbf{v}'\|$ and hence $\|\mathbf{v}\| \cdot \|\mathbf{v}'\| = \|\mathbf{v}\|^2$.

Since

$$\mathbf{v}' = \begin{bmatrix} v_1 \cos \alpha - v_2 \sin \alpha \\ v_1 \sin \alpha + v_2 \cos \alpha \end{bmatrix},$$

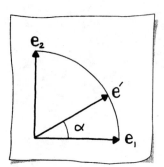

Sketch 4.5.

Rotating a unit vector.

 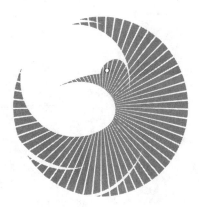

Figure 4.7.
Rotations: a rotation by 45°.

the dot product $\mathbf{v} \cdot \mathbf{v}'$ is given by

$$\begin{aligned}
\mathbf{v} \cdot \mathbf{v}' &= v_1^2 \cos\alpha - v_1 v_2 \sin\alpha + v_1 v_2 \sin\alpha + v_2^2 \cos\alpha \\
&= (v_1^2 + v_2^2)\cos\alpha \\
&= \|\mathbf{v}\|^2 \cos\alpha,
\end{aligned}$$

and all is shown! See Figure 4.7 for an illustration. There, $\alpha = 45°$, and the rotation matrix is thus given by

$$R = \begin{bmatrix} \sqrt{2}/2 & -\sqrt{2}/2 \\ \sqrt{2}/2 & \sqrt{2}/2 \end{bmatrix}.$$

Rotations are in a special class of transformations; these are called *rigid body motions*. See Section 5.5 for more details. Finally, it should come without saying that rotations do not change areas.

4.7   Shears

What map takes a rectangle to a parallelogram? Pictorially, one such map is shown in Sketch 4.6.

In this example, we have a map:

$$\mathbf{v} = \begin{bmatrix} 0 \\ 1 \end{bmatrix} \quad \longrightarrow \quad \mathbf{v}' = \begin{bmatrix} d_1 \\ 1 \end{bmatrix}.$$

In matrix form, this is realized by

$$\begin{bmatrix} d_1 \\ 1 \end{bmatrix} = \begin{bmatrix} 1 & d_1 \\ 0 & 1 \end{bmatrix} \begin{bmatrix} 0 \\ 1 \end{bmatrix}. \tag{4.17}$$

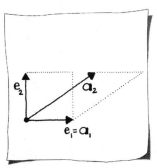

Sketch 4.6.
A special shear.

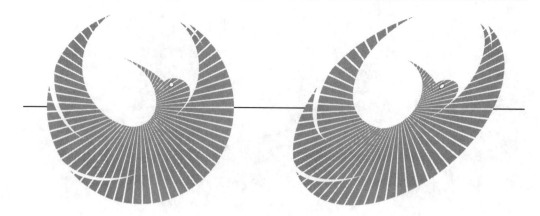

Figure 4.8.
Shears: shearing parallel to the $\mathbf{e}_1$-axis.

Verify! The 2×2 matrix in this equation is called a *shear matrix*. It is the kind of matrix that is used when you generate italic fonts from standard ones.

A shear matrix may be applied to arbitrary vectors. If $\mathbf{v}$ is an input vector, then a shear maps it to $\mathbf{v}'$:

$$\mathbf{v}' = \begin{bmatrix} 1 & d_1 \\ 0 & 1 \end{bmatrix} \begin{bmatrix} v_1 \\ v_2 \end{bmatrix} = \begin{bmatrix} v_1 + v_2 d_1 \\ v_2 \end{bmatrix},$$

as illustrated in Figure 4.8.

We have so far restricted ourselves to shears along the $\mathbf{e}_1$-axis; we may also shear along the $\mathbf{e}_2$-axis. Then we would have

$$\mathbf{v}' = \begin{bmatrix} 1 & 0 \\ d_2 & 1 \end{bmatrix} \begin{bmatrix} v_1 \\ v_2 \end{bmatrix} = \begin{bmatrix} v_1 \\ v_1 d_2 + v_2 \end{bmatrix},$$

as illustrated in Figure 4.9.

Since it will be needed later, we look at the following. What is the shear that achieves

$$\mathbf{v} = \begin{bmatrix} v_1 \\ v_2 \end{bmatrix} \quad \longrightarrow \quad \mathbf{v}' = \begin{bmatrix} v_1 \\ 0 \end{bmatrix}?$$

It is obviously a shear parallel to the $\mathbf{e}_2$-axis and is given by the map

$$\mathbf{v}' = \begin{bmatrix} v_1 \\ 0 \end{bmatrix} = \begin{bmatrix} 1 & 0 \\ -v_2/v_1 & 1 \end{bmatrix} \begin{bmatrix} v_1 \\ v_2 \end{bmatrix}. \tag{4.18}$$

Shears do not change areas. In Sketch 4.6, we see that the rectangle and its image, a parallelogram, have the same area: both have the same base and the same height.

Figure 4.9.
Shears: shearing parallel to the $\mathbf{e}_2$-axis.

4.8 Projections

Projections—parallel projections, for our purposes—act like sunlight casting shadows. In 2D, this is modeled as follows: take any vector $\mathbf{v}$ and "flatten it out" onto the $\mathbf{e}_1$-axis. This simply means: set the v_2-coordinate of the vector to zero. For example, if we project the vector

$$\mathbf{v} = \begin{bmatrix} 3 \\ 1 \end{bmatrix}$$

onto the $\mathbf{e}_1$-axis, it becomes

$$\mathbf{v}' = \begin{bmatrix} 3 \\ 0 \end{bmatrix},$$

as shown in Sketch 4.7.

What matrix achieves this map? That's easy:

$$\begin{bmatrix} 3 \\ 0 \end{bmatrix} = \begin{bmatrix} 1 & 0 \\ 0 & 0 \end{bmatrix} \begin{bmatrix} 3 \\ 1 \end{bmatrix}.$$

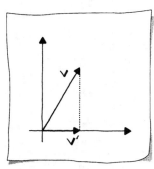

Sketch 4.7.
A projection.

This matrix will not only project the vector

$$\begin{bmatrix} 3 \\ 1 \end{bmatrix}$$

onto the e_1-axis, but in fact *every vector*! This is so since

$$\begin{bmatrix} 1 & 0 \\ 0 & 0 \end{bmatrix} \begin{bmatrix} v_1 \\ v_2 \end{bmatrix} = \begin{bmatrix} v_1 \\ 0 \end{bmatrix}.$$

While this is a somewhat trivial example of "real" projections, we see that this projection does indeed feature the main property of a projection: it *reduces dimensionality*. Every vector from 2D space is mapped into 1D space, namely onto the e_1-axis. Figure 4.10 illustrates this property. We also observe that the two columns of our projection matrix are linearly dependent (one column being the zero vector) and thus the matrix has rank one.

The analogous case, projecting onto the e_2-axis, is not more difficult; it is given by

$$\begin{bmatrix} 0 & 0 \\ 0 & 1 \end{bmatrix} \begin{bmatrix} v_1 \\ v_2 \end{bmatrix} = \begin{bmatrix} 0 \\ v_2 \end{bmatrix}.$$

Parallel projections are characterized by the fact that all vectors are projected in a parallel direction. In 2D, all vectors are projected onto a line. If that line is in the direction of a unit vector $\mathbf{a}$, then our matrix must be of the form

$$\begin{bmatrix} a_1 & ca_1 \\ a_2 & ca_2 \end{bmatrix}.$$

Figure 4.10.

Projections: all vectors are "flattened out" onto the e_1-axis.

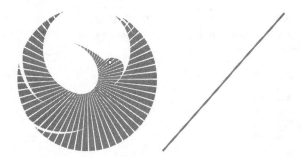

Figure 4.11.
Projections: all vectors are "flattened out" in one direction.

Figure 4.11 shows the effect of the matrix

$$\begin{bmatrix} 0.5 & 0.5 \\ 0.5 & 0.5 \end{bmatrix}.$$

Try sketching the projection of a few vectors yourself to get a feel for how this projection works. We will discuss a more practical approach to projections in a 3D context in Section 13.4.

We mention one additional property of projections: once a vector has been projected onto a line, application of the same projection will leave the result unchanged. Maps with that property are called *idempotent*; see more on these in Section 4.11.

As far as areas are concerned, projections take a lean approach: whatever an area was before the map, it is zero afterwards. We will revisit projection matrices in the context of homogeneous linear systems in Section 5.8.

4.9   The Kernel of a Projection

Let A be a projection matrix, i.e., A is of the form $A = [\mathbf{a}_1, c\mathbf{a}_1]$. We know that A maps every vector onto a multiple of $\mathbf{a}_1$. We might ask: which vectors $\mathbf{v}$ are mapped to the zero vector, i.e., for which $\mathbf{v}$ do we have

$$A\mathbf{v} = \mathbf{0}? \tag{4.19}$$

Since

$$A\mathbf{v} = (v_1 + cv_2)\mathbf{a}_1,$$

we find that the desired vectors $\mathbf{v}$ are characterized by $v_1 + cv_2 = 0$. This set of vectors depends on the factor c but is independent of $\mathbf{a}_1$.

It is also known as the *kernel* of the map A. We will revisit kernels several times in future chapters. Equation (4.19) is examined in more detail in Section 5.8.

Example 4.4

Let the matrix A be given by

$$A = \begin{bmatrix} 0.5 & 0.5 \\ 0.5 & 0.5 \end{bmatrix},$$

i.e., $c = 1$. Then all vectors of the form $v_1 = -v_2$ will be mapped to the zero vector, for example

$$\begin{bmatrix} 0.5 & 0.5 \\ 0.5 & 0.5 \end{bmatrix} \begin{bmatrix} 20 \\ -20 \end{bmatrix} = \begin{bmatrix} 0 \\ 0 \end{bmatrix}.$$

4.10   Areas and Linear Maps: Determinants

As you might have noticed, we discussed one particular aspect of linear maps for each type: how areas are changed. We will now discuss this aspect for an arbitrary linear map. Such a map takes the two vectors $[\mathbf{e}_1, \mathbf{e}_2]$ to the two vectors $[\mathbf{a}_1, \mathbf{a}_2]$. The area of the square spanned by $[\mathbf{e}_1, \mathbf{e}_2]$ is 1, that is area$(\mathbf{e}_1, \mathbf{e}_2) = 1$. If we knew the area of the parallelogram spanned by $[\mathbf{a}_1, \mathbf{a}_2]$, then we could say how the linear map affects areas.

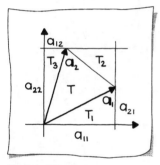

Sketch 4.8.

Area of a parallelogram.

How do we find the area P of a parallelogram spanned by two vectors $\mathbf{a}_1$ and $\mathbf{a}_2$? Referring to Sketch 4.8, let us first determine the area T of the triangle formed by $\mathbf{a}_1$ and $\mathbf{a}_2$. We see that

$$T = a_{1,1}a_{2,2} - T_1 - T_2 - T_3.$$

We then observe that

$$T_1 = \frac{1}{2}a_{1,1}a_{2,1}, \tag{4.20}$$

$$T_2 = \frac{1}{2}(a_{1,1} - a_{1,2})(a_{2,2} - a_{2,1}), \tag{4.21}$$

$$T_3 = \frac{1}{2}a_{1,2}a_{2,2}. \tag{4.22}$$

Working out the algebra, we arrive at

$$T = \frac{1}{2}a_{1,1}a_{2,2} - \frac{1}{2}a_{1,2}a_{2,1}.$$

Our aim was not really T, but the parallelogram area P. Clearly (see Sketch 4.9),

$$P = 2T,$$

and we have our desired area.

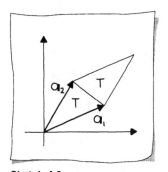

Sketch 4.9.
Parallelogram and triangles.

It is customary to use the term *determinant* for the (signed) area of the parallelogram spanned by $[\mathbf{a}_1, \mathbf{a}_2]$. Since the two vectors $\mathbf{a}_1$ and $\mathbf{a}_2$ form the columns of the matrix A, we also speak of the determinant of the matrix A, and denote it by $\det A$ or $|A|$:

$$|A| = \begin{vmatrix} a_{1,1} & a_{1,2} \\ a_{2,1} & a_{2,2} \end{vmatrix} = a_{1,1}a_{2,2} - a_{1,2}a_{2,1}. \tag{4.23}$$

Since A maps a square with area one onto a parallelogram with area $|A|$, the determinant of a matrix characterizes it as follows:

- If $|A| = 1$, then the linear map does not change areas.

- If $0 \leq |A| < 1$, then the linear map shrinks areas.

- If $|A| > 1$, then the linear map expands areas.

- If $|A| < 0$, then the linear map changes the orientation of objects. (We'll look at this closer after Example 4.5.) Areas may still contract or expanded depending on the magnitude of the determinant.

Example 4.5

We will look at a few examples. Let

$$A = \begin{bmatrix} 1 & 5 \\ 0 & 1 \end{bmatrix},$$

then

$$|A| = \begin{vmatrix} 1 & 5 \\ 0 & 1 \end{vmatrix} = 1.$$

Since A represents a *shear*, we see again that those maps do not change areas.

For another example, let

$$A = \begin{bmatrix} 1 & 0 \\ 0 & -1 \end{bmatrix}.$$

Then

$$|A| = \begin{vmatrix} 1 & 0 \\ 0 & -1 \end{vmatrix} = -1.$$

This matrix corresponds to a *reflection*, and it leaves areas unchanged, except for a *sign change*.

Finally, let

$$A = \begin{bmatrix} 0.5 & 0.5 \\ 0.5 & 0.5 \end{bmatrix}.$$

Then

$$|A| = \begin{vmatrix} 0.5 & 0.5 \\ 0.5 & 0.5 \end{vmatrix} = 0.$$

This matrix corresponds to the *projection* from Section 4.8. In that example, we saw that projections collapse any object onto a straight line, i.e., to a zero area.

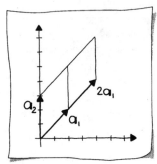

Sketch 4.10.

Resulting area after scaling one column of A.

There are some rules for working with determinants:
If $A = [\mathbf{a}_1, \mathbf{a}_2]$, then

$$|c\mathbf{a}_1, \mathbf{a}_2| = c|\mathbf{a}_1, \mathbf{a}_2| = c|A|.$$

In other words, if one of the columns of A is scaled by a factor c, then A's determinant is also scaled by c. Verify that this is true from the definition of the determinant of A! Sketch 4.10 illustrates this for the example, $c = 2$. If *both* columns of A are scaled by c, then the determinant is scaled by c^2:

$$|c\mathbf{a}_1, c\mathbf{a}_2| = c^2|\mathbf{a}_1, \mathbf{a}_2| = c^2|A|.$$

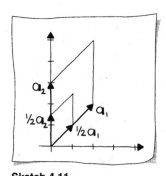

Sketch 4.11.

Resulting area after scaling both columns of A.

Sketch 4.11 illustrates for the example, $c = 1/2$.

If $|A|$ is positive and c is negative, then replacing $\mathbf{a}_1$ by $c\mathbf{a}_1$ will cause $c|A|$, the area formed by $c\mathbf{a}_1$ and $\mathbf{a}_2$, to become negative. The notion of a negative area is very useful computationally. Two 2D vectors whose determinant is positive are called *right-handed*. The standard example are the two vectors $\mathbf{e}_1$ and $\mathbf{e}_2$. If their determinant is negative, then they are called *left-handed*.[1] Sketch 4.12 shows

---

[1]The reason for this terminology will become apparent when we revisit these definitions for the 3D case (see Section 10.2).

a right-handed pair of vectors (top) and a pair of left-handed ones (bottom). Our definition of positive and negative area is not totally arbitrary: the triangle formed by vectors $\mathbf{a}_1$ and $\mathbf{a}_2$ has area $1/2 \times \sin(\alpha)\|\mathbf{a}_1\|\|\mathbf{a}_2\|$. Here, the angle α indicates how much we have to rotate $\mathbf{a}_1$ in order to line up with $\mathbf{a}_2$. If we interchange the two vectors, the sign of α and hence of $\sin(\alpha)$ also changes!

The area also changes sign when we *interchange* the columns of A:

$$|\mathbf{a}_1, \mathbf{a}_2| = -|\mathbf{a}_2, \mathbf{a}_1|. \qquad (4.24)$$

This fact is easily computed using the definition of a determinant:

$$|\mathbf{a}_2, \mathbf{a}_1| = a_{1,2}a_{2,1} - a_{2,2}a_{1,1}.$$

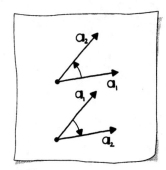

Sketch 4.12.
Right-handed and left-handed vectors.

4.11   Composing Linear Maps

Suppose you have mapped a vector $\mathbf{v}$ to $\mathbf{v}'$ using a matrix A. Next, you want to map $\mathbf{v}'$ to $\mathbf{v}''$ using a matrix B. We start out with

$$\mathbf{v}' = \begin{bmatrix} a_{1,1} & a_{1,2} \\ a_{2,1} & a_{2,2} \end{bmatrix} \begin{bmatrix} v_1 \\ v_2 \end{bmatrix} = \begin{bmatrix} a_{1,1}v_1 + a_{1,2}v_2 \\ a_{2,1}v_1 + a_{2,2}v_2 \end{bmatrix}.$$

Next, we have

$$\mathbf{v}'' = \begin{bmatrix} b_{1,1} & b_{1,2} \\ b_{2,1} & b_{2,2} \end{bmatrix} \begin{bmatrix} a_{1,1}v_1 + a_{1,2}v_2 \\ a_{2,1}v_1 + a_{2,2}v_2 \end{bmatrix} =$$

$$\begin{bmatrix} b_{1,1}(a_{1,1}v_1 + a_{1,2}v_2) + b_{1,2}(a_{2,1}v_1 + a_{2,2}v_2) \\ b_{2,1}(a_{1,1}v_1 + a_{1,2}v_2) + b_{2,2}(a_{2,1}v_1 + a_{2,2}v_2) \end{bmatrix}.$$

Collecting the terms in v_1 and v_2, we get

$$\mathbf{v}'' = \begin{bmatrix} b_{1,1}a_{1,1} + b_{1,2}a_{2,1} & b_{1,1}a_{1,2} + b_{1,2}a_{2,2} \\ b_{2,1}a_{1,1} + b_{2,2}a_{2,1} & b_{2,1}a_{1,2} + b_{2,2}a_{2,2} \end{bmatrix} \begin{bmatrix} v_1 \\ v_2 \end{bmatrix}.$$

The matrix that we have created here, let's call it C, is called the *product matrix* of B and A:

$$B \cdot A = C.$$

In more detail,

$$\begin{bmatrix} b_{1,1} & b_{1,2} \\ b_{2,1} & b_{2,2} \end{bmatrix} \begin{bmatrix} a_{1,1} & a_{1,2} \\ a_{2,1} & a_{2,2} \end{bmatrix} = \begin{bmatrix} b_{1,1}a_{1,1} + b_{1,2}a_{2,1} & b_{1,1}a_{1,2} + b_{1,2}a_{2,2} \\ b_{2,1}a_{1,1} + b_{2,2}a_{2,1} & b_{2,1}a_{1,2} + b_{2,2}a_{2,2} \end{bmatrix}.$$

$$(4.25)$$

This looks messy, but a simple rule puts order into chaos: the element $c_{i,j}$ is computed as the dot product of B's i^{th} row and A's j^{th} column.

We can use this product to describe the composite map:

$$\mathbf{v}'' = B\mathbf{v}' = B[A\mathbf{v}] = BA\mathbf{v}.$$

Example 4.6

Let

$$\mathbf{v} = \begin{bmatrix} 2 \\ -1 \end{bmatrix}, \quad A = \begin{bmatrix} -1 & 2 \\ 0 & 3 \end{bmatrix}, \quad B = \begin{bmatrix} 0 & -2 \\ -3 & 1 \end{bmatrix}.$$

Then

$$\mathbf{v}' = \begin{bmatrix} -1 & 2 \\ 0 & 3 \end{bmatrix} \begin{bmatrix} 2 \\ -1 \end{bmatrix} = \begin{bmatrix} -4 \\ -3 \end{bmatrix}$$

and

$$\mathbf{v}'' = \begin{bmatrix} 0 & -2 \\ -3 & 1 \end{bmatrix} \begin{bmatrix} -4 \\ -3 \end{bmatrix} = \begin{bmatrix} 6 \\ 9 \end{bmatrix}.$$

We can also compute $\mathbf{v}''$ using the matrix product BA:

$$C = BA = \begin{bmatrix} 0 & -2 \\ -3 & 1 \end{bmatrix} \begin{bmatrix} -1 & 2 \\ 0 & 3 \end{bmatrix} = \begin{bmatrix} 0 & -6 \\ 3 & -3 \end{bmatrix}.$$

Verify for yourself that $\mathbf{v}'' = C\mathbf{v}$!

There is a neat way to arrange two matrices when forming their product for manual computation (yes, that is still encountered!), analogous to the matrix/vector product from Section 4.2. Using the matrices of Example 4.6, and highlighting the computation of $c_{2,1}$, we write

$$\begin{array}{cc|cc}
 & & -\mathbf{1} & 2 \\
 & & \mathbf{0} & 3 \\
\hline
0 & -2 & & \\
-\mathbf{3} & \mathbf{1} & \mathbf{3} & \\
\end{array}$$

You see how $c_{2,1}$ is at the intersection of column one of the "top" matrix and row two of the "left" matrix.

The complete multiplication scheme is then arranged like this

$$\begin{array}{cc|cc}
 & & -1 & 2 \\
 & & 0 & 3 \\
\hline
0 & -2 & 0 & -6 \\
-3 & 1 & 3 & -3 \\
\end{array}$$

Figure 4.12.
Linear map composition is order dependent. Top: rotate by −120°, then reflect about the (rotated) $\mathbf{e}_2$-axis. Bottom: reflect, then rotate.

While we use the term "product" for BA, it is very important to realize that this kind of product differs significantly from products of real numbers: it is not *commutative*. That is, in general

$$AB \neq BA.$$

Matrix products correspond to linear map compositions—since the products are not commutative, it follows that it matters in which order we carry out linear maps. *Linear map composition is order dependent.* Figure 4.12 gives an example.

Example 4.7

Let us take two very simple matrices and demonstrate that the product is not commutative. This example is illustrated in Figure 4.12. A rotates by −120°, and B reflects about the $\mathbf{e}_2$-axis:

$$A = \begin{bmatrix} -0.5 & 0.866 \\ -0.866 & -0.5 \end{bmatrix}, \qquad B = \begin{bmatrix} 1 & 0 \\ 0 & -1 \end{bmatrix}.$$

We first form AB (reflect and then rotate),

$$AB = \begin{bmatrix} -0.5 & 0.866 \\ -0.866 & -0.5 \end{bmatrix} \begin{bmatrix} 1 & 0 \\ 0 & -1 \end{bmatrix} = \begin{bmatrix} -0.5 & -0.866 \\ -0.866 & 0.5 \end{bmatrix}.$$

Next, we form BA (rotate then reflect),

$$BA = \begin{bmatrix} 1 & 0 \\ 0 & -1 \end{bmatrix} \begin{bmatrix} -0.5 & 0.866 \\ -0.866 & -0.5 \end{bmatrix} = \begin{bmatrix} -0.5 & 0.866 \\ 0.866 & 0.5 \end{bmatrix}.$$

Clearly, these are not the same!

Of course, *some* maps *do* commute; for example, the rotations. It does not matter if we rotate by α first and then by β or the other way around. In either case, we have rotated by $\alpha + \beta$. In terms of matrices,

$$\begin{bmatrix} \cos\alpha & -\sin\alpha \\ \sin\alpha & \cos\alpha \end{bmatrix} \begin{bmatrix} \cos\beta & -\sin\beta \\ \sin\beta & \cos\beta \end{bmatrix} =$$

$$\begin{bmatrix} \cos\alpha\cos\beta - \sin\alpha\sin\beta & -\cos\alpha\sin\beta - \sin\alpha\cos\beta \\ \sin\alpha\cos\beta + \cos\alpha\sin\beta & -\sin\alpha\sin\beta + \cos\alpha\cos\beta \end{bmatrix}.$$

Check for yourself that the other alternative gives the same result! The product matrix must correspond to a rotation by $\alpha + \beta$, and thus it must equal to

$$\begin{bmatrix} \cos(\alpha + \beta) & -\sin(\alpha + \beta) \\ \sin(\alpha + \beta) & \cos(\alpha + \beta) \end{bmatrix}.$$

We have thus shown that

$$\begin{bmatrix} \cos(\alpha + \beta) & -\sin(\alpha + \beta) \\ \sin(\alpha + \beta) & \cos(\alpha + \beta) \end{bmatrix} =$$

$$\begin{bmatrix} \cos\alpha\cos\beta - \sin\alpha\sin\beta & -\cos\alpha\sin\beta - \sin\alpha\cos\beta \\ \sin\alpha\cos\beta + \cos\alpha\sin\beta & -\sin\alpha\sin\beta + \cos\alpha\cos\beta \end{bmatrix},$$

and so we have, without much effort, proved two "trig" identities!

One more example on composing linear maps, which we have seen in Section 4.8, is that projections are idempotent. If A is a projection matrix, then this means

$$A\mathbf{v} = AA\mathbf{v}$$

for any vector $\mathbf{v}$. Written out using only matrices, this becomes

$$A = AA \quad \text{or} \quad A = A^2. \qquad (4.26)$$

Verify this property for the projection matrix

$$\begin{bmatrix} 0.5 & 0.5 \\ 0.5 & 0.5 \end{bmatrix}.$$

4.12 More on Matrix Multiplication

Matrix multiplication is not limited to the product of 2×2 matrices. In fact, we are constantly using a different kind of matrix multiplication; when we multiply a matrix by a vector, we follow the rules of matrix multiplication! If $\mathbf{v}' = A\mathbf{v}$, then the first component of $\mathbf{v}'$ is the dot product of A's first row and $\mathbf{v}$; the second component of $\mathbf{v}'$ is the dot product of A's second row and $\mathbf{v}$.

We may even write the dot product of two vectors in the form of matrix multiplication, as an example should show:

$$\begin{bmatrix} 3 & 4 \end{bmatrix} \cdot \begin{bmatrix} -3 \\ 6 \end{bmatrix} = 15.$$

Usually, we write $\mathbf{u} \cdot \mathbf{v}$ for the dot product of $\mathbf{u}$ and $\mathbf{v}$, but sometimes the above form $\mathbf{u}^T \mathbf{v}$ is useful as well.

In Section 4.3, we introduced the transpose A^T of a matrix A. We saw that addition of matrices is "well-behaved" under transposition; matrix multiplication is not that straightforward. We have

$$[AB]^T = B^T A^T. \qquad (4.27)$$

To see why this is true, recall that each element of a product matrix is obtained as a dot product. How do dot products react to transposition? If we have a product $\mathbf{u}^T \mathbf{v}$, what is $[\mathbf{u}^T \mathbf{v}]^T$? This has an easy answer:

$$[\mathbf{u}^T \mathbf{v}]^T = \mathbf{v}^T \mathbf{u},$$

as an example will clarify.

Example 4.8

$$\begin{bmatrix} 3 & 4 \end{bmatrix} \begin{bmatrix} -3 \\ 6 \end{bmatrix} = \begin{bmatrix} -3 & 6 \end{bmatrix} \begin{bmatrix} 3 \\ 4 \end{bmatrix} = 15.$$

Since matrix multiplication is just the application of several dot products, we see that (4.27) does make sense. More on matrix multiplication appears in Chapter 12.

What is the *determinant* of a product matrix? If $C = AB$ denotes a matrix product, then we know that B scales objects by $|B|$, and A scales objects by $|A|$. It is clear then that the composition of the maps scales by the product of the individual scales

$$|AB| = |A| \cdot |B|. \tag{4.28}$$

Example 4.9

As a simple example, take two scalings

$$A = \begin{bmatrix} 1/2 & 0 \\ 0 & 1/2 \end{bmatrix}, \quad B = \begin{bmatrix} 4 & 0 \\ 0 & 4 \end{bmatrix}.$$

We have $|A| = 1/4$ and $|B| = 16$. Thus, A scales down, and B scales up, but the effect of B's scaling is greater than that of A's. The product

$$AB = \begin{bmatrix} 2 & 0 \\ 0 & 2 \end{bmatrix}$$

thus scales up: $|AB| = |A| \cdot |B| = 4$.

Just as for real numbers, we can define *exponents* for matrices:

$$A^r = \underbrace{A \cdot \ldots \cdots A}_{r \text{ times}}.$$

Here are some rules.

$$\begin{aligned} A^{r+s} &= A^r A^s \\ A^{rs} &= (A^r)^s \\ A^0 &= I \end{aligned}$$

For now, assume r and s are positive integers. See Sections 5.5 and 12.10 for a discussion of A^{-1}, the *inverse matrix*.

4.13   Working with Matrices

Yet more rules of matrix arithmetic! We encountered many of these rules throughout this chapter in terms of matrix and vector multiplication: a vector is simply a special matrix.

For 2×2 matrices A, B, C, the following rules hold.

Commutative Law for Addition
$$A + B = B + A$$

Associative Law for Addition
$$A + (B + C) = (A + B) + C$$

Associative Law for Multiplication
$$A(BC) = (AB)C$$

Distributive Law
$$A(B + C) = AB + AC$$

Distributive Law
$$(B + C)A = BA + CA$$

And some rules involving scalars:

$$a(B + C) = aB + aC$$
$$(a + b)C = aC + bC$$
$$(ab)C = a(bC)$$
$$a(BC) = (aB)C = B(aC).$$

Although the focus of this chapter is on 2×2 matrices, these rules apply to matrices of any size. Of course it is assumed in the rules above that the sizes of the matrices are such that the operations can be performed. See Chapters 12 and 14 for more information on matrices larger than 2×2.

- column space
- add matrices
- multiply matrices
- linear combinations
- transpose matrix
- transpose of a product or sum of matrices
- symmetric matrix
- rank of a matrix
- idempotent map
- construct the basic linear maps
- determinant
- right- and left-handed
- rigid body motions
- non-commutative property of matrix multiplication
- rules of matrix arithmetic
- kernel

4.14   Exercises

For the following exercises, let

$$A = \begin{bmatrix} 0 & -1 \\ 1 & 0 \end{bmatrix}, \quad B = \begin{bmatrix} 1 & -1 \\ -1 & 1/2 \end{bmatrix}, \quad \mathbf{v} = \begin{bmatrix} 2 \\ 3 \end{bmatrix}.$$

1. Describe geometrically the effect of A and B. (You may do this analytically or by suitably modifying the PostScript file `Rotate.ps` available from the downloads section on the book's website. The URL is given in the Preface.)

2. Compute $A\mathbf{v}$ and $B\mathbf{v}$.

3. Compute $A + B$. Show that $A\mathbf{v} + B\mathbf{v} = (A + B)\mathbf{v}$.

4. Compute $AB\mathbf{v}$ and $BA\mathbf{v}$.

5. Compute $B^{\mathrm{T}} A$.

6. What is the shear matrix that maps $\mathbf{v}$ onto the $\mathbf{e}_2$-axis?

7. What is the determinant of A?

8. What is the rank of B?

9. Are either A or B symmetric matrices?

10. What type of linear map is the matrix

$$\begin{bmatrix} -1 & 0 \\ 0 & -1 \end{bmatrix}?$$

11. Are either A or B a rigid body motion?

12. What is A^2?

13. What is the kernel of the matrix

$$A = \begin{bmatrix} 2 & 6 \\ 4 & 12 \end{bmatrix}?$$

14. Is the matrix from the previous exercise idempotent?

————

PS1 Modify the ps-file `Nocomm.ps` to show that $AB \neq BA$. The whole PostScript program may look somewhat involved—all you have to do is change the definition of the input matrices `mat1` and `mat2` and watch what happens!

PS2 Modify the ps-file `Rotate.ps` such that a rotation of 100 degrees is achieved.

5

2×2 Linear Systems

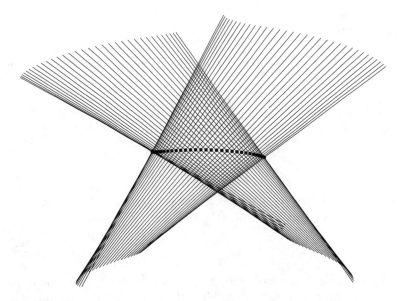

Figure 5.1.
Intersection of lines: two families of lines are shown; the intersections of corresponding line pairs are marked. For each intersection, a 2×2 linear system has to be solved.

Just about anybody can solve two equations in two unknowns by somehow manipulating the equations. In this chapter, we will develop a systematic way for finding the solution, simply by checking the underlying geometry. This approach will later enable us to solve much larger systems of equations. Figure 5.1 illustrates many instances of intersecting two lines: a problem that can be formulated as a 2 × 2 linear system.

5.1 Skew Target Boxes Revisited

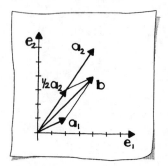

Sketch 5.1.

Geometry of a 2 × 2 system.

In our standard $[\mathbf{e}_1, \mathbf{e}_2]$-coordinate system, suppose we are given two vectors $\mathbf{a}_1$ and $\mathbf{a}_2$. In Section 4.1, we showed how these vectors define a skew target box with its lower-left corner at the origin. As illustrated in Sketch 5.1, suppose we also are given a vector $\mathbf{b}$ with respect to the $[\mathbf{e}_1, \mathbf{e}_2]$-system. Now the question arises, what are the components of $\mathbf{b}$ with respect to the $[\mathbf{a}_1, \mathbf{a}_2]$-system? In other words, we want to find a vector $\mathbf{u}$ with components u_1 and u_2, satisfying

$$u_1\mathbf{a}_1 + u_2\mathbf{a}_2 = \mathbf{b}.$$

Example 5.1

Before we proceed further, let's look at an example. Following Sketch 5.1, let

$$\mathbf{a}_1 = \begin{bmatrix} 2 \\ 1 \end{bmatrix}, \quad \mathbf{a}_2 = \begin{bmatrix} 4 \\ 6 \end{bmatrix}, \quad \mathbf{b} = \begin{bmatrix} 4 \\ 4 \end{bmatrix}.$$

Upon examining the sketch, we see that

$$1 \times \begin{bmatrix} 2 \\ 1 \end{bmatrix} + \frac{1}{2}\begin{bmatrix} 4 \\ 6 \end{bmatrix} = \begin{bmatrix} 4 \\ 4 \end{bmatrix}.$$

In the $[\mathbf{a}_1, \mathbf{a}_2]$-system, $\mathbf{b}$ has components $(1, 1/2)$. In the $[\mathbf{e}_1, \mathbf{e}_2]$-system, it has components $(4, 4)$.

What we have here are really two equations in the two unknowns u_1 and u_2, which we see by expanding the point/vector equations into

$$\begin{aligned} 2u_1 + 4u_2 &= 4 \\ u_1 + 6u_2 &= 4. \end{aligned} \tag{5.1}$$

And as we saw in Example 5.1, these two equations in two unknowns have the solution $u_1 = 1$ and $u_2 = 1/2$, as is seen by inserting these values for u_1 and u_2 into the equations.

Being able to solve two simultaneous sets of equations allows us to switch back and forth between different coordinate systems. The rest of this chapter is dedicated to a detailed discussion of how to solve these equations.

5.2   The Matrix Form

The two equations in (5.1) are also called a *linear system*. It can be written more compactly if we use matrix notation:

$$\begin{bmatrix} 2 & 4 \\ 1 & 6 \end{bmatrix} \begin{bmatrix} u_1 \\ u_2 \end{bmatrix} = \begin{bmatrix} 4 \\ 4 \end{bmatrix}. \tag{5.2}$$

In general, a 2×2 linear system looks like this:

$$\begin{bmatrix} a_{1,1} & a_{1,2} \\ a_{2,1} & a_{2,2} \end{bmatrix} \begin{bmatrix} u_1 \\ u_2 \end{bmatrix} = \begin{bmatrix} b_1 \\ b_2 \end{bmatrix}. \tag{5.3}$$

It is shorthand notation for the equations

$$a_{1,1} u_1 + a_{1,2} u_2 = b_1$$
$$a_{2,1} u_1 + a_{2,2} u_2 = b_2.$$

We sometimes write it even shorter, using a matrix A:

$$A\mathbf{u} = \mathbf{b}, \tag{5.4}$$

where

$$A = \begin{bmatrix} a_{1,1} & a_{1,2} \\ a_{2,1} & a_{2,2} \end{bmatrix}, \quad \mathbf{u} = \begin{bmatrix} u_1 \\ u_2 \end{bmatrix}, \quad \mathbf{b} = \begin{bmatrix} b_1 \\ b_1 \end{bmatrix}.$$

Both $\mathbf{u}$ and $\mathbf{b}$ represent vectors, not points! (See Sketch 5.1 for an illustration.) The vector $\mathbf{u}$ is called the *solution* of the linear system.

While the savings of this notation is not completely obvious in the 2×2 case, it will save a lot of work for more complicated cases with more equations and unknowns.

The columns of the matrix A correspond to the vectors $\mathbf{a}_1$ and $\mathbf{a}_2$. We could then rewrite our linear system as

$$u_1 \mathbf{a}_1 + u_2 \mathbf{a}_2 = \mathbf{b}.$$

Geometrically, we are trying to express the given vector $\mathbf{b}$ as a linear combination of the given vectors $\mathbf{a}_1$ and $\mathbf{a}_2$; we need to determine the factors u_1 and u_2. If we are able to find at least one solution, then the linear system is called *consistent*, otherwise it is called inconsistent. Three possibilities for our *solution space* exist.

1. There is exactly one solution vector **u**.

2. There is no solution, or in other words, the system is inconsistent. (See Section 5.6 for a geometric description.)

3. There are infinitely many solutions. (See Sections 5.7 and 5.8 for examples.)

5.3   A Direct Approach: Cramer's Rule

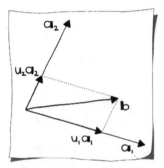

Sketch 5.2.

Cramer's rule.

Sketch 5.2 offers a direct solution to our linear system. By simply inspecting the areas of the parallelograms in the sketch, we see that

$$u_1 = \frac{\text{area}(\mathbf{b}, \mathbf{a}_2)}{\text{area}(\mathbf{a}_1, \mathbf{a}_2)},$$

$$u_2 = \frac{\text{area}(\mathbf{a}_1, \mathbf{b})}{\text{area}(\mathbf{a}_1, \mathbf{a}_2)}.$$

The area of a parallelogram is given by the determinant of the two vectors spanning it. Recall from Section 4.10 that this is a signed area. This method of solving for the solution of a linear system is called *Cramer's rule*.

Example 5.2

Applying Cramer's rule to the linear system in (5.2), we get

$$u_1 = \frac{\begin{vmatrix} 4 & 4 \\ 4 & 6 \end{vmatrix}}{\begin{vmatrix} 2 & 4 \\ 1 & 6 \end{vmatrix}} = \frac{8}{8},$$

$$u_2 = \frac{\begin{vmatrix} 2 & 4 \\ 1 & 4 \end{vmatrix}}{\begin{vmatrix} 2 & 4 \\ 1 & 6 \end{vmatrix}} = \frac{4}{8}.$$

Examining the determinant in the numerator, notice that **b** replaces **a**$_1$ in the solution for u_1 and then **b** replaces **a**$_2$ in the solution for u_2.

Notice that if the area spanned by $\mathbf{a}_1$ and $\mathbf{a}_2$ is zero, that is, the vectors are multiples of each other, then Cramer's rule will not result in a solution. (See Sections 5.6–5.8 for more information on this situation.)

Cramer's rule is primarily of theoretical importance. For larger systems, Cramer's rule is both expensive and numerically unstable. Hence, we now study a more effective method.

5.4 Gauss Elimination

Let's consider a special 2×2 linear system:

$$\begin{bmatrix} a_{1,1} & a_{1,2} \\ 0 & a_{2,2} \end{bmatrix} \mathbf{u} = \mathbf{b}. \tag{5.5}$$

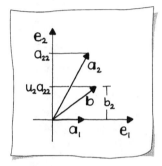

Sketch 5.3.

A special linear system.

This situation is shown in Sketch 5.3. This matrix is called *upper triangular* because all elements below the diagonal are zero, forming a triangle of numbers above the diagonal.

We can solve this system without much work. Examining the last equation, we see it is possible to solve for

$$u_2 = b_2/a_{2,2}.$$

With u_2 in hand, we can solve the first equation for

$$u_1 = \frac{1}{a_{1,1}}(b_1 - u_2 a_{1,2}).$$

This technique of solving the equations from the bottom up is called *back substitution.*

Notice that the process of back substitution requires divisions. Therefore, if the diagonal elements, a_{11} or a_{22}, equal zero then the algorithm will fail. This type of failure indicates that the columns of A are not linearly independent. (See Sections 5.6–5.8 for more information on this situation.)

In general, we will not be so lucky to encounter an upper triangular system as in (5.5). But any (almost any, at least) linear system may be *transformed* to this simple case, as we shall see by re-examining the system in (5.2). We write it as

$$u_1 \begin{bmatrix} 2 \\ 1 \end{bmatrix} + u_2 \begin{bmatrix} 4 \\ 6 \end{bmatrix} = \begin{bmatrix} 4 \\ 4 \end{bmatrix}.$$

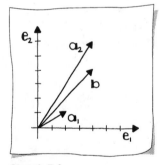

Sketch 5.4.

The geometry of a linear system.

This situation is shown in Sketch 5.4. Clearly, $\mathbf{a}_1$ is not on the $\mathbf{e}_1$-axis as we would like, but we can apply a stepwise procedure so that it will become just that. This systematic, stepwise procedure is called *Gauss elimination*.

Recall one key fact from Chapter 4: *linear maps do not change linear combinations*. That means if we apply the same linear map to all vectors in our system, then the factors u_1 and u_2 won't change. If the map is given by a matrix S, then

$$S \left[u_1 \begin{bmatrix} 2 \\ 1 \end{bmatrix} + u_2 \begin{bmatrix} 4 \\ 6 \end{bmatrix} \right] = S \begin{bmatrix} 4 \\ 4 \end{bmatrix}.$$

In order to get $\mathbf{a}_1$ to line up with the $\mathbf{e}_1$-axis, we will employ a *shear* parallel to the $\mathbf{e}_2$-axis, such that

$$\begin{bmatrix} 2 \\ 1 \end{bmatrix} \text{ is mapped to } \begin{bmatrix} 2 \\ 0 \end{bmatrix}.$$

That shear (see Section 4.7) is given by the matrix

$$S_1 = \begin{bmatrix} 1 & 0 \\ -1/2 & 1 \end{bmatrix}.$$

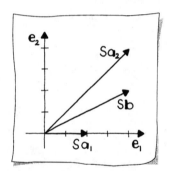

Sketch 5.5.

Shearing the vectors in a linear system.

We apply S_1 to all vectors involved in our system:

$$\begin{bmatrix} 1 & 0 \\ -1/2 & 1 \end{bmatrix} \begin{bmatrix} 2 \\ 1 \end{bmatrix} = \begin{bmatrix} 2 \\ 0 \end{bmatrix}, \quad \begin{bmatrix} 1 & 0 \\ -1/2 & 1 \end{bmatrix} \begin{bmatrix} 4 \\ 6 \end{bmatrix} = \begin{bmatrix} 4 \\ 4 \end{bmatrix},$$

$$\begin{bmatrix} 1 & 0 \\ -1/2 & 1 \end{bmatrix} \begin{bmatrix} 4 \\ 4 \end{bmatrix} = \begin{bmatrix} 4 \\ 2 \end{bmatrix}.$$

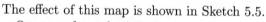

The effect of this map is shown in Sketch 5.5.

Our transformed system now reads

$$\begin{bmatrix} 2 & 4 \\ 0 & 4 \end{bmatrix} \begin{bmatrix} u_1 \\ u_2 \end{bmatrix} = \begin{bmatrix} 4 \\ 2 \end{bmatrix}.$$

Now we can employ *back substitution* to find

$$u_2 = 2/4 = 1/2,$$

$$u_1 = \frac{1}{2} \left(4 - 4 \times \frac{1}{2} \right) = 1.$$

For 2×2 linear systems there is only one matrix entry to zero in the Gauss elimination procedure. We will restate the procedure in a more algorithmic way in Chapter 14 when there is more work to do.

Example 5.3

We will look at one more example of Gauss elimination and back substitution. Let a linear system be given by

$$\begin{bmatrix} -1 & 4 \\ 2 & 2 \end{bmatrix} \begin{bmatrix} u_1 \\ u_2 \end{bmatrix} = \begin{bmatrix} 0 \\ 2 \end{bmatrix}.$$

The shear that takes $\mathbf{a}_1$ to the $\mathbf{e}_1$-axis is given by

$$S_1 = \begin{bmatrix} 1 & 0 \\ 2 & 1 \end{bmatrix},$$

and it transforms the system to

$$\begin{bmatrix} -1 & 4 \\ 0 & 10 \end{bmatrix} \begin{bmatrix} u_1 \\ u_2 \end{bmatrix} = \begin{bmatrix} 0 \\ 2 \end{bmatrix}.$$

Draw your own sketch to understand the geometry.

Using back substitution, the solution is now easily found as $u_1 = 8/10$ and $u_2 = 2/10$.

5.5   Undoing Maps: Inverse Matrices

In this section, we will see how to *undo* a linear map. Reconsider the linear system

$$A\mathbf{u} = \mathbf{b}.$$

The matrix A maps $\mathbf{u}$ to $\mathbf{b}$. Now that we know $\mathbf{u}$, what is the matrix B that maps $\mathbf{b}$ back to $\mathbf{u}$,

$$\mathbf{u} = B\mathbf{b}? \tag{5.6}$$

Defining B—the inverse map—is the purpose of this section.

In solving the original linear system, we applied shears to the column vectors of A and to $\mathbf{b}$. After the first shear, we had

$$S_1 A\mathbf{u} = S_1\mathbf{b}.$$

This demonstrated how shears can be used to zero elements of the matrix. Let's return to the example linear system in (5.2). After applying S_1 the system became

$$\begin{bmatrix} 2 & 4 \\ 0 & 4 \end{bmatrix} \begin{bmatrix} u_1 \\ u_2 \end{bmatrix} = \begin{bmatrix} 4 \\ 2 \end{bmatrix}.$$

Let's use another shear to zero the upper right element. Geometrically, this corresponds to constructing a shear that will map the new a_2 to the e_2-axis. It is given by the matrix

$$S_2 = \begin{bmatrix} 1 & -1 \\ 0 & 1 \end{bmatrix}.$$

Applying it to all vectors gives the new system

$$\begin{bmatrix} 2 & 0 \\ 0 & 4 \end{bmatrix} \begin{bmatrix} u_1 \\ u_2 \end{bmatrix} = \begin{bmatrix} 2 \\ 2 \end{bmatrix}.$$

After the second shear, our linear system has been changed to

$$S_2 S_1 A\mathbf{u} = S_2 S_1 \mathbf{b}.$$

We are not limited to applying shears—any affine map will do. Next, apply a non-uniform scaling S_3 in the e_1 and e_2 directions that will map the latest a_1 and a_2 onto the vectors e_1 and e_2. For our current example,

$$S_3 = \begin{bmatrix} 1/2 & 0 \\ 0 & 1/4 \end{bmatrix}.$$

The new system becomes

$$\begin{bmatrix} 1 & 0 \\ 0 & 1 \end{bmatrix} \begin{bmatrix} u_1 \\ u_2 \end{bmatrix} = \begin{bmatrix} 1 \\ 1/2 \end{bmatrix},$$

which corresponds to

$$S_3 S_2 S_1 A\mathbf{u} = S_3 S_2 S_1 \mathbf{b}.$$

This is a very special system. First of all, to solve for $\mathbf{u}$ is now trivial because A has been transformed into the *unit matrix* or *identity matrix* I,

$$I = \begin{bmatrix} 1 & 0 \\ 0 & 1 \end{bmatrix}. \tag{5.7}$$

This process of transforming A until it becomes the identity is theoretically equivalent to the back substitution process of Section 5.4. However, back substitution uses fewer operations and thus is the method of choice for solving linear systems.

Yet we have now found the matrix B in (5.6)! The two shears and scaling transformed A into the identity matrix I:

$$S_3 S_2 S_1 A = I; \tag{5.8}$$

thus, the solution of the system is

$$\mathbf{u} = S_3 S_2 S_1 \mathbf{b}. \tag{5.9}$$

This leads to the definition of the *inverse matrix* A^{-1} of a matrix A:

$$A^{-1} = S_3 S_2 S_1. \tag{5.10}$$

The matrix A^{-1} *undoes* the effect of the matrix A: the vector $\mathbf{u}$ was mapped to $\mathbf{b}$ by A, and $\mathbf{b}$ is mapped back to $\mathbf{u}$ by A^{-1}. Thus, we can now write (5.9) as

$$\mathbf{u} = A^{-1}\mathbf{b}.$$

If we combine (5.8) and (5.10), we immediately get

$$A^{-1}A = I. \tag{5.11}$$

This makes intuitive sense, since the actions of a map and its inverse should cancel out, i.e., not change anything—that is what I does! Figures 5.2 and 5.3 illustrate this.

The inverse of the identity is the identity

$$I^{-1} = I.$$

The inverse of a scaling is given by:

$$\begin{bmatrix} s & 0 \\ 0 & t \end{bmatrix}^{-1} = \begin{bmatrix} 1/s & 0 \\ 0 & 1/t \end{bmatrix}.$$

Multiply this out to convince yourself!

Figure 5.2 shows the effects of a matrix and its inverse for the scaling

$$\begin{bmatrix} 1 & 0 \\ 0 & 0.5 \end{bmatrix}.$$

Figure 5.3 shows the effects of a matrix and its inverse for the shear

$$\begin{bmatrix} 1 & 1 \\ 0 & 1 \end{bmatrix}.$$

We consider the inverse of a rotation as follows: if R_α rotates by α degrees counterclockwise, then $R_{-\alpha}$ rotates by α degrees clockwise, or

$$R_{-\alpha} = R_\alpha^{-1} = R_\alpha^{\mathrm{T}},$$

as you can see from the definition of a rotation matrix (4.16).

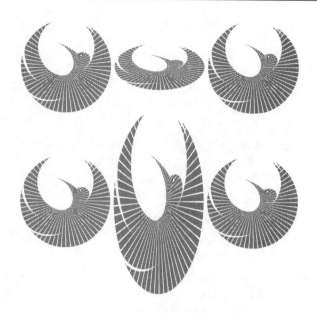

Figure 5.2.
Inverse matrices: illustrating scaling and its inverse, and that $AA^{-1} = A^{-1}A = I$. Top: the original Phoenix, the result of applying a scale, then the result of the inverse scale. Bottom: the original Phoenix, the result of applying the inverse scale, then the result of the original scale.

The rotation matrix is an example of an *orthogonal matrix*. An orthogonal matrix A is characterized by the fact that

$$A^{-1} = A^{\mathrm{T}}.$$

The column vectors $\mathbf{a}_1$ and $\mathbf{a}_2$ of an orthogonal matrix satisfy $\|\mathbf{a}_1\| = 1$, $\|\mathbf{a}_2\| = 1$ and $\mathbf{a}_1 \cdot \mathbf{a}_2 = 0$. In words, the column vectors are *orthonormal*. The row vectors are orthonormal as well. Those transformations that are described by orthogonal matrices are called *rigid body motions*. The determinant of an orthogonal matrix is ± 1.

We add without proof two fairly obvious identities:

$$A^{-1^{-1}} = A, \tag{5.12}$$

which should be obvious from Figures 5.2 and 5.3, and

$$A^{-1^{\mathrm{T}}} = A^{\mathrm{T}^{-1}}. \tag{5.13}$$

Figure 5.3.
Inverse matrices: illustrating a shear and its inverse, and that $AA^{-1} = A^{-1}A = I$. Top: the original Phoenix, the result of applying a shear, then the result of the inverse shear. Bottom: the original Phoenix, the result of applying the inverse shear, then the result of the original shear.

Figure 5.4 illustrates this for

$$A = \begin{bmatrix} 1 & 0 \\ 1 & 0.5 \end{bmatrix}.$$

Given a matrix A, how do we compute its inverse? Let us start with

$$AA^{-1} = I. \tag{5.14}$$

If we denote the two (unknown) columns of A^{-1} by $\bar{\mathbf{a}}_1$ and $\bar{\mathbf{a}}_2$, and those of I by $\mathbf{e}_1$ and $\mathbf{e}_2$, then (5.14) may be written as

$$A \begin{bmatrix} \bar{\mathbf{a}}_1 & \bar{\mathbf{a}}_2 \end{bmatrix} = \begin{bmatrix} \mathbf{e}_1 & \mathbf{e}_2 \end{bmatrix}.$$

This is really short for two linear systems

$$A\bar{\mathbf{a}}_1 = \mathbf{e}_1 \quad \text{and} \quad A\bar{\mathbf{a}}_2 = \mathbf{e}_2.$$

Both systems have the same matrix A and can thus be solved *simultaneously*. All we have to do is to apply the familiar shears and scale—those that transform A to I—to both $\mathbf{e}_1$ and $\mathbf{e}_2$.

Figure 5.4.
Inverse matrices: the top illustrates I, A^{-1}, $A^{-1}{}^T$ and the bottom illustrates I, A^T, $A^{T^{-1}}$.

Example 5.4

Let's revisit Example 5.3 with

$$A = \begin{bmatrix} -1 & 4 \\ 2 & 2 \end{bmatrix}.$$

Our two simultaneous systems are:

$$\begin{bmatrix} -1 & 4 \\ 2 & 2 \end{bmatrix} \begin{bmatrix} \bar{\mathbf{a}}_1 & \bar{\mathbf{a}}_2 \end{bmatrix} = \begin{bmatrix} 1 & 0 \\ 0 & 1 \end{bmatrix}.$$

The first shear takes this to

$$\begin{bmatrix} -1 & 4 \\ 0 & 10 \end{bmatrix} \begin{bmatrix} \bar{\mathbf{a}}_1 & \bar{\mathbf{a}}_2 \end{bmatrix} = \begin{bmatrix} 1 & 0 \\ 2 & 1 \end{bmatrix}.$$

The second shear yields

$$\begin{bmatrix} -1 & 0 \\ 0 & 10 \end{bmatrix} \begin{bmatrix} \bar{\mathbf{a}}_1 & \bar{\mathbf{a}}_2 \end{bmatrix} = \begin{bmatrix} 2/10 & -4/10 \\ 2 & 1 \end{bmatrix}.$$

Finally the scaling produces

$$\begin{bmatrix} 1 & 0 \\ 0 & 1 \end{bmatrix} \begin{bmatrix} \bar{\mathbf{a}}_1 & \bar{\mathbf{a}}_2 \end{bmatrix} = \begin{bmatrix} -2/10 & 4/10 \\ 2/10 & 1/10 \end{bmatrix}.$$

Thus the inverse matrix

$$A^{-1} = \begin{bmatrix} -2/10 & 4/10 \\ 2/10 & 1/10 \end{bmatrix}.$$

It can be the case that a matrix A does not have an inverse. For example, the matrix

$$\begin{bmatrix} 2 & 1 \\ 4 & 2 \end{bmatrix}$$

is not *invertible* because the columns are linearly dependent. A non-invertible matrix is also referred to as *singular*. If we try to compute the inverse by setting up two simultaneous systems,

$$\begin{bmatrix} 2 & 1 \\ 4 & 2 \end{bmatrix} \begin{bmatrix} \bar{\mathbf{a}}_1 & \bar{\mathbf{a}}_2 \end{bmatrix} = \begin{bmatrix} 1 & 0 \\ 0 & 1 \end{bmatrix},$$

then the first shear produces

$$\begin{bmatrix} 2 & 1 \\ 0 & 0 \end{bmatrix} \begin{bmatrix} \bar{\mathbf{a}}_1 & \bar{\mathbf{a}}_2 \end{bmatrix} = \begin{bmatrix} 1 & 0 \\ -2 & 1 \end{bmatrix}.$$

At this point it is clear that we will not be able to construct linear maps to achieve the identity matrix on the left side of the equation. Thus, a matrix with $|A| = 0$ is not invertible. In the next three sections we'll look at this type of matrix in more detail.

5.6 Unsolvable Systems

Consider the situation shown in Sketch 5.6. The two vectors $\mathbf{a}_1$ and $\mathbf{a}_2$ are multiples of each other. In other words, they are *linearly dependent*.

The corresponding linear system is

$$\begin{bmatrix} 2 & 1 \\ 4 & 2 \end{bmatrix} \begin{bmatrix} u_1 \\ u_2 \end{bmatrix} = \begin{bmatrix} 1 \\ 1 \end{bmatrix}.$$

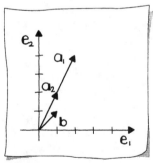

Sketch 5.6.
An unsolvable linear system.

It is obvious from the sketch that we have a problem here, but let's just blindly apply Gauss elimination; apply a shear such that $\mathbf{a}_1$ is mapped to the $\mathbf{e}_1$-axis. The resulting system is

$$\begin{bmatrix} 2 & 1 \\ 0 & 0 \end{bmatrix} \begin{bmatrix} u_1 \\ u_2 \end{bmatrix} = \begin{bmatrix} 1 \\ -1 \end{bmatrix}.$$

But the last equation reads $0 = -1$, and now we really are in trouble! This means that our system is *inconsistent*, and therefore does not have a solution.

5.7   Underdetermined Systems

Consider the system

$$\begin{bmatrix} 2 & 1 \\ 4 & 2 \end{bmatrix} \begin{bmatrix} u_1 \\ u_2 \end{bmatrix} = \begin{bmatrix} 3 \\ 6 \end{bmatrix},$$

shown in Sketch 5.7.

Again, we shear $\mathbf{a}_1$ onto the $\mathbf{e}_1$-axis, and obtain

$$\begin{bmatrix} 1 & 3 \\ 0 & 0 \end{bmatrix} \begin{bmatrix} u_1 \\ u_2 \end{bmatrix} = \begin{bmatrix} 0 \\ 0 \end{bmatrix}.$$

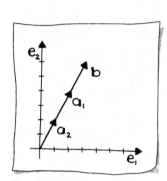

Sketch 5.7.
An underdetermined linear system.

Now the last equation reads $0 = 0$—true, but a bit trivial! In reality, our system is just one equation written down twice in slightly different forms. This is also clear from the sketch: $\mathbf{b}$ may be written as a multiple of either $\mathbf{a}_1$ or $\mathbf{a}_2$. This type of system is *consistent* because at least one solution exists.

5.8   Homogeneous Systems

A system of the form

$$A\mathbf{u} = \mathbf{0}, \tag{5.15}$$

i.e., one where the right-hand side consists of the zero vector, is called *homogeneous*. If it has a solution $\mathbf{u}$, then clearly all multiples $c\mathbf{u}$ are also solutions: we multiply both sides of the equations by a common factor c. In other words, the system has an *infinite number of solutions*. One obvious solution is the zero vector itself; this is called the *trivial solution* and is usually of little interest.

Not all homogeneous systems do have a nontrivial solution, however. Equation (5.15) may be read as follows: What vector $\mathbf{u}$, when mapped by A, has the zero vector as its image? The only maps capable of achieving this are *projections*, which have rank 1. They are characterized by the fact that their two columns $\mathbf{a}_1$ and $\mathbf{a}_2$ are parallel, or linearly dependent.

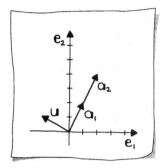

Example 5.5

An example, illustrated in Sketch 5.8, should help. Let our homogeneous system be

$$\begin{bmatrix} 1 & 2 \\ 2 & 4 \end{bmatrix} \mathbf{u} = \begin{bmatrix} 0 \\ 0 \end{bmatrix}.$$

Sketch 5.8.
A homogeneous system.

Clearly, $\mathbf{a}_2 = 2\mathbf{a}_1$; the matrix A maps all vectors onto the line defined by $\mathbf{a}_1$ and the origin. In this example, any vector $\mathbf{u}$ which is perpendicular to $\mathbf{a}_1$ will be projected to the zero vector:

$$A[c\mathbf{u}] = \mathbf{0}.$$

An easy check reveals that

$$\mathbf{u} = \begin{bmatrix} -2 \\ 1 \end{bmatrix}$$

is a solution to the system; so is any multiple of it. Also check that $\mathbf{a}_1 \cdot \mathbf{u} = 0$, so they are in fact perpendicular. Recall from Section 4.9 that all vectors $\mathbf{u}$ that satisfy a homogeneous system make up the *kernel* of the matrix.

Example 5.6

We now consider an example of a homogeneous system that has only the trivial solution:

$$\begin{bmatrix} 1 & 2 \\ 2 & 1 \end{bmatrix} \mathbf{u} = \begin{bmatrix} 0 \\ 0 \end{bmatrix}.$$

The two columns of A are linearly independent; therefore, A does not reduce dimensionality. Then it cannot map any (nonzero) vector $\mathbf{u}$ to the zero vector!

In general, we may state that a homogeneous system has nontrivial solutions (and therefore, infinitely many solutions) only if the columns of the matrix are linearly dependent.

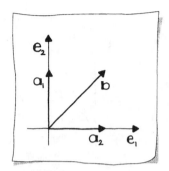

Sketch 5.9.

A linear system that needs pivoting.

5.9 Numerical Strategies: Pivoting

Consider the system

$$\begin{bmatrix} 0 & 1 \\ 1 & 0 \end{bmatrix} \begin{bmatrix} u_1 \\ u_2 \end{bmatrix} = \begin{bmatrix} 1 \\ 1 \end{bmatrix},$$

illustrated in Sketch 5.9.

Our standard approach, shearing $\mathbf{a}_1$ onto the $\mathbf{e}_1$-axis, will not work here; there is no shear that takes

$$\begin{bmatrix} 0 \\ 1 \end{bmatrix}$$

onto the $\mathbf{e}_1$-axis. However, there is no problem if we simply interchange the two equations! Then we have

$$\begin{bmatrix} 1 & 0 \\ 0 & 1 \end{bmatrix} \begin{bmatrix} u_1 \\ u_2 \end{bmatrix} = \begin{bmatrix} 1 \\ 1 \end{bmatrix},$$

and thus $u_1 = u_2 = 1$. So we cannot blindly apply a shear to $\mathbf{a}_1$; we must first check that one exists. If it does not—i.e., if $a_{1,1} = 0$—interchange the equations.

As a rule of thumb, if a method fails because some number equals zero, then it will work poorly if that number is small. It is thus advisable to interchange the two equations anytime we have $|a_{1,1}| < |a_{1,2}|$. The absolute value is used here since we are interested in the magnitude of the involved numbers, not their sign. The process of interchanging equations is called *pivoting*, and it is used to improve numerical stability.

Example 5.7

Let's study a somewhat realistic example taken from [14]:

$$\begin{bmatrix} 0.0001 & 1 \\ 1 & 1 \end{bmatrix} \begin{bmatrix} u_1 \\ u_2 \end{bmatrix} = \begin{bmatrix} 1 \\ 2 \end{bmatrix},$$

If we shear $\mathbf{a}_1$ onto the $\mathbf{e}_1$-axis, the new system reads

$$\begin{bmatrix} 0.0001 & 1 \\ 0 & -9999 \end{bmatrix} \begin{bmatrix} u_1 \\ u_2 \end{bmatrix} = \begin{bmatrix} 1 \\ -9998 \end{bmatrix},$$

Note how "far out" the new $\mathbf{a}_2$ and $\mathbf{b}$ are relative to $\mathbf{a}_1$! This is the type of behavior that causes numerical problems. Numbers that differ greatly in magnitude tend to need more digits for calculation. On a finite precision machine these extra digits are not always available, thus round-off can take us far from the true solution.

Suppose we have a machine which only stores three digits, although it calculates with six digits. Due to round-off, the system above would be stored as

$$\begin{bmatrix} 0.0001 & 1 \\ 0 & -10000 \end{bmatrix} \begin{bmatrix} u_1 \\ u_2 \end{bmatrix} = \begin{bmatrix} 1 \\ -10000 \end{bmatrix},$$

which would result in a solution of $u_2 = 1$ and $u_1 = 0$, which is not close to the true solution (rounded to five digits) of $u_2 = 1.0001$ and $u_1 = 0.99990$.

Luckily, pivoting is a tool to damper the effects of round-off. Now employ pivoting by interchanging the rows, yielding the system

$$\begin{bmatrix} 1 & 1 \\ 0.0001 & 1 \end{bmatrix} \begin{bmatrix} u_1 \\ u_2 \end{bmatrix} = \begin{bmatrix} 2 \\ 1 \end{bmatrix}.$$

Shear $\mathbf{a}_1$ onto the $\mathbf{e}_1$-axis, and the new system reads

$$\begin{bmatrix} 1 & 1 \\ 0 & 0.9999 \end{bmatrix} \begin{bmatrix} u_1 \\ u_2 \end{bmatrix} = \begin{bmatrix} 2 \\ 0.9998 \end{bmatrix}.$$

Notice that the vectors are all within the same range. Even with the three digit machine, this system will allow us to compute a result which is "close" to the true solution, because the effects of round-off have been minimized.

5.10   Defining a Map

Matrices map vectors to vectors. If we know the result of such a map, namely that two vectors $\mathbf{v}_1$ and $\mathbf{v}_2$ were mapped to $\mathbf{v}_1'$ and $\mathbf{v}_2'$, can we find the matrix that did it?

Suppose some matrix A was responsible for the map. We would then have the two equations

$$A\mathbf{v}_1 = \mathbf{v}_1' \quad \text{and} \quad A\mathbf{v}_2 = \mathbf{v}_2'.$$

Combining them, we can write

$$A[\mathbf{v}_1, \mathbf{v}_2] = [\mathbf{v}_1', \mathbf{v}_2'],$$

or, even shorter,

$$AV = V'. \tag{5.16}$$

To define A, we simply use shears, S_1 and S_2, and a scaling S_3 to transform V into the identity matrix as follows:

$$AVS_1S_2S_3 = V'S_1S_2S_3$$
$$AVV^{-1} = V'V^{-1}$$
$$A = V'V^{-1}.$$

- solution spaces
- consistent linear system
- Cramer's rule
- upper triangular
- Gauss elimination
- back substitution
- linear combination
- inverse
- orthogonal matrix

- orthonormal
- rigid body motion
- inconsistent system of equations
- underdetermined system of equations
- homogeneous system
- kernel
- pivoting
- linear map construction

5.11 Exercises

1. Using the matrix form, write down the linear system to find

$$\begin{bmatrix} 6 \\ 3 \end{bmatrix}$$

in terms of the local coordinate system defined by the origin,

$$\mathbf{a}_1 = \begin{bmatrix} 2 \\ -3 \end{bmatrix} \quad \text{and} \quad \mathbf{a}_2 = \begin{bmatrix} 6 \\ 0 \end{bmatrix}.$$

2. Use Cramer's rule to solve the system in Exercise 1.

3. Give an example of an upper triangular matrix.

4. Use Gauss elimination and back substitution to solve the system in Exercise 1.

5. Find the inverse of the matrix in Exercise 1.

6. What is the inverse of the matrix

$$\begin{bmatrix} 10 & 0 \\ 0 & 0.5 \end{bmatrix}?$$

7. What is the inverse of the matrix

$$\begin{bmatrix} \cos 30° & -\sin 30° \\ \sin 30° & \cos 30° \end{bmatrix}?$$

8. Give an example along with a sketch of an unsolvable system. Do the same for an underdetermined system.

9. Under what conditions can a nontrivial solution be found to a homogeneous system?

10. Resolve the system in Exercise 1 with Gauss elimination with pivoting.

11. Define the matrix A that maps

$$\begin{bmatrix} 1 \\ 0 \end{bmatrix} \rightarrow \begin{bmatrix} 1 \\ 0 \end{bmatrix} \quad \text{and} \quad \begin{bmatrix} 1 \\ 1 \end{bmatrix} \rightarrow \begin{bmatrix} 1 \\ -1 \end{bmatrix}.$$

12. What is the kernel of the matrix

$$A = \begin{bmatrix} 2 & 6 \\ 4 & 12 \end{bmatrix}?$$

6

Moving Things Around:
Affine Maps in 2D

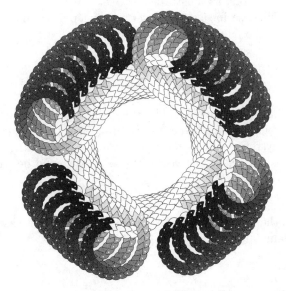

Figure 6.1.
Moving things around: affine maps in 2D applied to a familiar video game character.

Imagine playing a video game. As you press a button, figures and objects on the screen start moving around; they shift their position, they rotate, they zoom in or out. As you see this kind of motion,

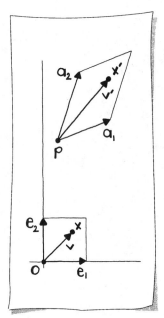

Sketch 6.1.

A skew target box.

the video software must carry out quite a few computations. Such computations have been applied to the familiar face in Figure 6.1. These computations are implementations of *affine maps*, the subject of this chapter.

6.1   Coordinate Transformations

In Section 4.1 the focus was on constructing a linear map which takes a vector $\mathbf{v}$ in the $[\mathbf{e}_1, \mathbf{e}_2]$-system,

$$\mathbf{v} = v_1\mathbf{e}_1 + v_2\mathbf{e}_2,$$

to a vector $\mathbf{v}'$ in the $[\mathbf{a}_1, \mathbf{a}_2]$-system

$$\mathbf{v}' = v_1\mathbf{a}_1 + v_2\mathbf{a}_2.$$

Recall that this latter system describes a parallelogram (skew) target box with lower-left corner at the origin. In this chapter, we want to construct this skew target box anywhere, and we want to map points rather than vectors, as illustrated by Sketch 6.1.

Now we will describe a skew target box by a point $\mathbf{p}$ and two vectors $\mathbf{a}_1, \mathbf{a}_2$. A point $\mathbf{x}$ is mapped to a point $\mathbf{x}'$ by

$$\mathbf{x}' = \mathbf{p} + x_1\mathbf{a}_1 + x_2\mathbf{a}_2, \tag{6.1}$$
$$= \mathbf{p} + A\mathbf{x} \tag{6.2}$$

as illustrated by Sketch 6.1. This simply states that we duplicate the $[\mathbf{e}_1, \mathbf{e}_2]$-geometry in the $[\mathbf{a}_1, \mathbf{a}_2]$-system: $\mathbf{x}'$ has the same coordinates in the new system as $\mathbf{x}$ did in the old one. Technically, the linear map A in (6.2) is applied to the vector $\mathbf{x} - \mathbf{o}$, so it should be written as

$$\mathbf{x}' = \mathbf{p} + A(\mathbf{x} - \mathbf{o}), \tag{6.3}$$

where $\mathbf{o}$ is the origin of $\mathbf{x}$'s coordinate system. In most cases, we will have the familiar

$$\mathbf{o} = \begin{bmatrix} 0 \\ 0 \end{bmatrix},$$

and then we will simply drop the "$-\mathbf{o}$" part, as in (6.2).

Affine maps are the basic tool to move and orient objects. All are of the form given in (6.3) and thus have two parts: a *translation*, given by $\mathbf{p}$ and a *linear map*, given by A.

Let's try representing the coordinate transformation of Section 1.1 as an affine map. The point $\mathbf{u}$ lives in the $[\mathbf{e}_1, \mathbf{e}_2]$-system, and we wish to find $\mathbf{x}$ in the $[\mathbf{a}_1, \mathbf{a}_2]$-system. Recall that the extents of the target box defined $\Delta_1 = \max_1 - \min_1$ and $\Delta_2 = \max_2 - \min_2$, so we set

$$\mathbf{p} = \begin{bmatrix} \min_1 \\ \min_2 \end{bmatrix}, \qquad \mathbf{a}_1 = \begin{bmatrix} \Delta_1 \\ 0 \end{bmatrix}, \qquad \mathbf{a}_2 = \begin{bmatrix} 0 \\ \Delta_2 \end{bmatrix}.$$

The affine map is defined as

$$\begin{bmatrix} x_1 \\ x_2 \end{bmatrix} = \begin{bmatrix} \min_1 \\ \min_2 \end{bmatrix} + \begin{bmatrix} \Delta_1 & 0 \\ 0 & \Delta_2 \end{bmatrix} \begin{bmatrix} u_1 \\ u_2 \end{bmatrix},$$

and we recover (1.3) and (1.4).

Of course we are not restricted to target boxes that are parallel to the $[\mathbf{e}_1, \mathbf{e}_2]$-coordinate axes. Let's look at such an example.

Example 6.1

Let

$$\mathbf{p} = \begin{bmatrix} 2 \\ 2 \end{bmatrix}, \qquad \mathbf{a}_1 = \begin{bmatrix} 2 \\ 1 \end{bmatrix}, \qquad \mathbf{a}_2 = \begin{bmatrix} -2 \\ 4 \end{bmatrix}$$

define a new coordinate system, and let

$$\mathbf{x} = \begin{bmatrix} 2 \\ 1/2 \end{bmatrix}$$

be a point in the $[\mathbf{e}_1, \mathbf{e}_2]$-system. In the new coordinate system, the $[\mathbf{a}_1, \mathbf{a}_2]$-system, the coordinates of $\mathbf{x}$ define a new point $\mathbf{x}'$. What is this point with respect to the $[\mathbf{e}_1, \mathbf{e}_2]$-system? The solution:

$$\mathbf{x}' = \begin{bmatrix} 2 \\ 2 \end{bmatrix} + \begin{bmatrix} 2 & -2 \\ 1 & 4 \end{bmatrix} \begin{bmatrix} 2 \\ 1/2 \end{bmatrix} = \begin{bmatrix} 5 \\ 6 \end{bmatrix}. \tag{6.4}$$

Thus, $\mathbf{x}'$ has coordinates

$$\begin{bmatrix} 2 \\ 1/2 \end{bmatrix}$$

with respect to the $[\mathbf{a}_1, \mathbf{a}_2]$-system; with respect to the $[\mathbf{e}_1, \mathbf{e}_2]$-system, it has coordinates

$$\begin{bmatrix} 5 \\ 6 \end{bmatrix}.$$

(See Sketch 6.2 for an illustration.)

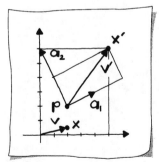

Sketch 6.2.
Mapping a point and a vector.

And now an example of a skew target box. Let's revisit the Example from Section 5.1, and add an affine aspect to it by translating our target box.

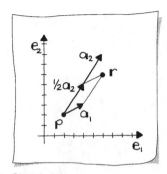

Sketch 6.3.

A new coordinate system.

Example 6.2

Sketch 6.3 illustrates the given geometry,

$$\mathbf{p} = \begin{bmatrix} 2 \\ 2 \end{bmatrix}, \quad \mathbf{a}_1 = \begin{bmatrix} 2 \\ 1 \end{bmatrix}, \quad \mathbf{a}_2 = \begin{bmatrix} 4 \\ 6 \end{bmatrix},$$

and the point

$$\mathbf{r} = \begin{bmatrix} 6 \\ 6 \end{bmatrix}$$

with respect to the $[\mathbf{e}_1, \mathbf{e}_2]$-system. We may ask, what are the coordinates of $\mathbf{r}$ with respect to the $[\mathbf{a}_1, \mathbf{a}_2]$-system? This was the topic of Chapter 5; we simply set up the linear system, $A\mathbf{u} = (\mathbf{r} - \mathbf{p})$, or

$$\begin{bmatrix} 2 & 4 \\ 1 & 6 \end{bmatrix} \begin{bmatrix} u_1 \\ u_2 \end{bmatrix} = \begin{bmatrix} 4 \\ 4 \end{bmatrix}.$$

Using the methods of Chapter 5, we find that $\mathbf{u} = \begin{bmatrix} 1 \\ 1/2 \end{bmatrix}$.

6.2 Affine and Linear Maps

A map of the form $\mathbf{v}' = A\mathbf{v}$ is called a *linear map* because it preserves linear combinations of vectors. This idea is expressed in (4.8) and illustrated in Sketch 4.4. A very fundamental property of linear maps has to do with *ratios*, which are defined in Section 2.5. What happens to the ratio of three collinear points when we map them by an affine map? The answer to this question is fairly fundamental to all of geometry, and it is: nothing. In other words, affine maps leave ratios unchanged, or *invariant*. To see this, let

$$\mathbf{p}_2 = (1 - t)\mathbf{p}_1 + t\mathbf{p}_3$$

and let an affine map be defined by

$$\mathbf{x}' = A\mathbf{x} + \mathbf{p}.$$

We now have

$$\begin{aligned}
\mathbf{p}_2' &= A((1 - t)\mathbf{p}_1 + t\mathbf{p}_3) + \mathbf{p} \\
&= (1 - t)A\mathbf{p}_1 + tA\mathbf{p}_3 + [(1 - t) + t]\mathbf{p} \\
&= (1 - t)[A\mathbf{p}_1 + \mathbf{p}] + t[A\mathbf{p}_3 + \mathbf{p}] \\
&= (1 - t)\mathbf{p}_1' + t\mathbf{p}_3'.
\end{aligned}$$

The step from the first to the second equation may seem a bit contrived; yet it is the one that makes crucial use of the fact that we are combining points using *barycentric combinations*: $(1-t)+t=1$.

The last equation shows that the linear $(1-t), t$ relationship between three points is not changed by affine maps—meaning that their ratio is invariant, as is illustrated in Sketch 6.4. In particular, the midpoint of two points will be mapped to the midpoint of the image points.

The other basic property of affine maps is this: they map parallel lines to parallel lines. If two lines do not intersect before they are mapped, then they will not intersect afterwards either. Conversely, two lines that intersect before the map will also do so afterwards. Figure 6.2 shows how two families of parallel lines are mapped to two families of parallel lines. The two families intersect before and after the affine map. The map uses the matrix

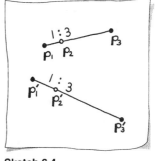

Sketch 6.4.

Ratios are invariant under affine maps.

$$A = \begin{bmatrix} 1 & 2 \\ 2 & 1 \end{bmatrix}.$$

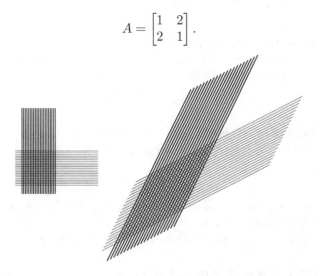

Figure 6.2.
Affine maps: parallel lines are mapped to parallel lines.

6.3 Translations

If an object is moved without changing its orientation, then it is *translated*. See Figure 6.3 for several translations of the letter **D**.[1]

---

[1] We rendered the **D** darker as it was translated more.

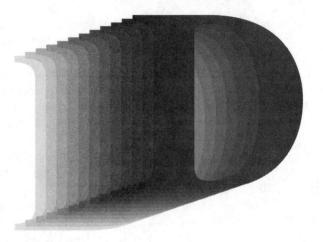

Figure 6.3.
Translations: the letter **D** is moved several times.

How is this action covered by the general affine map in (6.3)? Recall the *identity matrix* from Section 5.5 which has no effect whatsoever on any vector: we always have

$$I\mathbf{x} = \mathbf{x},$$

which you should be able to verify without effort.

A translation is thus written in the context of (6.3) as

$$\mathbf{x}' = \mathbf{p} + I\mathbf{x}.$$

One property of translations is that they do not change areas; all **D**'s in Figure 6.3 have the same area. A translation causes a *rigid body motion*. Recall that rotations are also of this type.

6.4   More General Affine Maps

In this section, we present two very common geometric problems which are affine maps. It is one thing to say "every affine map is of the form $A\mathbf{x} + \mathbf{p}$," but it is not always clear what A and $\mathbf{p}$ should be for a given problem. Sometimes a more constructive approach is called for, as is the case with Problem 2 in this section.

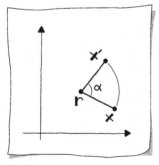

Sketch 6.5.
Rotating a point about another point.

Problem 1: Let $\mathbf{r}$ be some point around which you would like to *rotate* some other point $\mathbf{x}$ by α degrees, as shown in Sketch 6.5. Let $\mathbf{x}'$ be the rotated point.

Rotations have only been defined around the origin, not around arbitrary points. Hence, we translate our given geometry (the two points $\mathbf{r}$ and $\mathbf{x}$) such that $\mathbf{r}$ moves to the origin. This is easy:

$$\bar{\mathbf{r}} = \mathbf{r} - \mathbf{r} = \mathbf{0}, \quad \bar{\mathbf{x}} = \mathbf{x} - \mathbf{r}.$$

Now we rotate the vector $\bar{\mathbf{x}}$ around the origin by α degrees:

$$\bar{\bar{\mathbf{x}}} = A\bar{\mathbf{x}}.$$

The matrix A would be taken directly from (4.16). Finally, we translate $\bar{\bar{\mathbf{x}}}$ back to the center $\mathbf{r}$ of rotation:

$$\mathbf{x}' = A\bar{\mathbf{x}} + \mathbf{r}.$$

Let's reformulate this in terms of the given information. This is achieved by replacing $\bar{\mathbf{x}}$ by its definition:

$$\mathbf{x}' = A(\mathbf{x} - \mathbf{r}) + \mathbf{r}. \tag{6.5}$$

Example 6.3

Let

$$\mathbf{r} = \begin{bmatrix} 2 \\ 1 \end{bmatrix}, \quad \mathbf{x} = \begin{bmatrix} 3 \\ 0 \end{bmatrix}$$

and $\alpha = 90°$. We obtain

$$\mathbf{x}' = \begin{bmatrix} 0 & -1 \\ 1 & 0 \end{bmatrix} \begin{bmatrix} 1 \\ -1 \end{bmatrix} + \begin{bmatrix} 2 \\ 1 \end{bmatrix} = \begin{bmatrix} 3 \\ 2 \end{bmatrix}.$$

See Sketch 6.6 for an illustration.

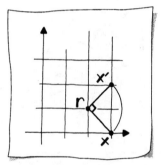

Sketch 6.6.
Rotate $\mathbf{x}$ 90° around $\mathbf{r}$.

Problem 2: Let l be a line and $\mathbf{x}$ be a point. You want to *reflect* $\mathbf{x}$ across l, with result $\mathbf{x}'$, as shown in Sketch 6.7. This problem could be solved using the following affine maps. Find the intersection $\mathbf{r}$ of l with the $\mathbf{e}_1$-axis. Find the cosine of the angle between l and $\mathbf{e}_1$. Rotate $\mathbf{x}$ around $\mathbf{r}$ such that l is mapped to the $\mathbf{e}_1$-axis. Reflect the rotated $\mathbf{x}$ across the $\mathbf{e}_1$-axis, and finally undo the rotation. Complicated!

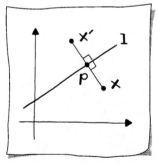

Sketch 6.7.
Reflect a point across a line.

It is much easier to employ the 'foot of a point' algorithm which finds the closest point $\mathbf{p}$ on a line l to a point $\mathbf{x}$, which was developed in Section 3.7. Then $\mathbf{p}$ must be the midpoint of $\mathbf{x}$ and $\mathbf{x}'$:

$$\mathbf{p} = \frac{1}{2}\mathbf{x} + \frac{1}{2}\mathbf{x}',$$

from which we conclude

$$\mathbf{x}' = 2\mathbf{p} - \mathbf{x}. \tag{6.6}$$

While this does not have the standard affine map form, it is equivalent to it, yet computationally much less complex.

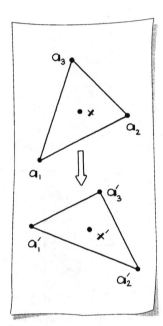

Sketch 6.8.

Two triangles define an affine map.

6.5   Mapping Triangles to Triangles

Affine maps may be viewed as combinations of linear maps and translations. Another flavor of affine maps is described in this section; it draws from concepts in Chapter 5.

This other flavor arises like this: given a (source) triangle T with vertices $\mathbf{a}_1, \mathbf{a}_2, \mathbf{a}_3$, and a (target) triangle T' with vertices $\mathbf{a}_1', \mathbf{a}_2', \mathbf{a}_3'$, what affine maps takes T to T'? More precisely, if $\mathbf{x}$ is a point inside T, it will be mapped to a point $\mathbf{x}'$ inside T': how do we find $\mathbf{x}'$? For starters, see Sketch 6.8.

Our desired affine map will be of the form

$$\mathbf{x}' = A[\mathbf{x} - \mathbf{a}_1] + \mathbf{a}_1',$$

thus we need to find the matrix A. (We have chosen $\mathbf{a}_1$ and $\mathbf{a}_1'$ as the origins in the two coordinate systems.) We define (see Sketch 6.8)

$$\mathbf{v}_2 = \mathbf{a}_2 - \mathbf{a}_1, \qquad \mathbf{v}_3 = \mathbf{a}_3 - \mathbf{a}_1$$

and

$$\mathbf{v}_2' = \mathbf{a}_2' - \mathbf{a}_1', \qquad \mathbf{v}_3' = \mathbf{a}_3' - \mathbf{a}_1'.$$

We know

$$A\mathbf{v}_2 = \mathbf{v}_2',$$
$$A\mathbf{v}_3 = \mathbf{v}_3'.$$

These two vector equations may be combined into one matrix equation:

$$A\begin{bmatrix}\mathbf{v}_2 & \mathbf{v}_3\end{bmatrix} = \begin{bmatrix}\mathbf{v}_2' & \mathbf{v}_3'\end{bmatrix},$$

which we abbreviate as

$$AV = V'.$$

We multiply both sides of this equation by V's inverse V^{-1}, see Chapter 5, and obtain A as

$$A = V'V^{-1}.$$

Example 6.4

Triangle T is defined by the vertices

$$\mathbf{a}_1 = \begin{bmatrix} 0 \\ 1 \end{bmatrix}, \quad \mathbf{a}_2 = \begin{bmatrix} -1 \\ -1 \end{bmatrix}, \quad \mathbf{a}_3 = \begin{bmatrix} 1 \\ -1 \end{bmatrix},$$

and triangle T' is defined by the vertices

$$\mathbf{a}_1' = \begin{bmatrix} 0 \\ 1 \end{bmatrix}, \quad \mathbf{a}_2' = \begin{bmatrix} 1 \\ 3 \end{bmatrix}, \quad \mathbf{a}_3' = \begin{bmatrix} -1 \\ 3 \end{bmatrix}.$$

The matrices V and V' are then defined as

$$V = \begin{bmatrix} -1 & 1 \\ -2 & -2 \end{bmatrix}, \quad V' = \begin{bmatrix} 1 & -1 \\ 2 & 2 \end{bmatrix}.$$

The inverse of the matrix V is

$$V^{-1} = \begin{bmatrix} -1/2 & -1/4 \\ 1/2 & -1/4 \end{bmatrix},$$

thus the linear map A is defined as

$$A = \begin{bmatrix} 1 & 0 \\ 0 & -1 \end{bmatrix}.$$

Do you recognize the map?

Let's try a sample point

$$\mathbf{x} = \begin{bmatrix} 0 \\ -1/3 \end{bmatrix}$$

in T. This point is mapped to

$$\mathbf{x}' = \begin{bmatrix} 1 & 0 \\ 0 & -1 \end{bmatrix} \left[\begin{bmatrix} 0 \\ -1/3 \end{bmatrix} - \begin{bmatrix} 0 \\ 1 \end{bmatrix} \right] + \begin{bmatrix} 0 \\ 1 \end{bmatrix} = \begin{bmatrix} 0 \\ 7/3 \end{bmatrix}$$

in T'.

Note that V's inverse V^{-1} might not exist; this is the case when $\mathbf{v}_2$ and $\mathbf{v}_3$ are linearly dependent and thus $|V| = 0$.

6.6   Composing Affine Maps

Linear maps are an important theoretical tool, but ultimately we are interested in affine maps; they map objects which are defined by points to other such objects.

If an affine map is given by

$$\mathbf{x}' = \mathbf{p} + A(\mathbf{x} - \mathbf{o}),$$

nothing keeps us from applying it twice, resulting in $\mathbf{x}''$:

$$\mathbf{x}'' = \mathbf{p} + A(\mathbf{x}' - \mathbf{o}).$$

This may be repeated several times—for interesting choices of A and $\mathbf{p}$, interesting images will result. (We added extra gray scales to our figures to help distinguish between the individual maps.)

Our test objects will be the letters $\mathbf{D}$ and $\mathbf{S}$. If we apply a sequence of scalings, we get a result as in Figure 6.4. As the iteration of maps proceeds, we darkened the corresponding letters. We have moved the origin to $\mathbf{D}$'s center. The affine map is defined by

$$A = \begin{bmatrix} 1.1 & 0 \\ 0 & 1.1 \end{bmatrix} \quad \text{and} \quad \mathbf{p} = \begin{bmatrix} 0 \\ 0 \end{bmatrix}.$$

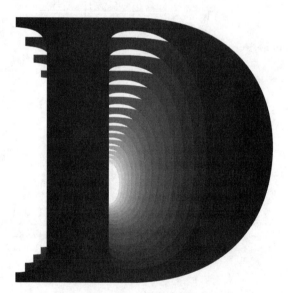

Figure 6.4.
Scaling: the letter D is scaled several times; the origin is at its center.

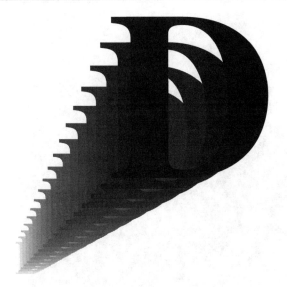

Figure 6.5.

Scaling: the letter D is scaled several times; a translation was applied at each step as well.

In affine space, we can introduce a translation, and obtain Figure 6.5. This was achieved by setting

$$\mathbf{p} = \begin{bmatrix} 2 \\ 2 \end{bmatrix}.$$

Rotations can also be made more interesting. In Figure 6.6, you see the letter **S** rotated several times around the origin, which is near the lower left of the letter.

Adding scaling and rotation results in Figure 6.7. The basic affine map for this case is given by

$$\mathbf{x}' = S[R\mathbf{x} + \mathbf{p}]$$

where R rotates by $-20°$, S scales nonuniformly, and $\mathbf{p}$ translates:

$$R = \begin{bmatrix} \cos(-20) & -\sin(-20) \\ \sin(-20) & \cos(-20) \end{bmatrix}, \quad S = \begin{bmatrix} 1.25 & 0 \\ 0 & 1.1 \end{bmatrix}, \quad \mathbf{p} = \begin{bmatrix} 5 \\ 5 \end{bmatrix}.$$

Figure 6.6.
Rotations: the letter S is rotated several times; the origin is at the lower left of the letter.

Figure 6.7.
Rotations: the letter S is rotated several times; scalings and translations are also applied.

Figure 6.8.
M.C. Escher: Magic Mirror (1949).

We finish this chapter with Figure 6.8 by the Dutch artist, M.C. Escher [5], who in a very unique way mixed complex geometric issues with a unique style.

Figure 6.8 is itself a 2D object, and so may be subjected to affine maps. Figure 6.9 gives an example. The matrix used here is

$$A = \begin{bmatrix} 1 & 0.5 \\ -0.2 & 0.7 \end{bmatrix}. \tag{6.7}$$

Figure 6.9.
M.C. Escher: Magic Mirror (1949); affine map applied.

- linear map
- rigid body motion
- affine map
- translation
- identity matrix

- invariant ratios
- rotate a point about another point
- reflect a point about a line
- three points mapped to three points

6.7 Exercises

For Exercises 1 and 2 let

$$A = \begin{bmatrix} 2 & 1 \\ 1 & 2 \end{bmatrix} \quad \text{and} \quad \mathbf{p} = \begin{bmatrix} 2 \\ 2 \end{bmatrix}.$$

1. Let

$$\mathbf{r} = \begin{bmatrix} 0 \\ 1 \end{bmatrix}, \qquad \mathbf{s} = \begin{bmatrix} 1 \\ 3/2 \end{bmatrix},$$

and $\mathbf{q} = 1/3\mathbf{r} + 2/3\mathbf{s}$. Compute $\mathbf{r}', \mathbf{s}', \mathbf{q}'$; e.g., $\mathbf{r}' = A\mathbf{r} + \mathbf{p}$. Show that $\mathbf{q}' = 1/3\mathbf{r}' + 2/3\mathbf{s}'$.

2. Let

$$\mathbf{t} = \begin{bmatrix} 0 \\ 1 \end{bmatrix} \quad \text{and} \quad \mathbf{m} = \begin{bmatrix} 2 \\ 1 \end{bmatrix}.$$

Compute $\mathbf{t}'$ and $\mathbf{m}'$. Sketch the lines defined by $\mathbf{t}, \mathbf{m}$ and $\mathbf{t}', \mathbf{m}'$. Do the same for $\mathbf{r}$ and $\mathbf{s}$ from the previous exercise. What does this illustrate?

3. Rotate the point

$$\mathbf{x} = \begin{bmatrix} -2 \\ -2 \end{bmatrix}$$

by $90°$ around the point

$$\mathbf{r} = \begin{bmatrix} -2 \\ 2 \end{bmatrix}.$$

Define A and $\mathbf{p}$ of the affine map.

4. Reflect the point $\mathbf{x} = \begin{bmatrix} 0 \\ 2 \end{bmatrix}$ about the line $\mathbf{l}(t) = \begin{bmatrix} 0 \\ 0 \end{bmatrix} + t \begin{bmatrix} 1 \\ 2 \end{bmatrix}$.

5. Given a triangle T with vertices

$$\mathbf{a}_1 = \begin{bmatrix} 2 \\ 0 \end{bmatrix} \qquad \mathbf{a}_2 = \begin{bmatrix} 0 \\ 1 \end{bmatrix} \qquad \mathbf{a}_3 = \begin{bmatrix} -2 \\ 0 \end{bmatrix},$$

and T' with vertices

$$\mathbf{a}_1' = \begin{bmatrix} 2 \\ 0 \end{bmatrix} \qquad \mathbf{a}_2' = \begin{bmatrix} 0 \\ -1 \end{bmatrix} \qquad \mathbf{a}_3' = \begin{bmatrix} -2 \\ 0 \end{bmatrix},$$

suppose that the triangle T has been mapped to T' via an affine map. What are the coordinates of the point $\mathbf{x}'$ corresponding to

$$\mathbf{x} = \begin{bmatrix} 0 \\ 0 \end{bmatrix}?$$

6. Let's revisit the coordinate transformation from Exercise 5 in Chapter 1. Construct the affine map which takes a 2D point $\mathbf{x}$ in *NDC coordinates* to the 2D point $\mathbf{x}'$ in a *viewport*. Recall that the extents of the NDC system are defined by a the lower-left and upper-right points

$$\mathbf{l}_n = \begin{bmatrix} -1 \\ -1 \end{bmatrix} \quad \text{and} \quad \mathbf{u}_n = \begin{bmatrix} 1 \\ 1 \end{bmatrix},$$

respectively. Suppose we want to map to a viewport with extents

$$\mathbf{l}_v = \begin{bmatrix} 10 \\ 10 \end{bmatrix} \quad \text{and} \quad \mathbf{u}_v = \begin{bmatrix} 30 \\ 20 \end{bmatrix}.$$

After constructing the affine map, find the points in the viewport associated with the NDC points

$$\mathbf{x}_1 = \begin{bmatrix} -1 \\ -1 \end{bmatrix}, \quad \mathbf{x}_2 = \begin{bmatrix} 1 \\ 1 \end{bmatrix}, \quad \mathbf{x}_3 = \begin{bmatrix} -1/2 \\ 1/2 \end{bmatrix}.$$

7. Affine maps transform parallel lines to parallel lines. Do affine maps transform perpendicular lines to perpendicular lines?

8. Which affine maps are rigid body motions?

9. The solution to the problem of reflecting a point across a line is given by (6.6). Why is this a valid combination of points?

PS1 Experiment with the file S_rotran.ps by changing some of the parameters in the for loop.

PS2 Experiment with the file Escher_aff.ps. Just before the unreadable part, you see the line

/matrix[1 −.2 0.5 0.7 0 0]def.

This is PostScript's way of defining the matrix from (6.7). The last two zeroes are meaningless here. Change some of the other elements and see what happens.

7

Eigen Things

Figure 7.1.
The Tacoma Narrows bridge: a view from the approach shortly before collapsing.

A linear map is described by a matrix, but that does not say much about its geometric properties. When you look at the 2D linear map figures from Chapter 4, you see that they all map a circle[1] to some

---

[1]The circle is formed from the wings of the Phoenix.

Figure 7.2.
The Tacoma Narrows bridge: a view from shore shortly before collapsing.

ellipse, thereby stretching and rotating the circle. This stretching and rotating is the geometry of a linear map; it is captured by its eigenvectors and eigenvalues, the subject of this chapter.

Eigenvalues and eigenvectors play an important role in the analysis of mechanical structures. If a bridge starts to sway because of strong winds, then this may be described in terms of certain eigenvalues associated with the bridge's mathematical model. Figures 7.1 and 7.2 shows how the Tacoma Narrows bridge swayed violently during mere 42-mile-per-hour winds on November 7, 1940. It collapsed seconds later. Today, a careful eigenvalue analysis is carried-out before any bridge is built! For more images, see http://www.nwwf.com /wa003a.htm.

The essentials of all eigen-theory are already present in the humble 2D case, the subject of this chapter. A discussion of the higher-dimensional case is given in Section 15.5.

7.1 Fixed Directions

Consider Figure 4.2. You see that the e_1-axis is mapped to itself; so is the e_2-axis. This means that any vector of the form ce_1 or de_2 is mapped to some multiple of itself. Similarly, in Figure 4.8, you see that all vectors of the form $c\mathbf{e}_1$ are mapped to multiples of each other.

The directions defined by those vectors are called *fixed directions*, for the reason that those directions are not changed by the map. All vectors in the fixed directions change only in length. The fixed directions need not be the coordinate axes.

If a matrix A takes a (nonzero) vector $\mathbf{r}$ to a multiple of itself, then this may be written as

$$A\mathbf{r} = \lambda\mathbf{r} \qquad (7.1)$$

with some real number λ. The value of λ will determine if $\mathbf{r}$ will dilate, contract, or reverse direction. Furthermore, A will treat any multiple of $\mathbf{r}$ in this way as well. Given a matrix A, one might then ask which vectors it treats in this special way. It turns out that there are at most two directions (in 2D), and when we study symmetric matrices (e.g., a scaling) in Section 7.5, we'll see in that case they are orthogonal to each other. These special vectors are called the *eigenvectors* of A, from the German word "*eigen*," meaning special or proper. An eigenvector is mapped to a multiple of itself, and the corresponding factor λ is called its *eigenvalue*. The eigenvalues and eigenvectors of a matrix are the key to understanding its geometry.

7.2   Eigenvalues

We now develop a way to find the eigenvalues of a matrix A. First, we rewrite (7.1) as

$$A\mathbf{r} = \lambda I \mathbf{r},$$

with I being the identity matrix. We may change this to

$$[A - \lambda I]\mathbf{r} = \mathbf{0}. \qquad (7.2)$$

This means that the matrix $[A - \lambda I]$ maps a nonzero vector $\mathbf{r}$ to the zero vector; $[A-\lambda I]$ must be a projection. Then $[A-\lambda I]$'s determinant vanishes:

$$\det[A - \lambda I] = 0. \qquad (7.3)$$

This, as you will see, is a quadratic equation in λ, called the *characteristic equation* of A. The left-hand side of the equation is called the *characteristic polynomial*.

Example 7.1

Before we proceed further, we will look at an example. Let

$$A = \begin{bmatrix} 2 & 1 \\ 1 & 2 \end{bmatrix}.$$

Its action is shown in Figure 7.3.

So let's write out (7.3). It is

$$\begin{vmatrix} 2 - \lambda & 1 \\ 1 & 2 - \lambda \end{vmatrix} = 0.$$

If we expand the determinant, we get the simple expression—the characteristic equation—

$$(2 - \lambda)^2 - 1 = 0.$$

Expanding and gathering terms, we have a quadratic equation in λ:

$$\lambda^2 - 4\lambda + 3 = 0$$

with the solutions[2]

$$\lambda_1 = 3, \qquad \lambda_2 = 1.$$

Thus, the eigenvalues of a 2×2 matrix are nothing but the zeroes of a quadratic equation.

The product of the eigenvalues equals the determinant of A, so for a 2×2 matrix,

$$|A| = \lambda_1 \cdot \lambda_2.$$

Check this in Example 7.1. This makes intuitive sense if we consider the determinant as a measure of the change in area of the unit square as it is mapped by A to a parallelogram. The eigenvalues indicate a scaling of certain fixed directions defined by A. Now let's look at how to compute these fixed directions, or eigenvectors.

---

[2]Recall that a quadratic equation $a\lambda^2 + b\lambda + c = 0$ has the solutions $\lambda_1 = \frac{-b+\sqrt{b^2-4ac}}{2a}$ and $\lambda_2 = \frac{-b-\sqrt{b^2-4ac}}{2a}$.

Figure 7.3.
Action of a matrix: behavior of the matrix from Example 7.1.

7.3 Eigenvectors

Continuing with Example 7.1, we would still like to know the corresponding eigenvectors. We know that one of them will be mapped to three times itself, the other one to itself. Let's call the corresponding eigenvectors $\mathbf{r}_1$ and $\mathbf{r}_2$. The eigenvector $\mathbf{r}_1$ satisfies

$$\begin{bmatrix} 2-3 & 1 \\ 1 & 2-3 \end{bmatrix} \mathbf{r}_1 = \mathbf{0},$$

or

$$\begin{bmatrix} -1 & 1 \\ 1 & -1 \end{bmatrix} \mathbf{r}_1 = \mathbf{0}.$$

This is a *homogeneous* system, as discussed in Section 5.8. Such systems either have none or infinitely many solutions. In our case, since the matrix has rank 1, there are infinitely many solutions. Any vector of the form

$$\mathbf{r}_1 = \begin{bmatrix} c \\ c \end{bmatrix}$$

will do. And indeed, Figure 7.4 indicates that

$$\begin{bmatrix} 1 \\ 1 \end{bmatrix}$$

Figure 7.4.
Eigenvectors: the action of the matrix from Exercise 7.1 and its eigenvectors, scaled
by their corresponding eigenvalues.

is stretched by a factor of three, that is $A\mathbf{r}_1 = 3\mathbf{r}_1$. Of course

$$\begin{bmatrix} -1 \\ -1 \end{bmatrix}$$

is also stretched by a factor of three.

Next, we determine $\mathbf{r}_2$. We get the linear system

$$\begin{bmatrix} 1 & 1 \\ 1 & 1 \end{bmatrix} \mathbf{r}_2 = \mathbf{0}.$$

Again, we have a homogeneous system with infinitely many solutions.
They are all of the form

$$\mathbf{r}_2 = \begin{bmatrix} c \\ -c \end{bmatrix}.$$

Now recheck Figure 7.4; you see that the vector

$$\begin{bmatrix} 1 \\ -1 \end{bmatrix}$$

is not stretched, and indeed it is mapped to itself!

Typically, eigenvectors are normalized to achieve a degree of uniqueness, and we then have

$$\mathbf{r}_1 = \frac{1}{\sqrt{2}} \begin{bmatrix} 1 \\ 1 \end{bmatrix} \qquad \mathbf{r}_2 = \frac{1}{\sqrt{2}} \begin{bmatrix} 1 \\ -1 \end{bmatrix}.$$

Let us return to the general case. The expression $\det[A - \lambda I] = 0$ is a quadratic polynomial in λ, and its zeroes λ_1 and λ_2 are A's eigenvalues. To find the corresponding eigenvectors, we set up the linear systems $[A - \lambda_1 I]\mathbf{r}_1 = \mathbf{0}$ and $[A - \lambda_2 I]\mathbf{r}_2 = \mathbf{0}$. Both are homogeneous linear systems with infinitely many solutions, corresponding to the eigenvectors $\mathbf{r}_1$ and $\mathbf{r}_2$.

Example 7.2

Let's look at another matrix, namely

$$A = \begin{bmatrix} 1 & 2 \\ 0 & 2 \end{bmatrix}.$$

The characteristic equation, $(1 - \lambda)(2 - \lambda) = 0$ results in eigenvalues $\lambda_1 = 1$ and $\lambda_2 = 2$.

The homogeneous systems

$$\begin{bmatrix} 0 & 2 \\ 0 & 1 \end{bmatrix} \mathbf{r}_1 = \mathbf{0} \quad \text{and} \quad \begin{bmatrix} -1 & 2 \\ 0 & 0 \end{bmatrix} \mathbf{r}_2 = \mathbf{0},$$

result in eigenvectors

$$\mathbf{r}_1 = \begin{bmatrix} 1 \\ 0 \end{bmatrix} \quad \text{and} \quad \mathbf{r}_2 = \frac{1}{\sqrt{5}} \begin{bmatrix} 2 \\ 1 \end{bmatrix},$$

which unlike the eigenvectors of Example 7.1, are not orthogonal. (The matrix in Example 7.1 is symmetric.) We can confirm that $A\mathbf{r}_1 = \mathbf{r}_1$ and $A\mathbf{r}_2 = 2\mathbf{r}_2$.

7.4 Special Cases

Not all quadratic polynomials have zeroes which are real. As you might recall from calculus, there are either no, one, or two real zeroes of a quadratic polynomial,[3] as illustrated in Figure 7.5. If there are no zeroes, then the corresponding matrix A has no fixed directions. We know one example—rotations. They rotate every vector, leaving no direction unchanged. Let's look at a rotation by $-90°$, given by

$$\begin{bmatrix} 0 & 1 \\ -1 & 0 \end{bmatrix}.$$

---

[3] Actually, every quadratic polynomial has two zeroes, but they may be complex numbers.

Figure 7.5.
Quadratic polynomials: from left to right, no zero, one zero, two zeroes.

Its characteristic equation is

$$\begin{bmatrix} -\lambda & 1 \\ -1 & -\lambda \end{bmatrix} = 0$$

or

$$\lambda^2 + 1 = 0.$$

This has no real solutions, as expected.

A quadratic equation may also have one double root; then there is only one fixed direction. A shear in the $\mathbf{e}_1$-direction provides an example—it maps all vectors in the $\mathbf{e}_1$-direction to themselves. An example is

$$A = \begin{bmatrix} 1 & 1/2 \\ 0 & 1 \end{bmatrix}.$$

The action of this shear is illustrated in Figure 4.8. You clearly see that the $\mathbf{e}_1$-axis is not changed.

The characteristic equation for A is

$$\begin{vmatrix} 1 - \lambda & 1/2 \\ 0 & 1 - \lambda \end{vmatrix} = 0$$

or

$$(1 - \lambda)^2 = 0.$$

It has the double root $\lambda_1 = \lambda_2 = 1$. For the corresponding eigenvector, we have to solve

$$\begin{bmatrix} 0 & 1/2 \\ 0 & 0 \end{bmatrix} \mathbf{r} = \mathbf{0}.$$

While this may look strange, it is nothing but a homogeneous system. All vectors of the form

$$\mathbf{r} = \begin{bmatrix} c \\ 0 \end{bmatrix}$$

are solutions. This is quite as expected; those vectors line up along the $\mathbf{e}_1$-direction.

Is there an easy way to decide if a matrix has real eigenvalues or not? In general, no. But there is one important special case: *every symmetric matrix has real eigenvalues.* We will skip the proof but note that these matrices do arise quite often in "real life." In Section 7.5 we'll take a closer look at symmetric matrices.

The last special case to be covered is that of a *zero eigenvalue*.

Example 7.3

Take the matrix from Figure 4.11. It was given by

$$A = \begin{bmatrix} 0.5 & 0.5 \\ 0.5 & 0.5 \end{bmatrix}.$$

The characteristic equation is

$$(0.5 - \lambda)^2 - 0.25 = 0,$$

resulting in $\lambda_1 = 1$ and $\lambda_2 = 0$. The eigenvector corresponding to λ_2 is found by solving

$$\begin{bmatrix} 0.5 & 0.5 \\ 0.5 & 0.5 \end{bmatrix} \mathbf{r}_1 = \begin{bmatrix} 0 \\ 0 \end{bmatrix},$$

yet another homogeneous linear system. Its solutions are of the form

$$\mathbf{r}_1 = c \begin{bmatrix} -1 \\ 1 \end{bmatrix},$$

as you should convince yourself! Since this matrix maps nonzero vectors (multiples $\mathbf{r}_1$) to the zero vector, it reduces dimensionality, and thus has rank one. Note that the eigenvector corresponding to the zero eigenvalue is the kernel of the matrix!

The matrix from Example 7.3 illustrates a defining property of projections. We know that such rank one matrices are idempotent, i.e., $A^2 = A$. One eigenvalue is zero; let λ be the nonzero one, with corresponding eigenvector $\mathbf{r}$. Then

$$A^2\mathbf{v} = A\mathbf{v}, \tag{7.4}$$

resulting in

$$\lambda^2\mathbf{v} = \lambda\mathbf{v},$$

and hence $\lambda = 1$. Thus, a 2D projection matrix always has eigenvalues 0 and 1.

As a general statement, we may say that a 2×2 matrix with one zero eigenvalue has rank one. A matrix with two zero eigenvalues has rank zero, and thus must be the zero matrix.

7.5 The Geometry of Symmetric Matrices

Symmetric matrices arise often in practical problems, and two important examples are addressed in this book:

- conics in Chapter 9 and

- least squares approximation in Section 14.4.

However, many more practical examples exist, coming from fields such as classical mechanics, elasticity theory, quantum mechanics, and thermodynamics. One nice thing about symmetric matrices: we don't have to worry about complex eigenvalues. And another nice thing: they have an interesting geometric interpretation, as we will see below.

We know the two basic equations for eigenvalues and eigenvectors of a symmetric 2×2 matrix A:

$$A\mathbf{r}_1 = \lambda_1 \mathbf{r}_1, \qquad (7.5)$$

$$A\mathbf{r}_2 = \lambda_2 \mathbf{r}_2. \qquad (7.6)$$

Since A is symmetric, we may use (7.5) to write the following:

$$\mathbf{r}_1^{\mathrm{T}} A \mathbf{r}_2 = \lambda_1 \mathbf{r}_1^{\mathrm{T}} \mathbf{r}_2. \qquad (7.7)$$

Using (7.6), we obtain

$$\mathbf{r}_1^{\mathrm{T}} A \mathbf{r}_2 = \lambda_2 \mathbf{r}_1^{\mathrm{T}} \mathbf{r}_2 \qquad (7.8)$$

and thus

$$\lambda_1 \mathbf{r}_1^{\mathrm{T}} \mathbf{r}_2 = \lambda_2 \mathbf{r}_1^{\mathrm{T}} \mathbf{r}_2$$

or

$$(\lambda_1 - \lambda_2)\mathbf{r}_1^{\mathrm{T}} \mathbf{r}_2 = 0.$$

If $\lambda_1 \neq \lambda_2$ (the standard case), then we conclude that $\mathbf{r}_1^{\mathrm{T}} \mathbf{r}_2 = 0$; in other words, A's two eigenvectors are *orthogonal*. Check this for Example 7.1, started in Section 7.2 and continued in Section 7.3!

We may condense (7.5) and (7.6) into one matrix equation,

$$\begin{bmatrix} A\mathbf{r}_1 & A\mathbf{r}_2 \end{bmatrix} = \begin{bmatrix} \lambda_1 \mathbf{r}_1 & \lambda_2 \mathbf{r}_2 \end{bmatrix}. \qquad (7.9)$$

If we define (using the capital Greek Lambda: Λ)

$$\Lambda = \begin{bmatrix} \lambda_1 & 0 \\ 0 & \lambda_2 \end{bmatrix} \quad \text{and} \quad R = \begin{bmatrix} \mathbf{r}_1 & \mathbf{r}_2 \end{bmatrix},$$

then (7.9) becomes

$$AR = R\Lambda. \tag{7.10}$$

Example 7.4

For Example 7.1, $AR = R\Lambda$ becomes

$$\begin{bmatrix} 2 & 1 \\ 1 & 2 \end{bmatrix} \begin{bmatrix} 1/\sqrt{2} & 1/\sqrt{2} \\ 1/\sqrt{2} & -1/\sqrt{2} \end{bmatrix} = \begin{bmatrix} 1/\sqrt{2} & 1/\sqrt{2} \\ 1/\sqrt{2} & -1/\sqrt{2} \end{bmatrix} \begin{bmatrix} 3 & 0 \\ 0 & 1 \end{bmatrix}.$$

Verify this identity!

We know $\mathbf{r}_1^T \mathbf{r}_1 = 1$ and $\mathbf{r}_2^T \mathbf{r}_2 = 1$ (since we assume eigenvectors are normalized) and also $\mathbf{r}_1^T \mathbf{r}_2 = \mathbf{r}_2^T \mathbf{r}_1 = 0$ (since they are orthogonal). These four equations may also be written in matrix form

$$R^T R = I, \tag{7.11}$$

with I the identity matrix. Thus,

$$R^{-1} = R^T$$

and R is an orthogonal matrix. Now (7.10) becomes

$$A = R\Lambda R^T, \tag{7.12}$$

and A is said to be *diagonalizable* because it possible to transform A to the diagonal matrix Λ.

What does this mean geometrically? Since R is an orthogonal matrix, it is a *rotation*, a *reflection*, or a combination of the two. Recall that these linear maps preserve lengths and angles. Its inverse, R^T, is the same type of linear map as R, but a reversal of the action of R. In Example 7.4, R is a rotation ($\hat{R}$) of 45° and a reflection (S) about the $x_1 = x_2$ line, or

$$R = S\hat{R} = \begin{bmatrix} 0 & 1 \\ 1 & 0 \end{bmatrix} \begin{bmatrix} 1/\sqrt{2} & -1/\sqrt{2} \\ 1/\sqrt{2} & 1/\sqrt{2} \end{bmatrix}. \tag{7.13}$$

Figure 7.6.

Symmetric matrices: the action of the matrix from Example 7.1 and its decomposition into rotations, reflections, and a scaling. Top: I, A. Bottom: I, S, $\hat{R}^TS$, $\Lambda\hat{R}^TS$, $\hat{R}\Lambda\hat{R}^TS$, $S\hat{R}\Lambda\hat{R}^TS$.

Therefore, for this set of eigenvectors, (7.12) can be expanded to

$$A = S\hat{R}\Lambda(S\hat{R})^T = S\hat{R}\Lambda\hat{R}^TS, \tag{7.14}$$

with the observation that $S^T = S$. The diagonal matrix Λ is a scaling along each of the coordinate axes. Figure 7.6 illustrates (7.14). Notice that the reflection really isn't necessary, and if we use the degree of freedom available in selecting the direction of the eigenvalues properly, we can construct R to be simply a rotation, or $R = \hat{R}$ from (7.13). Figure 7.7 illustrates this. The column vectors of $\hat{R}$ are eigenvectors of A and form a rotation matrix. As a conclusion, (7.12), in its simplest form, states that the action of every symmetric matrix may be obtained by applying a rotation, then a scaling, and then undoing the rotation.

7.5.1 Positive Definite Symmetric Matrices

Positive definite matrices are a special class of matrices that arise in a number of applications, and they lend themselves to numerically stable and efficient algorithms. A real matrix is positive definite if

$$\mathbf{x}^TA\mathbf{x} > 0$$

for any nonzero vector $\mathbf{x} \in \mathbb{R}^2$. Geometrically we can get a handle on this condition by first considering only unit vectors. Then, this

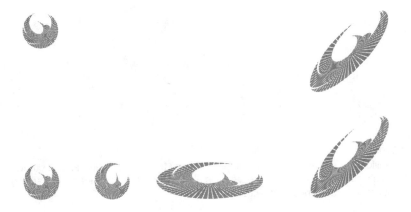

Figure 7.7.
Symmetric matrices: the action of the matrix from Example 7.1 and its decomposition into rotations and a scaling. Top: I, A. Bottom: I, R^T, ΛR^T, $R\Lambda R^T$.

condition states that the angle between $\mathbf{x}$ and $A\mathbf{x}$ is between $-90°$ and $90°$, indicating that A is somehow constrained in its action on $\mathbf{x}$. It isn't sufficient to only consider unit vectors, though. Therefore, for a general matrix, this is a difficult condition to verify. However, for symmetric matrices, the condition takes on a form that is easier to verify.

First, we need a definition: the *symmetric part A_s* of a real matrix is defined as

$$A_s = \frac{1}{2}(A + A^T).$$

A real symmetric matrix is positive definite if and only if the eigenvalues of A_s are positive.[4] Additionally, the determinant of a positive definite matrix is always positive, and therefore the matrix is always nonsingular. Of course these concepts apply to $n \times n$ matrices, which are discussed in more detail in Chapter 14.

7.6 Repeating Maps

When we studied matrices, we saw that they always map the unit circle to an ellipse. Nothing keeps us from now mapping the ellipse again using the same map. We can then repeat again, and so on. Figures 7.8 and 7.9 show two such examples.

---

[4]The phrase *if and only if* communicates that positive definite implies positive eigenvalues *and* positive eigenvalues imply positive definite.

Figure 7.8.
Repetitions: a symmetric matrix is applied several times. One eigenvalue is greater than one, causing stretching in one direction. One eigenvalue is less than one, causing compaction in the opposing direction.

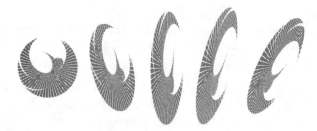

Figure 7.9.
Repetitions: a matrix is applied several times. The eigenvalues are not real, therefore the phoenixes do not line-up along fixed directions.

Figure 7.8 corresponds to the matrix

$$A = \begin{bmatrix} 1 & 0.3 \\ 0.3 & 1 \end{bmatrix}.$$

Being symmetric, it has two real eigenvalues and orthogonal eigenvectors. As the map is repeated several times, the resulting ellipses become more and more stretched: they are elongated in the direction $\mathbf{r}_1$ by $\lambda_1 = 1.3$ and compacted in the direction of $\mathbf{r}_2$ by a factor of $\lambda_2 = 0.7$, with

$$\mathbf{r}_1 = \begin{bmatrix} 1 \\ 1 \end{bmatrix}, \quad \mathbf{r}_2 = \begin{bmatrix} -1 \\ 1 \end{bmatrix}.$$

To get some more insight into this phenomenon, consider applying A (now a generic matrix) twice to $\mathbf{r}_1$. We get

$$AA\mathbf{r}_1 = A\lambda_1\mathbf{r}_1 = \lambda_1^2\mathbf{r}_1.$$

In general,

$$A^n \mathbf{r}_1 = \lambda_1^n \mathbf{r}_1. \qquad (7.15)$$

The same holds for $\mathbf{r}_2$ and λ_2, of course. So you see that once a matrix has real eigenvectors, they play a more and more prominent role as the matrix is applied repeatedly.

By contrast, the matrix corresponding to Figure 7.9 is given by

$$A = \begin{bmatrix} 0.7 & 0.3 \\ -1 & 1 \end{bmatrix}.$$

As you should verify for yourself, this matrix does not have real eigenvalues. In that sense, it is related to a rotation matrix. If you study Figure 7.9, you will notice a rotational component as we progress—the figures do not line up along any (real) fixed directions.

7.7   The Condition of a Map

In most of our figures about affine maps, we have mapped a circle (formed by many unit vectors) to an ellipse. This ellipse is evidently closely related to the geometry of the map, and indeed to its eigenvalues.

A unit circle is given by the equation

$$r_1^2 + r_2^2 = 1.$$

Using matrix notation, we may write this as

$$\begin{bmatrix} r_1 & r_2 \end{bmatrix} \begin{bmatrix} r_1 \\ r_2 \end{bmatrix} = 1,$$

or

$$\mathbf{r}^{\mathrm{T}} \mathbf{r} = 1. \qquad (7.16)$$

If we apply a linear map to the vector $\mathbf{r}$, we have $\mathbf{r}' = A\mathbf{r}$. Inserting this into (7.16), we get

$$\mathbf{r}'^{\mathrm{T}} \mathbf{r}' = [A\mathbf{r}]^{\mathrm{T}} [A\mathbf{r}] = c$$

for some scalar value c. Using the rules for transpose matrices (see Section 4.3), this becomes

$$\mathbf{r}^{\mathrm{T}} A^{\mathrm{T}} A \mathbf{r} = c. \qquad (7.17)$$

This is therefore the equation of an ellipse! (See Chapter 9 for more details.)

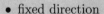

Figure 7.10.

Condition number: The action of $A^T A$ with $\lambda'_1/\lambda'_2 = 900$.

The matrix $A^T A$ is *symmetric* and *positive definite*, and thus has real and positive eigenvalues λ'_1 and λ'_2. Order them so that $\lambda'_1 \geq \lambda'_2$. If λ'_1 is very large and λ'_2 is very small, then the ellipse (7.17) will be very elongated (see Figure 7.10). The ratio λ'_1/λ'_2 of is called the *condition number* c_A of the original matrix A. Notice that $c_A \geq 1$. If a matrix has a condition number close to one, it is called *well-conditioned*. The larger c_A, the "worse" A distorts, and the matrix is called *ill-conditioned*. The eigenvalues λ'_1 and λ'_2 of $A^T A$ are called A's *singular values*.[5]

If you solve a linear system $A\mathbf{x} = \mathbf{b}$, you will be in trouble if c_A becomes large. In that case, a small change in one of the eigenvalues will result in a great change in the shape of the ellipse of (7.17). Practically speaking, a large condition number means that the solution to the linear system is numerically very sensitive to small changes in A or $\mathbf{b}$. Alternatively, we can say that we can confidently calculate the inverse of a well-conditioned matrix.

- fixed direction
- eigenvalue
- characteristic equation
- eigenvector
- homogeneous system
- orthogonal matrix
- eigenvectors of a symmetric matrix

- fixed directions
- repeated linear map
- condition number
- matrix with real eigenvalues
- diagonalizable matrix
- singular values

7.8 Exercises

1. Find the eigenvalues and eigenvectors of the matrix

$$A = \begin{bmatrix} 1 & -2 \\ -2 & 1 \end{bmatrix}.$$

[5]The singular values are typically computed using a method called *Singular Value Decomposition*, or *SVD*, [9].

2. Find the eigenvalues and eigenvectors of

$$A = \begin{bmatrix} 1 & -2 \\ -2 & 0 \end{bmatrix}.$$

3. What is the condition number of the matrix

$$A = \begin{bmatrix} 0.7 & 0.3 \\ -1 & 1 \end{bmatrix}$$

   which generated Figure 7.9?

4. What can you say about the condition number of a rotation matrix?

5. If all eigenvalues of a matrix have absolute value less than one, what will happen as you keep repeating the map?

6. Let

$$A = \begin{bmatrix} 1 & -1.5 \\ -1.5 & 0 \end{bmatrix}.$$

   Modify file `Matdecomp.ps` to show how A's action can be broken down into two rotations and a scaling. Don't try to understand the details of the PostScript program—it's fairly involved. Just replace the entries of `mat1` through `mat4` as described in the comments.

7. What is the condition number of the matrix

$$\begin{bmatrix} 1 & 0 \\ 0 & 0 \end{bmatrix}?$$

   What type of matrix is it, and is it invertible?

8. For the matrix A in Exercise 1, what are the eigenvalues and eigenvectors for A^2 and A^3?

9. For the matrix in Example 7.1, identify the four sets of eigenvectors that may be constructed (by using the degree of freedom available in choosing the direction). Consider each set as column vectors of a matrix R, and sketch the action of the matrix. Which sets involve a reflection?

8

Breaking It Up: Triangles

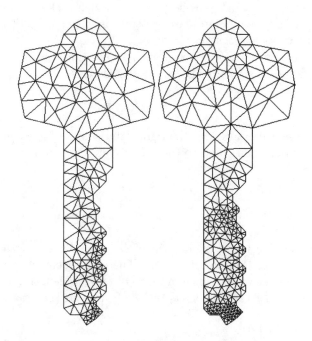

Figure 8.1.
2D FEM: Refinement of a triangulation based on stress and strain calculations.
(Source: J. Shewchuk, http://www.cs.cmu.edu/~quake/triangle.html)

Triangles are as old as geometry. They were of interest to the ancient Greeks, and in fact the roots of trigonometry can be found in their study. Triangles also became an indispensable tool in computer graphics and advanced disciplines such as *finite element analysis* (*FEM*). In graphics, objects are broken down into triangular facets for display purposes; in FEM, 2D shapes are broken down into triangles in order to facilitate complicated algorithms. Figure 8.1 illustrates a refinement procedure based on stress and strain calculations.

8.1 Barycentric Coordinates

A *triangle* T is given by three points, its *vertices*, $\mathbf{p}_1$, $\mathbf{p}_2$, and $\mathbf{p}_3$. The vertices may live in 2D or 3D. Three points define a plane, thus a triangle is a 2D element. We use the convention of labeling the $\mathbf{p}_i$ in a counterclockwise sense. The edge, or side, opposite point $\mathbf{p}_i$ is labeled $\mathbf{s}_i$. (See Sketch 8.1.)

When we study properties of this triangle, it is more convenient to work in terms of a local coordinate system which is closely tied to the triangle. This type of coordinate system was invented by F. Moebius and is known as *barycentric coordinates*.

Let $\mathbf{p}$ be an arbitrary point inside T. Our aim is to write it as a combination of the vertices $\mathbf{p}_i$, in a form like this:

$$\mathbf{p} = u\mathbf{p}_1 + v\mathbf{p}_2 + w\mathbf{p}_3. \tag{8.1}$$

We know one thing already: the right-hand side of this equation is a combination of points, and so the coefficients must sum to one:

$$u + v + w = 1.$$

Otherwise, we would not have a barycentric combination! (See Sketch 8.2.)

Before we give the explicit form of (u, v, w), let us revisit linear interpolation (see Section 2.1) briefly. There, barycentric coordinates on a line segment were defined in terms of ratios of lengths. It sounds reasonable to try the analogous ratios of areas in the triangle case, and so we get:

$$u = \frac{\text{area}(\mathbf{p}, \mathbf{p}_2, \mathbf{p}_3)}{\text{area}(\mathbf{p}_1, \mathbf{p}_2, \mathbf{p}_3)}, \tag{8.2}$$

$$v = \frac{\text{area}(\mathbf{p}, \mathbf{p}_3, \mathbf{p}_1)}{\text{area}(\mathbf{p}_1, \mathbf{p}_2, \mathbf{p}_3)}, \tag{8.3}$$

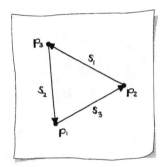

Sketch 8.1.

Vertices and edges of a triangle.

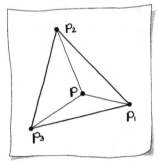

Sketch 8.2.

Barycentric coordinates.

$$w = \frac{\text{area}(\mathbf{p}, \mathbf{p}_1, \mathbf{p}_2)}{\text{area}(\mathbf{p}_1, \mathbf{p}_2, \mathbf{p}_3)}. \qquad (8.4)$$

Recall that areas are easily computed using determinants (see Section 4.10) or as we will learn later (see Section 10.2) using cross products.

Let's see why this works. First, we observe that (u, v, w) do indeed sum to one. Next, let $\mathbf{p} = \mathbf{p}_2$. Now (8.2)–(8.4) tell us that $v = 1$ and $u = w = 0$, just as expected. One more check: if $\mathbf{p}$ is on the edge $\mathbf{s}_1$, say, then $u = 0$, again as expected. Try for yourself that the remaining vertices and edges work the same way!

We call (u, v, w) *barycentric coordinates* and denote them by boldface: $\mathbf{u} = (u, v, w)$. Although they are not independent of each other (we may set $w = 1 - u - v$), they behave much like "normal" coordinates: if $\mathbf{p}$ is given, then we can find $\mathbf{u}$ from (8.2)–(8.4). If $\mathbf{u}$ is given, then we can find $\mathbf{p}$ from (8.1).

The three vertices of the triangle have barycentric coordinates

$$\mathbf{p}_1 \cong (1, 0, 0),$$
$$\mathbf{p}_2 \cong (0, 1, 0),$$
$$\mathbf{p}_3 \cong (0, 0, 1).$$

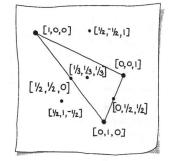

Sketch 8.3.

Examples of barycentric coordinates.

The $\cong$ symbol will be used to indicate the barycentric coordinates of a point. These and several other examples are shown in Sketch 8.3.

As you see, even points outside of T can be given barycentric coordinates! This works since the areas involved in (8.2)–(8.4) are *signed*. So points inside T have positive barycentric coordinates, and those outside have mixed signs.[1]

This observation is the basis for one of the most frequent uses of barycentric coordinates: the *triangle inclusion test*. If a triangle T and a point $\mathbf{p}$ are given, how do we determine if $\mathbf{p}$ is inside T or not? We simply compute $\mathbf{p}$'s barycentric coordinates and check their signs! If they are all of the same sign, inside—else, outside. Theoretically, one or two of the barycentric coordinates could be zero, indicating that $\mathbf{p}$ is on one of the edges. In "real" situations, you are not likely to encounter values which are *exactly* equal to zero; be sure not to test for a barycentric coordinate to be *equal* to zero! Instead, use a tolerance ϵ, and flag a point as being on an edge if one of its barycentric coordinates is less than ϵ in absolute value. A good value for ϵ? Obviously, this is application dependent, but something like $1.0E - 6$ should work for most cases.

---

[1]This assumes the triangle to be oriented counterclockwise. If it is oriented clockwise, then the points inside have all negative barycentric coordinates, and the outside ones still have mixed signs.

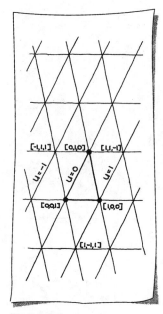

Sketch 8.4.

Barycentric coordinates
coordinate lines.

Finally, Sketch 8.4 shows how we may think of the whole plane as being covered by a grid of coordinate lines. Note that the plane is divided into seven regions by the (extended) edges of T!

Example 8.1

Let's work with a simple example that is easy for you to sketch. Suppose the three triangle vertices are given by

$$\mathbf{p}_1 = \begin{bmatrix} 0 \\ 0 \end{bmatrix}, \quad \mathbf{p}_2 = \begin{bmatrix} 1 \\ 0 \end{bmatrix}, \quad \mathbf{p}_3 = \begin{bmatrix} 0 \\ 1 \end{bmatrix}.$$

The points $\mathbf{q}$, $\mathbf{r}$, $\mathbf{s}$ with barycentric coordinates

$$\mathbf{q} \cong \left(0, \frac{1}{2}, \frac{1}{2} \right), \quad \mathbf{r} \cong (-1, 1, 1), \quad \mathbf{s} \cong \left(\frac{1}{3}, \frac{1}{3}, \frac{1}{3} \right),$$

have the following coordinates in the plane,

$$\mathbf{q} = 0 \times \mathbf{p}_1 + \frac{1}{2} \times \mathbf{p}_2 + \frac{1}{2} \times \mathbf{p}_3 = \begin{bmatrix} 1/2 \\ 1/2 \end{bmatrix},$$

$$\mathbf{r} = -1 \times \mathbf{p}_1 + 1 \times \mathbf{p}_2 + 1 \times \mathbf{p}_3 = \begin{bmatrix} 1 \\ 1 \end{bmatrix},$$

$$\mathbf{s} = \frac{1}{3} \times \mathbf{p}_1 + \frac{1}{3} \times \mathbf{p}_2 + \frac{1}{3} \times \mathbf{p}_3 = \begin{bmatrix} 1/3 \\ 1/3 \end{bmatrix}.$$

8.2   Affine Invariance

In this short section, we will discuss the statement: *barycentric coordinates are affinely invariant.*

Let $\hat{T}$ be an affine image of T, having vertices $\hat{\mathbf{p}}_1, \hat{\mathbf{p}}_2, \hat{\mathbf{p}}_3$. Let $\mathbf{p}$ be a point with barycentric coordinates $\mathbf{u}$ relative to T. We may apply the affine map to $\mathbf{p}$ also, and then we ask: What are the barycentric coordinates of $\hat{\mathbf{p}}$ with respect to $\hat{T}$?

While at first sight this looks like a daunting task, simple geometry yields the answer quickly. Note that in (8.2)–(8.4), we employ *ratios of areas*. These are, as introduced in Section 6.2, unchanged by affine maps! So while the individual areas in (8.2)–(8.4) do change, their quotients do not. Thus, $\hat{\mathbf{p}}$ also has barycentric coordinates $\mathbf{u}$ with respect to $\hat{T}$.

This fact, namely that affine maps do not change barycentric co-ordinates, is what is meant by the statement at the beginning of this section. (See Sketch 8.5 for an illustration.)

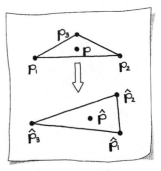

Example 8.2

Let's revisit the simple triangle in Example 8.1 and look at the affine invariance of barycentric coordinates. Suppose we apply a 90° rota-tion,

$$R = \begin{bmatrix} 0 & -1 \\ 1 & 0 \end{bmatrix}$$

Sketch 8.5.
Affine invariance of barycentric coordinates.

to the triangle vertices, resulting in $\hat{\mathbf{p}}_i = R\mathbf{p}_i$. Apply this rotation to $\mathbf{s}$ from Example 8.1,

$$\hat{\mathbf{s}} = R\mathbf{s} = \begin{bmatrix} -1/3 \\ 1/3 \end{bmatrix}.$$

Due to the affine invariance of barycentric coordinates, we could have found the coordinates of $\hat{\mathbf{s}}$ as

$$\hat{\mathbf{s}} = \frac{1}{3} \times \hat{\mathbf{p}}_1 + \frac{1}{3} \times \hat{\mathbf{p}}_2 + \frac{1}{3} \times \hat{\mathbf{p}}_3 = \begin{bmatrix} -1/3 \\ 1/3 \end{bmatrix}.$$

8.3   Some Special Points

In classical geometry, many special points relative to a triangle have been discovered, but for our purposes, just three will do: the centroid, the incenter, and the circumcenter. They are used for a multitude of geometric computations.

The *centroid* $\mathbf{c}$ of a triangle is given by the intersection of the three medians. (A median is the connection of an edge midpoint to the opposite vertex.) Its barycentric coordinates (see Sketch 8.6) are given by

$$\mathbf{c} \cong \left(\frac{1}{3}, \frac{1}{3}, \frac{1}{3} \right). \tag{8.5}$$

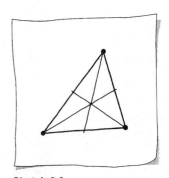

Sketch 8.6.
The centroid.

We verify this by writing

$$\left(\frac{1}{3}, \frac{1}{3}, \frac{1}{3} \right) = \frac{1}{3}(0, 1, 0) + \frac{2}{3}\left(\frac{1}{2}, 0, \frac{1}{2} \right),$$

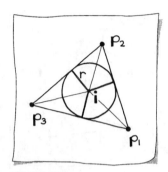

Sketch 8.7.

The incenter.

thus asserting that $\left(\frac{1}{3}, \frac{1}{3}, \frac{1}{3}\right)$ lies on the median associated with $\mathbf{p}_2$. In the same way, we show that it is also on the remaining two medians. We also observe that a triangle and its centroid are related in an affinely invariant way.

The *incenter* $\mathbf{i}$ of a triangle is the intersection of the three angle bisectors (see Sketch 8.7). There is a circle, called the *incircle*, that has $\mathbf{i}$ as its center and touches all three triangle edges. Let s_i be the length of the triangle edge opposite vertex $\mathbf{p}_i$. Let r be the radius of the incircle—there is a formula for it, but we won't need it here.

If the barycentric coordinates of $\mathbf{i}$ are (i_1, i_2, i_3), then we see that

$$i_1 = \frac{\text{area}(\mathbf{i}, \mathbf{p}_2, \mathbf{p}_3)}{\text{area}(\mathbf{p}_1, \mathbf{p}_2, \mathbf{p}_3)}.$$

This may be rewritten as

$$i_1 = \frac{rs_1}{rs_1 + rs_2 + rs_3},$$

using the "1/2 base times height" rule for triangle areas.

Simplifying, we obtain

$$i_1 = s_1/c,$$
$$i_2 = s_2/c,$$
$$i_3 = s_3/c,$$

where $c = s_1 + s_2 + s_3$ is the circumference of T. A triangle is not affinely related to its incenter—affine maps change the barycentric coordinates of $\mathbf{i}$.

The *circumcenter* $\mathbf{cc}$ of a triangle is the center of the circle through its vertices. It is obtained as the intersection of the edge bisectors. (See Sketch 8.8.) Notice that the circumcenter might not be inside the triangle. This circle is called the *circumcircle* and we will refer to its radius as R.

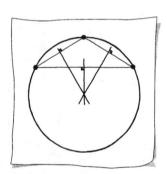

Sketch 8.8.

The circumcenter.

The barycentric coordinates (cc_1, cc_2, cc_3) of the circumcenter are

$$cc_1 = d_1(d_2 + d_3)/D$$
$$cc_2 = d_2(d_1 + d_3)/D$$
$$cc_3 = d_3(d_1 + d_2)/D$$

where

$$d_1 = (\mathbf{p}_2 - \mathbf{p}_1) \cdot (\mathbf{p}_3 - \mathbf{p}_1)$$
$$d_2 = (\mathbf{p}_1 - \mathbf{p}_2) \cdot (\mathbf{p}_3 - \mathbf{p}_2)$$

$$d_3 = (\mathbf{p}_1 - \mathbf{p}_3) \cdot (\mathbf{p}_2 - \mathbf{p}_3)$$
$$D = 2(d_1 d_2 + d_2 d_3 + d_3 d_1).$$

Furthermore,

$$R = \frac{1}{2}\sqrt{\frac{(d_1 + d_2)(d_2 + d_3)(d_3 + d_1)}{D/2}}.$$

These formulas are due to [8]. Confirming our observation in Sketch 8.8, that the circumcircle might be outside of the triangle, note that some of the cc_i may be negative. If T has an angle close to 180°, then the corresponding cc_i will be *very* negative, leading to serious numerical problems! As a result, the circumcenter will be far away from the vertices, and thus not be of practical use. As with the incenter, affine maps of the triangle change the barycentric coordinates of the circumcenter.

Example 8.3

Yet again, let's visit the simple triangle in Example 8.1. Be sure to make a sketch to check the results of this example. Let's compute the incenter. The lengths of the edges of the triangle are $s_1 = \sqrt{2}$, $s_2 = 1$, and $s_3 = 1$. The circumference of the triangle is $c = 2 + \sqrt{2}$. The barycentric coordinates of the incenter are then

$$\mathbf{i} \cong \left(\frac{\sqrt{2}}{2 + \sqrt{2}}, \frac{1}{2 + \sqrt{2}}, \frac{1}{2 + \sqrt{2}} \right).$$

(These barycentric coordinates are approximately, $(0.41, 0.29, 0.29)$.) The coordinates of the incenter are

$$\mathbf{i} = 0.41 \times \mathbf{p}_1 + 0.29 \times \mathbf{p}_2 + 0.29 \times \mathbf{p}_3 = \begin{bmatrix} 0.29 \\ 0.29 \end{bmatrix}.$$

The circumcircle's circumcenter is easily calculated, too. First compute $d_1 = 0$, $d_2 = 1$, $d_3 = 1$, and $D = 2$. Then the barycentric coordinates of the circumcenter are $\mathbf{c} \cong (0, 1/2, 1/2)$. This is the midpoint of the "diagonal" edge of the triangle.

Now the radius of the circumcircle is easily computed with the equation above, $R = \sqrt{2}/2$.

8.4   2D Triangulations

The study of *one* triangle is the realm of classical geometry; in modern applications, one often encounters millions of triangles. Typically, they are connected in some well-defined way; the most basic one being the 2D *triangulation*. Triangulations have been used in surveying for centuries; more modern applications rely on satellite data which are collected in triangulations called *TINS* (*Triangular Irregular Networks*).

Here is the formal definition of a 2D triangulation. A triangulation of a set of 2D points $\{\mathbf{p}_i\}_{i=1}^N$ is a connected set of triangles meeting the following criteria:

1. The vertices of the triangles consist of the given points.

2. The interiors of any two triangles do not intersect.

3. If two triangles are not disjoint, then they share a vertex or have coinciding edges.

4. The union of all triangles equals the convex hull of the $\mathbf{p}_i$.

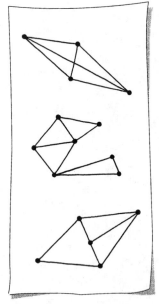

Sketch 8.9.

Illegal examples of triangulations.

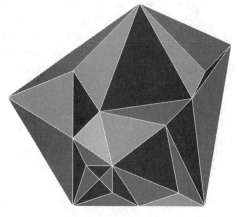

Figure 8.2.

Triangulation: a valid triangulation of the convex hull.

These rules sound abstract, but some examples will shed light on them. In Figure 8.2 you see a triangulation which satisfies the 2D triangulation definition. Evident from this example: the number of triangles surrounding a vertex, or *valence*, varies from vertex to vertex. These triangles make up the *star* of a vertex. On the other hand, in Sketch 8.9 you see three illegal triangulations, violating the above rules. The top example involves overlapping triangles. In the middle example, the boundary of the triangulation is not the convex hull of the point set. (A lot more on convex hulls may be found in [4].) The bottom example violates condition 3.

If we are given a point set, is there a unique triangulation? Certainly not, as Sketch 8.10 shows. Among the many possible triangulations, there is one that is most commonly agreed to be the "best." This is the *Delaunay triangulation*. Describing the details of this method is beyond the scope of this text, however a wealth of information can be found on the Web. Search or visit the sites listed in the introduction of this chapter or see [4].

8.5   A Data Structure

What is the best data structure for storing a triangulation? The factors which determine the best structure include storage requirements and accessibility. Let's build the "best" structure based on the point set and triangulation illustrated in Sketch 8.11.

In order to minimize storage, it is an accepted practice to store each point once only. Since these are floating point values, they take up the most space. Thus, a basic triangulation structure would be a listing of the point set followed by the triangulation information. This constitutes pointers into the point set, indicating which points are joined to form a triangle. Store the triangles in a counterclockwise orientation! This is the data structure for the triangulation in Sketch 8.11:

```
5              (number of points)
0.0   0.0      (point #1)
1.0   0.0
0.0   1.0
0.25  0.3
0.5   0.3
5              (number of triangles)
1 2 5          (first triangle - connects points #1,2,5)
2 3 5
4 5 3
1 5 4
1 4 3
```

We can improve this structure. We will encounter applications which require a knowledge of the connectivity of the triangulation, as described in Section 8.6. To facilitate this, it is not uncommon to also

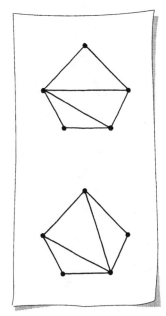

Sketch 8.10.
Non-uniqueness of triangulations.

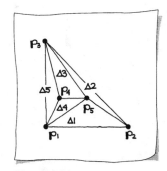

Sketch 8.11.
A sample triangulation.

see the *neighbor information* of the triangulation stored. This means
that for each triangle, the indices of the triangles surrounding it are
stored. For example, in Sketch 8.11, triangle 1 defined by points 1,2,5
is surrounded by triangles 2,4,–1. The neighboring triangles are listed
corresponding to the point across from the shared edge. Triangle –1
indicates that there is not a neighboring triangle across this edge.
Immediately, we see that this gives us a fast method for determining
the boundary of the triangulation! Listing the neighbor information
after each triangle, the final data structure is as follows.

```
5                    (number of points)
0.0  0.0             (point #1)
1.0  0.0
0.0  1.0
0.25 0.3
0.5  0.3
5                    (number of triangles)
1 2 5   2 4 -1       (first triangle and neighbors)
2 3 5   3 1 -1
4 5 3   2 5  4
1 5 4   3 5  1
1 4 3   3 -1 4
```

This is but one of many possible data structures for a triangulation. Based on the needs of particular applications, researchers have
developed a variety of structures to optimize searches. One such
structure that has proved to be popular is called the *winged-edge
data structure* [19].

8.6   Point Location

Given a triangulation of points $\mathbf{p}_i$, assume we are given a point $\mathbf{p}$
which has not been used in building the triangulation. Question:
Which triangle is $\mathbf{p}$ in, if any? The easiest way is to compute $\mathbf{p}$'s
barycentric coordinates with respect to all triangles; if all of them are
positive with respect to some triangle, then that is the desired one,
else, $\mathbf{p}$ is in none of the triangles.

While simple, this algorithm is expensive. In the worst case, every
triangle has to be considered; on average, half of all triangles have to
be considered. A much more efficient algorithm may be based upon
the following observation. Suppose $\mathbf{p}$ is not in a particular triangle
T. Then at least one of its barycentric coordinates with respect to

T must be negative; let's assume it is u. We then know that $\mathbf{p}$ has no chance of being inside T's two neighbors along edges $\mathbf{s}_2$ or $\mathbf{s}_3$ (see Sketch 8.12).

So a likely candidate to check is the neighbor along $\mathbf{s}_1$—recall that we have stored the neighboring information in a data structure. In this way—always searching in the direction of the currently most negative barycentric coordinate—we create a path from a starting triangle to the one that actually contains $\mathbf{p}$.

8.6.1 Point Location Algorithm

Input: Triangulation and neighbor information, plus one point $\mathbf{p}$.

Output: Triangle that $\mathbf{p}$ is in.

Step 0: Set the "current triangle" to be the first triangle in the triangulation.

Step 1: Perform the *triangle inclusion test* (see Section 8.1) for $\mathbf{p}$ and the current triangle. If all barycentric coordinates are positive, output the current triangle. If the barycentric coordinates are mixed in sign, then determine the barycentric coordinate of $\mathbf{p}$ with respect to the current triangle that has the most negative value. Set the current triangle to be the corresponding neighbor and repeat Step 1.

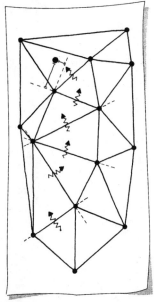

Sketch 8.12.
Neighbor check.

Notes:

- Try improving the speed of this algorithm by not completing the division for determining the barycentric coordinates in (8.2)–(8.4). This division does not change the sign. Keep in mind the test for which triangle to move to changes.

- Suppose the algorithm is to be executed for more than one point. Consider using the triangle that was output from the previous run as input, rather than always using the first triangle. Many times a data set has some *coherence*, and the output for the next run might be the same triangle, or one very near to the triangle from the previous run.

8.7 3D Triangulations

In computer applications, one often encounters millions of triangles, connected in some well-defined way, describing a geometric object. In particular, shading algorithms require this type of structure.

Figure 8.3.
3D triangulated chalice: a wireframe rendering.

Figure 8.4.
3D triangulated chalice: a shaded rendering.

The rules for 3D triangulations are the same as for 2D. Additionally, the data structure is the same, except that now each point has three instead of two coordinates.

Figure 8.3 shows a 3D object that is composed of triangles, and Figure 8.4 is the same object shaded. Another example is provided in Figure 12.1: all objects in that image were broken down into triangles before display. Shading requires a 3D unit vector, called a *normal*, to be associated with each triangle or vertex. A normal is perpendicular to object's surface at a particular point. This normal is used to calculate how light is reflected, and in turn the illumination of the object. (See [19] for details on such illumination methods.) We'll investigate just how to calculate normals in Section 10.5, after we have introduced tools for 3D geometry.

- barycentric coordinates
- triangle inclusion test
- affine invariance of barycentric coordinates
- centroid, barycenter
- incenter
- circumcenter
- 2D triangulation criteria
- star

- valence
- Delaunay triangulation
- triangulation data structure
- point location algorithm
- 3D triangulation criteria
- 3D triangulation data structure
- normal

8.8   Exercises

Let a triangle T_1 be given by the vertices

$$\mathbf{p}_1 = \begin{bmatrix} 1 \\ 1 \end{bmatrix}, \quad \mathbf{p}_2 = \begin{bmatrix} 2 \\ 2 \end{bmatrix}, \quad \mathbf{p}_3 = \begin{bmatrix} -1 \\ 2 \end{bmatrix}.$$

Let a triangle T_2 be given by the vertices

$$\mathbf{q}_1 = \begin{bmatrix} 0 \\ 0 \end{bmatrix}, \quad \mathbf{q}_2 = \begin{bmatrix} 0 \\ -1 \end{bmatrix}, \quad \mathbf{q}_3 = \begin{bmatrix} -1 \\ 0 \end{bmatrix}.$$

1. Using T_1,

    (a) What are the barycentric coordinates of $\mathbf{p} = \begin{bmatrix} 0 \\ 1.5 \end{bmatrix}$?

    (b) What are the barycentric coordinates of $\mathbf{p} = \begin{bmatrix} 0 \\ 0 \end{bmatrix}$?

    (c) Find the triangle's incenter.

    (d) Find the triangle's circumcenter.

    (e) Find the centroid of the triangle.

2. Repeat the above for T_2.

3. What are the areas of T_1 and T_2?

4. Let an affine map be given by

$$\mathbf{x}' = \begin{bmatrix} 1 & 2 \\ -1 & 2 \end{bmatrix} \mathbf{x} + \begin{bmatrix} -1 \\ 0 \end{bmatrix}.$$

What are the areas of mapped triangles T_1' and T_2'? Compare the ratios

$$\frac{T_1}{T_2} \quad \text{and} \quad \frac{T_1'}{T_2'}.$$

5. Using the file `sample.tri` in the downloads section of the book's website, write a program that determines the triangle that an input point $\mathbf{p}$ is in. Test with $\mathbf{p}$ being the barycenter of the data. Starting with triangle 1 in the data structure, print the numbers of all triangles encountered in your search. (Note: this file gives the first point the index 0 and the first triangle the index 0.) This triangulation is illustrated in Figure 8.2.

9

Conics

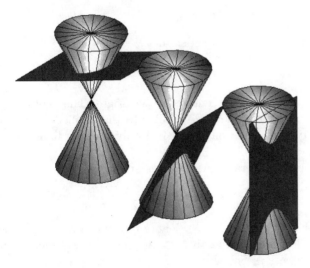

Figure 9.1.
Conic sections: three types of curves formed by the intersection of a plane and a cone.
From left to right: ellipse, parabola, and hyperbola.

Take a flashlight and shine it straight onto a wall. You will see a circle. Tilt the light, and the circle will turn into an ellipse. Tilt further, and the ellipse will become more and more elongated, and will become a parabola eventually. Tilt a little more, and you will have a hyperbola—actually one branch of it. The beam of your flashlight is

155

a *cone*, and the image it generates on the wall is the intersection of that cone with a *plane* (i.e., the wall). Thus, we have the name *conic section* for curves that are the intersections of cones and planes. (See Figure 9.1.)

The three curves, ellipses, parabolas, and hyperbolas, arise in many situations and are the subject of this chapter. The basic tools for handling them are nothing but the matrix theory developed earlier.

Before we delve into the theory of conic sections, we list some "real-life" occurances.

Sketch 9.1.

A pencil with hyperbolic arcs.

- The paths of the planets around the sun are ellipses.

- If you sharpen a pencil, you generate hyperbolas (see Sketch 9.1).

- If you water your lawn, the water leaving the hose traces a parabolic arc.

9.1   The General Conic

We know that all points $\mathbf{x}$ satisfying

$$x_1^2 + x_2^2 = r^2 \tag{9.1}$$

are on a *circle* of radius r, centered at the origin. This type of equation is called an *implicit equation*. Similar to the implicit equation for a line, this type of equation is satisfied only for coordinate pairs that lie on the circle.

A little more generality will give us an *ellipse*:

$$\lambda_1 x_1^2 + \lambda_2 x_2^2 = c. \tag{9.2}$$

The positive factors λ_1 and λ_2 denote how much the ellipse deviates from a circle. For example, if $\lambda_1 > \lambda_2$, the ellipse is more elongated in the x_2-direction. See Sketch 9.2 for the example

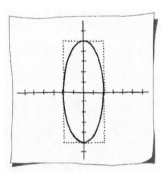

Sketch 9.2.

An ellipse with $\lambda_1 = 1/4$, $\lambda_2 = 1/25$, and c = 1.

$$\frac{1}{4}x_1^2 + \frac{1}{25}x_2^2 = 1.$$

An ellipse in this form is said to be in *standard position*, because its *minor and major axes* are coincident with the coordinate axes, and the *center* is at the origin. The ellipse is symmetric about both axes, and it lives in the rectangle with x_1 extents $[-\sqrt{c/\lambda_1}, \sqrt{c/\lambda_1}]$ and x_2 extents $[-\sqrt{c/\lambda_2}, \sqrt{c/\lambda_2}]$. For a derivation of (9.2), see an analytic geometry text such as [21].

We will now rewrite (9.2) in a much more complicated form:

$$\begin{bmatrix} x_1 & x_2 \end{bmatrix} \begin{bmatrix} \lambda_1 & 0 \\ 0 & \lambda_2 \end{bmatrix} \begin{bmatrix} x_1 \\ x_2 \end{bmatrix} - c = 0. \qquad (9.3)$$

You will see the wisdom of this in a short while. This equation allows for significant compaction:

$$\mathbf{x}^T D \mathbf{x} - c = 0. \qquad (9.4)$$

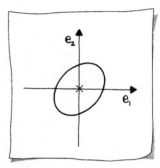

Suppose we encounter an ellipse whose minor and major axes are not aligned with the coordinate axes, however, the center is at the origin, as is the case with Sketch 9.3. What is the equation of such an ellipse? Points $\hat{\mathbf{x}}$ on this ellipse are mapped from an ellipse in standard position via a rotation, $\hat{\mathbf{x}} = R\mathbf{x}$. Using the fact that a rotation matrix is orthogonal, we replace $\mathbf{x}$ by $R^T\hat{\mathbf{x}}$ in (9.4), and the rotated conic takes the form

Sketch 9.3.
A rotated ellipse.

$$[R^T\hat{\mathbf{x}}]^T D[R^T\hat{\mathbf{x}}] - c = 0,$$

which becomes

$$\hat{\mathbf{x}}^T R D R^T \hat{\mathbf{x}} - c = 0. \qquad (9.5)$$

The ellipse of Sketch 9.3, rotated into standard form, is illustrated in Sketch 9.4.

Now suppose we encounter an ellipse as in Sketch 9.5, that is *rotated and translated* out of standard position. What is the equation of this ellipse? Points $\hat{\mathbf{x}}$ on this ellipse are mapped from an ellipse in standard position via a rotation and then a translation, $\hat{\mathbf{x}} = R\mathbf{x} + \mathbf{v}$. Again using the fact that a rotation matrix is orthogonal, we replace $\mathbf{x}$ by $R^T(\hat{\mathbf{x}} - \mathbf{v})$ in (9.4), and the rotated conic takes the form

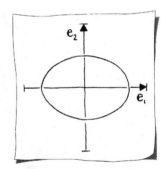

Sketch 9.4.
An ellipse in standard form with $\lambda_1 = 2$, $\lambda_2 = 4$, and c = 1.

$$[R^T(\hat{\mathbf{x}} - \mathbf{v})]^T D[R^T(\hat{\mathbf{x}} - \mathbf{v})] - c = 0,$$

or

$$[\hat{\mathbf{x}}^T - \mathbf{v}^T]R D R^T[\hat{\mathbf{x}} - \mathbf{v}] - c = 0.$$

An abbreviation of $A = R D R^T$ shortens this equation to

$$[\hat{\mathbf{x}}^T - \mathbf{v}^T]A[\hat{\mathbf{x}} - \mathbf{v}] - c = 0.$$

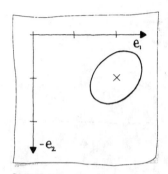

Since $\hat{\mathbf{x}}$ is simply a variable, we may drop the "hat" notation. Further, recognize that A is a symmetric matrix, and it is of the same form as we encountered in Section 7.5. This allows for the equality: $\mathbf{x}^{\mathrm{T}} A \mathbf{v} = \mathbf{v}^{\mathrm{T}} A \mathbf{x}$, and we obtain

$$\mathbf{x}^{\mathrm{T}} A \mathbf{x} - 2\mathbf{x}^{\mathrm{T}} A \mathbf{v} + \mathbf{v}^{\mathrm{T}} A \mathbf{v} - c = 0. \tag{9.6}$$

This denotes an ellipse in general position and it may be slightly abbreviated as

$$\mathbf{x}^{\mathrm{T}} A \mathbf{x} - 2\mathbf{x}^{\mathrm{T}} \mathbf{b} + d = 0, \tag{9.7}$$

with $\mathbf{b} = A\mathbf{v}$ and $d = \mathbf{v}^{\mathrm{T}} A \mathbf{v} - c$.

Example 9.1

Let's start with the ellipse $2x_1^2 + 4x_2^2 - 1 = 0$. In matrix form, corresponding to (9.3), we have

$$\begin{bmatrix} x_1 & x_2 \end{bmatrix} \begin{bmatrix} 2 & 0 \\ 0 & 4 \end{bmatrix} \begin{bmatrix} x_1 \\ x_2 \end{bmatrix} - 1 = 0.$$

This ellipse is shown in Sketch 9.4.

Now we rotate by $45°$, using the rotation matrix

$$R = \begin{bmatrix} s & -s \\ s & s \end{bmatrix}$$

with $s = \sin 45° = \cos 45° = 1/\sqrt{2}$. The matrix $A = RDR^{\mathrm{T}}$ becomes

$$A = \begin{bmatrix} 3 & -1 \\ -1 & 3 \end{bmatrix}.$$

This ellipse is illustrated in Sketch 9.3.

If we now translate by a vector

$$\mathbf{v} = \begin{bmatrix} 2 \\ -1 \end{bmatrix},$$

then (9.7) becomes

$$\mathbf{x}^{\mathrm{T}} \begin{bmatrix} 3 & -1 \\ -1 & 3 \end{bmatrix} \mathbf{x} - 2\mathbf{x}^{\mathrm{T}} \begin{bmatrix} 7 \\ -5 \end{bmatrix} + 18 = 0.$$

Expanding the previous equation, the conic is

$$3x_1^2 - 2x_1x_2 + 3x_2^2 - 14x_1 + 10x_2 + 18 = 0.$$

This ellipse is illustrated in Sketch 9.5.

This was a lot of work just to find the general form of an ellipse! However, as we shall see, a lot more has been achieved here; the form (9.7) does not just represent ellipses, but *any* conic. To see that (9.7) represents any conic, let's examine two remaining conics: a hyperbola and a parabola.

Example 9.2

Sketch 9.6 illustrates the conic

$$x_2 = \frac{1}{x_1}$$

which is a hyperbola. This may be written as

$$\begin{bmatrix} x_1 & x_2 \end{bmatrix} \begin{bmatrix} 0 & \frac{1}{2} \\ \frac{1}{2} & 0 \end{bmatrix} \begin{bmatrix} x_1 \\ x_2 \end{bmatrix} - 1 = 0,$$

which is clearly of the form (9.7).

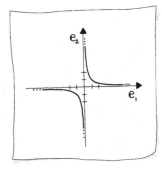

Sketch 9.6.
A hyperbola.

Example 9.3

A parabola is illustrated in Sketch 9.7,

$$x_2 = x_1^2.$$

This may be written as

$$\begin{bmatrix} x_1 & x_2 \end{bmatrix} \begin{bmatrix} -1 & 0 \\ 0 & 0 \end{bmatrix} \begin{bmatrix} x_1 \\ x_2 \end{bmatrix} + \begin{bmatrix} x_1 & x_2 \end{bmatrix} \begin{bmatrix} 0 \\ 1 \end{bmatrix} = 0,$$

and thus is also of the form (9.7)!

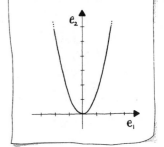

Sketch 9.7.
A parabola.

If we relabel the elements of $A, \mathbf{b}, d$ from (9.7), the equation takes the form

$$\begin{bmatrix} x_1 & x_2 \end{bmatrix} \begin{bmatrix} c_1 & \frac{1}{2}c_3 \\ \frac{1}{2}c_3 & c_2 \end{bmatrix} \begin{bmatrix} x_1 \\ x_2 \end{bmatrix} - 2 \begin{bmatrix} x_1 & x_2 \end{bmatrix} \begin{bmatrix} -\frac{1}{2}c_4 \\ -\frac{1}{2}c_5 \end{bmatrix} + c_6 = 0. \quad (9.8)$$

Expanding this, we arrive at a familiar equation of a conic,

$$c_1 x_1^2 + c_2 x_2^2 + c_3 x_1 x_2 + c_4 x_1 + c_5 x_2 + c_6 = 0. \quad (9.9)$$

In fact, many texts simply start out by using (9.9) as the initial *definition* of a conic.

9.2 Analyzing Conics

If you are given the equation of a conic as in (9.9), how can you tell which of the three basic types it is? Upon examining the matrix entries in (9.8), we observe it is a symmetric matrix, and therefore can be written as RDR^{T}, with a diagonal matrix D and a rotation matrix R. With this *decomposition* of A, we can tell what kind of conic we have.

- If D is the zero matrix, then the conic is degenerate and simply consists of a straight line.

- If D's diagonal has only one nonzero entry, then the conic is a parabola.

- If D's diagonal has two nonzero entries of the same sign, then the conic is an ellipse.

- If D's diagonal has two nonzero entries with opposite sign, then the conic is a hyperbola.

The problem of finding the diagonal matrix D has already been solved earlier in this book, namely in Section 7.5. All we have to do is find the eigenvalues of A; they determine the diagonal matrix D, and it in turn determines what kind of conic we have.

Example 9.4

Let's revisit the conic from Example 9.1. The conic was given by

$$3x_1^2 + 3x_2^2 - 2x_1 x_2 - 14x_1 + 10x_2 + 18 = 0. \quad (9.10)$$

We first rewrite this in the form of (9.8),

$$\mathbf{x}^T \begin{bmatrix} 3 & -1 \\ -1 & 3 \end{bmatrix} \mathbf{x} - 2\mathbf{x}^T \begin{bmatrix} 7 \\ -5 \end{bmatrix} + 18 = 0.$$

The eigenvalues of the 2×2 matrix are the solution of the quadratic equation

$$(3 - \lambda)^2 - 1 = 0,$$

and thus are $\lambda_1 = 2$ and $\lambda_2 = 4$.

Our desired diagonal matrix is:

$$D = \begin{bmatrix} 2 & 0 \\ 0 & 4 \end{bmatrix},$$

and thus this conic is an *ellipse*. It is illustrated in Sketch 9.5.

Example 9.5

Revisit Example 9.2, and find the eigenvalues of the matrix

$$\begin{bmatrix} 0 & \frac{1}{2} \\ \frac{1}{2} & 0 \end{bmatrix}.$$

The characteristic equation is $\lambda^2 - 1/4 = 0$, which results in

$$D = \begin{bmatrix} \frac{1}{2} & 0 \\ 0 & -\frac{1}{2} \end{bmatrix},$$

and this matches the condition that this conic is a hyperbola.

Example 9.6

Revisit Example 9.3, notice that the matrix

$$\begin{bmatrix} -1 & 0 \\ 0 & 0 \end{bmatrix}.$$

is already in diagonal form, and this matches the condition that this conic is a parabola.

In this section and the last, we have derived the general conic and folded this into a tool to determine its type. What might not be obvious: we found that affine maps take a particular type of conic to another one of the same type. The conic type is determined by D: it is unchanged by affine maps.

9.3   The Position of a Conic

If we are given a conic in the general equation format (9.9), how do we determine its position? If we know how it has been rotated and translated out of standard position then we know the location of the center and the direction of the minor and major axes. The solution to this problem is simple: we just work through Section 9.1 in reverse.

Let's work with a familiar example, the conic

$$3x_1^2 - 2x_1x_2 + 3x_2^2 - 14x_1 + 10x_2 + 18 = 0$$

from Example 9.1 and illustrated in Sketch 9.5. Upon converting this equation to the form in (9.8), we have

$$\mathbf{x}^{\mathrm{T}} \begin{bmatrix} 3 & -1 \\ -1 & 3 \end{bmatrix} \mathbf{x} - 2\mathbf{x}^{\mathrm{T}} \begin{bmatrix} 7 \\ -5 \end{bmatrix} + 18 = 0.$$

Breaking down this equation into the elements of (9.7), we have

$$A = \begin{bmatrix} 3 & -1 \\ -1 & 3 \end{bmatrix} \quad \text{and} \quad \mathbf{b} = \begin{bmatrix} 7 \\ -5 \end{bmatrix}.$$

This means that the translation $\mathbf{v}$ may be found by solving the 2×2 linear system

$$A\mathbf{v} = \mathbf{b},$$

which in this case is

$$\mathbf{v} = \begin{bmatrix} 2 \\ -1 \end{bmatrix}.$$

This linear system may be solved if A has full rank. This is equivalent to A having two nonzero eigenvalues, and so the given conic is either an ellipse or a hyperbola.

We can remove the translation from the conic, which moves the center to the origin. By calculating $c = d - \mathbf{v}^T A\mathbf{v}$ from (9.7), the conic without the translation is

$$\mathbf{x}^{\mathrm{T}} \begin{bmatrix} 3 & -1 \\ -1 & 3 \end{bmatrix} \mathbf{x} - 1 = 0.$$

The rotation that was applied to the ellipse in standard form is identified by finding the eigenvectors associated with A, following the method of Section 7.3. Recall that we found eigenvalues $\lambda_1 = 2$ and $\lambda_2 = 4$ in Section 9.2, so the corresponding rotation matrix is

$$\hat{R} = \frac{1}{\sqrt{2}} \begin{bmatrix} 1 & 1 \\ -1 & 1 \end{bmatrix}.$$

(Using the degree of freedom available, we chose the column vectors of $\hat{R}$ so that it represents a rotation instead of a rotation combined with a reflection.) This matrix represents the rotation necessary to align the conic with the coordinate axes, and in fact, $\hat{R} = R^T$ with respect to the decomposition of A into RAR^T. Points $\hat{\mathbf{x}}$ on the co-ordinate axes aligned conic must be mapped via $\hat{\mathbf{x}} = R^T\mathbf{x}$, so $\mathbf{x}$ is replaced by $R\hat{\mathbf{x}}$,

$$(R\hat{\mathbf{x}})^T R D R^T R\hat{\mathbf{x}} - c = 0.$$

Simplifying this equation results in,

$$\hat{\mathbf{x}}^T D \hat{\mathbf{x}} - c = 0.$$

In conclusion, the conic in standard form was rotated by $45°$ via R and then translated by

$$\mathbf{v} = \begin{bmatrix} 2 \\ -1 \end{bmatrix}.$$

- conic section
- implicit equation
- minor axis
- major axis
- center
- standard form

- ellipse
- hyperbola
- parabola
- eigenvalues
- eigenvectors
- conic type
- affine invariance

9.4 Exercises

1. Let $x_1^2 - 2x_1x_2 - 4 = 0$ be the equation of a conic section. What type is it?

2. What affine map takes the circle

$$(x_1 - 3)^2 + (x_2 + 1)^2 - 4 = 0$$

to the ellipse

$$2x_1^2 + 4x_2^2 - 1 = 0?$$

3. How many intersections does a straight line have with a conic? Given a conic in the form (9.2) and a parametric form of a line l(t), what are the t-values of the intersection points? Explain any singularities.

4. If the shear
$$\begin{bmatrix} 1 & 1 \\ 1/2 & 0 \end{bmatrix}$$
is applied to the conic of Example 9.1, what is the type of the resulting conic?

5. Let a conic be given by
$$3x_2^2 + 2x_1x_2 + 3x_2^2 + 10x_1 - 2x_2 + 10 = 0.$$

What type of conic is it, and what is the rotation and translation that took it out of standard form?

10

3D Geometry

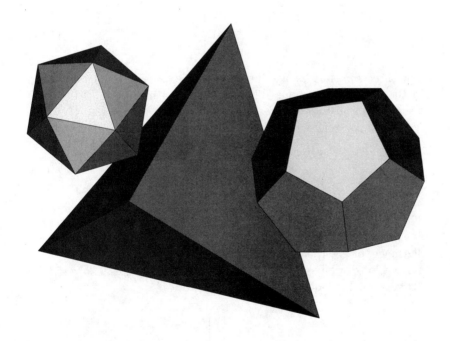

Figure 10.1.
3D objects: Planar facets joined to form 3D objects.

   This chapter introduces the essential building blocks of 3D geometry by first extending the 2D tools from Chapters 2 and 3 to 3D. But
beyond that, we will also encounter some concepts that are "truly"
3D, i.e., those that do not have 2D counterparts. With the geome-

try presented in this chapter, we will be ready to create and analyze simple 3D objects, such as those illustrated in Figure 10.1.

10.1   From 2D to 3D

Moving from 2D to 3D geometry requires a coordinate system with one more dimension. Sketch 10.1 illustrates the $[\mathbf{e}_1, \mathbf{e}_2, \mathbf{e}_3]$-system which consists of the vectors

$$\mathbf{e}_1 = \begin{bmatrix} 1 \\ 0 \\ 0 \end{bmatrix}, \quad \mathbf{e}_2 = \begin{bmatrix} 0 \\ 1 \\ 0 \end{bmatrix}, \quad \text{and} \quad \mathbf{e}_3 = \begin{bmatrix} 0 \\ 0 \\ 1 \end{bmatrix}.$$

Thus, a *vector* in 3D is given as

$$\mathbf{v} = \begin{bmatrix} v_1 \\ v_2 \\ v_3 \end{bmatrix}. \tag{10.1}$$

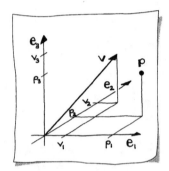

Sketch 10.1.

The $[\mathbf{e}_1, \mathbf{e}_2, \mathbf{e}_3]$-axes, a point, and a vector.

The three *components* of $\mathbf{v}$ indicate the displacement along each axis in the $[\mathbf{e}_1, \mathbf{e}_2, \mathbf{e}_3]$-system. This is illustrated in Sketch 10.1. A 3D vector $\mathbf{v}$ is said to live in real 3D space, or $\mathbb{R}^3$, that is $\mathbf{v} \in \mathbb{R}^3$.

A *point* is a reference to a *location*. Points in 3D are given as

$$\mathbf{p} = \begin{bmatrix} p_1 \\ p_2 \\ p_3 \end{bmatrix}. \tag{10.2}$$

The *coordinates* indicate the point's location in the $[\mathbf{e}_1, \mathbf{e}_2, \mathbf{e}_3]$-system, as illustrated in Sketch 10.1. A point $\mathbf{p}$ is said to live in Euclidean 3D-space, or $\mathbb{E}^3$, that is $\mathbf{p} \in \mathbb{E}^3$.

Let's look briefly at some basic 3D vector properties, as we did for 2D vectors. First of all, the 3D *zero vector*:

$$\mathbf{0} = \begin{bmatrix} 0 \\ 0 \\ 0 \end{bmatrix}.$$

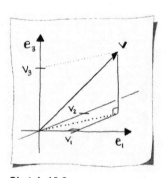

Sketch 10.2.

Length of a 3D vector.

Sketch 10.2 illustrates a 3D vector $\mathbf{v}$ along with its components. Notice the two right triangles. Applying the *Pythagorean theorem* twice, the *length* of $\mathbf{v}$, denoted as $\|\mathbf{v}\|$, is

$$\|\mathbf{v}\| = \sqrt{v_1^2 + v_2^2 + v_3^2}. \tag{10.3}$$

The length or magnitude of a 3D vector can be interpreted as distance, speed, or force.

Scaling a vector by an amount k yields $\|k\mathbf{v}\| = k\|\mathbf{v}\|$. Also, a *normalized vector* has unit length, $\|\mathbf{v}\| = 1$.

Example 10.1

We will get some practice working with 3D vectors. The first task is to normalize the vector

$$\mathbf{v} = \begin{bmatrix} 1 \\ 2 \\ 3 \end{bmatrix}.$$

First calculate the length of $\mathbf{v}$ as

$$\|\mathbf{v}\| = \sqrt{1^2 + 2^2 + 3^2} = \sqrt{14},$$

then the normalized vector $\mathbf{w}$ is

$$\mathbf{w} = \frac{\mathbf{v}}{\|\mathbf{v}\|} = \frac{1}{\sqrt{14}} \begin{bmatrix} 1 \\ 2 \\ 3 \end{bmatrix} \approx \begin{bmatrix} 0.27 \\ 0.53 \\ 0.80 \end{bmatrix}.$$

Check for yourself that $\|\mathbf{w}\| = 1$.

Scale $\mathbf{v}$ by $k = 2$:

$$2\mathbf{v} = \begin{bmatrix} 2 \\ 4 \\ 6 \end{bmatrix}.$$

Now calculate

$$\|2\mathbf{v}\| = \sqrt{2^2 + 4^2 + 6^2} = 2\sqrt{14}.$$

Thus we verified that $\|2\mathbf{v}\| = 2\|\mathbf{v}\|$.

There are infinitely many 3D unit vectors. In Sketch 10.3 a few of these are drawn emanating from the origin. The sketch is a sphere of radius one.

All the rules for combining points and vectors in 2D from Section 2.2 carry over to 3D. The *dot product* of two 3D vectors, $\mathbf{v}$ and $\mathbf{w}$, becomes

$$\mathbf{v} \cdot \mathbf{w} = v_1 w_1 + v_2 w_2 + v_3 w_3.$$

The cosine of the angle θ between the two vectors can be determined as

$$\cos \theta = \frac{\mathbf{v} \cdot \mathbf{w}}{\|\mathbf{v}\|\|\mathbf{w}\|}. \tag{10.4}$$

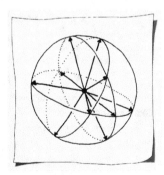

Sketch 10.3.
All 3D unit vectors define a sphere.

10.2 Cross Product

The dot product is a type of multiplication for two vectors which reveals geometric information, namely the angle between them. However, this does not reveal information about their orientation in relation to $\mathbb{R}^3$. Two vectors define a plane—which is a *subspace* of $\mathbb{R}^3$. Thus, it would be useful to have yet another vector in order to create a 3D coordinate system which is *embedded* in the $[\mathbf{e}_1, \mathbf{e}_2, \mathbf{e}_3]$-system. This is the purpose of another form of vector multiplication called the *cross product*.

In other words, the cross product of $\mathbf{v}$ and $\mathbf{w}$, written as

$$\mathbf{u} = \mathbf{v} \wedge \mathbf{w},$$

produces the vector $\mathbf{u}$ which satisfies the following:

1. The vector $\mathbf{u}$ is perpendicular to $\mathbf{v}$ and $\mathbf{w}$, that is

$$\mathbf{u} \cdot \mathbf{v} = 0 \quad \text{and} \quad \mathbf{u} \cdot \mathbf{w} = 0.$$

Sketch 10.4.

Characteristics of the cross product

2. The orientation of the vector $\mathbf{u}$ follows the *right-hand rule*. This means that if you curl the fingers of your right hand from $\mathbf{v}$ to $\mathbf{w}$, your thumb will point in the direction of $\mathbf{u}$.

3. The magnitude of $\mathbf{u}$ is the area of the parallelogram defined by $\mathbf{v}$ and $\mathbf{w}$.

These items are illustrated in Sketch 10.4. Because the cross product produces a vector, it is also called a *vector product*.

Items 1 and 2 determine the direction of $\mathbf{u}$ and item 3 determines the length of $\mathbf{u}$. The cross product is defined as

$$\mathbf{v} \wedge \mathbf{w} = \begin{bmatrix} v_2 w_3 - w_2 v_3 \\ v_3 w_1 - w_3 v_1 \\ v_1 w_2 - w_1 v_2 \end{bmatrix}. \tag{10.5}$$

Item 3 ensures that the cross product of orthogonal and unit length vectors $\mathbf{v}$ and $\mathbf{w}$ results in a vector $\mathbf{u}$ such that $\mathbf{u}, \mathbf{v}, \mathbf{w}$ are *orthonormal*. In other words, $\mathbf{u}$ is unit length and perpendicular to $\mathbf{v}$ and $\mathbf{w}$. This fact results in computation savings when constructing an orthonormal coordinate frame, as we will see in Section 11.8.

Example 10.2

Compute the cross product of

$$\mathbf{v} = \begin{bmatrix} 1 \\ 0 \\ 2 \end{bmatrix} \quad \text{and} \quad \mathbf{w} = \begin{bmatrix} 0 \\ 3 \\ 4 \end{bmatrix}.$$

The cross product is

$$\mathbf{u} = \mathbf{v} \wedge \mathbf{w} = \begin{bmatrix} 0 \times 4 - 3 \times 2 \\ 2 \times 0 - 4 \times 1 \\ 1 \times 3 - 0 \times 0 \end{bmatrix} = \begin{bmatrix} -6 \\ -4 \\ 3 \end{bmatrix}. \tag{10.6}$$

Section 4.10 described why the 2×2 determinant, formed from two 2D vectors, is equal to the area P of the parallelogram defined by these two vectors. The analogous result for two vectors in 3D is

$$P = \|\mathbf{v} \wedge \mathbf{w}\|. \tag{10.7}$$

Recall that P is also defined by measuring a height and side length of the parallelogram, as illustrated in Sketch 10.5. The height h is

$$h = \|\mathbf{w}\| \sin \theta,$$

and the side length is $\|\mathbf{v}\|$, which makes

$$P = \|\mathbf{v}\| \|\mathbf{w}\| \sin \theta. \tag{10.8}$$

Equating (10.7) and (10.8) results in

$$\|\mathbf{v} \wedge \mathbf{w}\| = \|\mathbf{v}\| \|\mathbf{w}\| \sin \theta. \tag{10.9}$$

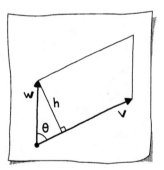

Sketch 10.5.
Area of a parallelogram.

Example 10.3

Compute the area of the parallelogram formed by

$$\mathbf{v} = \begin{bmatrix} 2 \\ 2 \\ 0 \end{bmatrix} \quad \text{and} \quad \mathbf{w} = \begin{bmatrix} 0 \\ 0 \\ 1 \end{bmatrix}.$$

Set up the cross product

$$\mathbf{v} \wedge \mathbf{w} = \begin{bmatrix} 2 \\ -2 \\ 0 \end{bmatrix}.$$

Then the area is

$$P = \|\mathbf{v} \wedge \mathbf{w}\| = 2\sqrt{2}.$$

Since the parallelogram is a rectangle, the area is the product of the edge lengths, so this is the correct result. Verifying (10.9),

$$P = 2\sqrt{2}\sin 90° = 2\sqrt{2}.$$

In order to derive another useful expression in terms of the cross product, square both sides of (10.9). Thus, we have

$$\begin{aligned} \|\mathbf{v} \wedge \mathbf{w}\|^2 &= \|\mathbf{v}\|^2 \|\mathbf{w}\|^2 \sin^2 \theta \\ &= \|\mathbf{v}\|^2 \|\mathbf{w}\|^2 (1 - \cos^2 \theta) \\ &= \|\mathbf{v}\|^2 \|\mathbf{w}\|^2 - \|\mathbf{v}\|^2 \|\mathbf{w}\|^2 \cos^2 \theta \\ &= \|\mathbf{v}\|^2 \|\mathbf{w}\|^2 - (\mathbf{v} \cdot \mathbf{w})^2. \end{aligned} \qquad (10.10)$$

The last line is referred to as *Lagrange's identity.*

To get a better feeling for the behavior of the cross product, let's look at some of its properties.

- Parallel vectors result in the zero vector: $\mathbf{v} \wedge c\mathbf{v} = \mathbf{0}$.

- Homogeneous: $c\mathbf{v} \wedge \mathbf{w} = c(\mathbf{v} \wedge \mathbf{w})$.

- Anti-symmetric: $\mathbf{v} \wedge \mathbf{w} = -(\mathbf{w} \wedge \mathbf{v})$.

- Non-associative: $\mathbf{u} \wedge (\mathbf{v} \wedge \mathbf{w}) \neq (\mathbf{u} \wedge \mathbf{v}) \wedge \mathbf{w}$, in general.

- Distributive: $\mathbf{u} \wedge (\mathbf{v} + \mathbf{w}) = \mathbf{u} \wedge \mathbf{v} + \mathbf{u} \wedge \mathbf{w}$.

- Right-hand rule:

$$\begin{aligned} \mathbf{e}_1 \wedge \mathbf{e}_2 &= \mathbf{e}_3 \\ \mathbf{e}_2 \wedge \mathbf{e}_3 &= \mathbf{e}_1 \\ \mathbf{e}_3 \wedge \mathbf{e}_1 &= \mathbf{e}_2 \end{aligned}$$

Example 10.4

Let's test these properties of the cross product with

$$\mathbf{u} = \begin{bmatrix} 1 \\ 1 \\ 1 \end{bmatrix} \quad \mathbf{v} = \begin{bmatrix} 2 \\ 0 \\ 0 \end{bmatrix} \quad \mathbf{w} = \begin{bmatrix} 0 \\ 3 \\ 0 \end{bmatrix}.$$

Make your own sketches and don't forget the right-hand rule to guess the resulting vector direction.

Parallel vectors:

$$\mathbf{v} \wedge 3\mathbf{v} = \begin{bmatrix} 0 \times 0 - 0 \times 0 \\ 0 \times 6 - 0 \times 2 \\ 2 \times 0 - 6 \times 0 \end{bmatrix} = \mathbf{0}.$$

Homogeneous:

$$4\mathbf{v} \wedge \mathbf{w} = \begin{bmatrix} 0 \times 0 - 3 \times 0 \\ 0 \times 0 - 0 \times 8 \\ 8 \times 3 - 0 \times 0 \end{bmatrix} = \begin{bmatrix} 0 \\ 0 \\ 24 \end{bmatrix},$$

and

$$4(\mathbf{v} \wedge \mathbf{w}) = 4 \begin{bmatrix} 0 \times 0 - 3 \times 0 \\ 0 \times 0 - 0 \times 2 \\ 2 \times 3 - 0 \times 0 \end{bmatrix} = 4 \begin{bmatrix} 0 \\ 0 \\ 6 \end{bmatrix} = \begin{bmatrix} 0 \\ 0 \\ 24 \end{bmatrix}.$$

Anti-symmetric:

$$\mathbf{v} \wedge \mathbf{w} = \begin{bmatrix} 0 \\ 0 \\ 6 \end{bmatrix} \quad \text{and} \quad -(\mathbf{w} \wedge \mathbf{v}) = -\left(\begin{bmatrix} 0 \\ 0 \\ -6 \end{bmatrix} \right).$$

Non-associative:

$$\mathbf{u} \wedge (\mathbf{v} \wedge \mathbf{w}) = \begin{bmatrix} 1 \times 6 - 0 \times 1 \\ 1 \times 0 - 6 \times 1 \\ 1 \times 0 - 0 \times 1 \end{bmatrix} = \begin{bmatrix} 6 \\ -6 \\ 0 \end{bmatrix},$$

which is not the same as

$$(\mathbf{u} \wedge \mathbf{v}) \wedge \mathbf{w} = \begin{bmatrix} 0 \\ 2 \\ -2 \end{bmatrix} \wedge \begin{bmatrix} 0 \\ 3 \\ 0 \end{bmatrix} = \begin{bmatrix} 6 \\ 0 \\ 0 \end{bmatrix}.$$

Distributive:

$$\mathbf{u} \wedge (\mathbf{v} + \mathbf{w}) = \begin{bmatrix} 1 \\ 1 \\ 1 \end{bmatrix} \wedge \begin{bmatrix} 2 \\ 3 \\ 0 \end{bmatrix} = \begin{bmatrix} -3 \\ 2 \\ 1 \end{bmatrix},$$

which is equal to

$$(\mathbf{u} \wedge \mathbf{v}) + (\mathbf{u} \wedge \mathbf{w}) = \begin{bmatrix} 0 \\ 2 \\ -2 \end{bmatrix} + \begin{bmatrix} -3 \\ 0 \\ 3 \end{bmatrix} = \begin{bmatrix} -3 \\ 2 \\ 1 \end{bmatrix}.$$

The cross product is an invaluable tool for engineering. One reason: it facilitates the construction of a coordinate independent frame of reference.

10.3 Lines

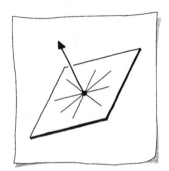

Sketch 10.6.

Point and perpendicular don't define a line.

Specifying a line with 3D geometry differs a bit from 2D. In terms of points and vectors, two pieces of information define a line; however, we are restricted to specifying

- two points or

- a point and a vector parallel to the line.

The 2D geometry item

- a point and a vector perpendicular to the line,

no longer works. It isn't specific enough. (See Sketch 10.6.) In other words, an entire family of lines satisfies this specification; this family lies in a plane. (More on planes in Section 10.4.) As a consequence, the concept of a *normal* to a 3D line does not exist.

Let's look at the mathematical representations of a 3D line. Clearly, from the discussion above, there cannot be an *implicit form*.

The *parametric form* of a 3D line does not differ from the 2D line except for the fact that the given information lives in 3D. A line $\mathbf{l}(t)$ has the form

$$\mathbf{l}(t) = \mathbf{p} + t\mathbf{v}, \tag{10.11}$$

where $\mathbf{p} \in \mathbb{E}^3$ and $\mathbf{v} \in \mathbb{R}^3$. Points are generated on the line as the parameter t varies.

In 2D, two lines either intersect or they are parallel. In 3D this is not the case; a third possibility is that the lines are *skew*. Sketch 10.7 illustrates skew lines using a cube as a reference frame.

Because lines in 3D can be skew, the intersection of two lines might not have a solution. Revisiting the problem of the intersection of

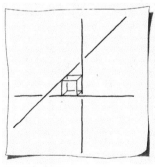

Sketch 10.7.

Skew lines.

two lines given in parametric form from Section 3.8, we can see the algebraic truth in this statement. Now the two lines are

$$l_1 : \quad l_1(t) = \mathbf{p} + t\mathbf{v}$$
$$l_2 : \quad l_2(s) = \mathbf{q} + s\mathbf{w}$$

where $\mathbf{p}, \mathbf{q} \in \mathbb{E}^3$ and $\mathbf{v}, \mathbf{w} \in \mathbb{R}^3$. To find the intersection point, we solve for t or s. Repeating (3.15), we have the linear system

$$\hat{t}\mathbf{v} - \hat{s}\mathbf{w} = \mathbf{q} - \mathbf{p}.$$

However, now there are three equations and still only two unknowns. Thus, the system is *overdetermined*; more information on this type of system is given in Section 14.4. No solution exists when the lines are skew. In many applications it is important to know the closest point on a line to another line. This problem is solved in Section 11.2.

We still have the concepts of perpendicular and parallel lines in 3D.

10.4 Planes

While exploring the possibility of a 3D implicit line, we encountered a plane. We'll essentially repeat that here, however, with a little change in notation. Suppose we are given a point $\mathbf{p}$ and a vector $\mathbf{n}$ bound to $\mathbf{p}$. The locus of all points $\mathbf{x}$ which satisfy the equation

$$\mathbf{n} \cdot (\mathbf{x} - \mathbf{p}) = 0 \qquad (10.12)$$

defines the *implicit form* of a plane. This is illustrated in Sketch 10.8. The vector $\mathbf{n}$ is called the *normal* to the plane if $\|\mathbf{n}\| = 1$. If this is the case, then (10.12) is called the *point normal plane equation*.

Expanding (10.12), we have

$$n_1 x_1 + n_2 x_2 + n_3 x_3 - (n_1 p_1 + n_2 p_2 + n_3 p_3) = 0.$$

Typically, this is written as

$$Ax_1 + Bx_2 + Cx_3 + D = 0, \qquad (10.13)$$

where

$$A = n_1$$
$$B = n_2$$
$$C = n_3$$
$$D = -(n_1 p_1 + n_2 p_2 + n_3 p_3).$$

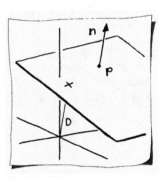

Sketch 10.8.
Point normal plane equation.

Example 10.5

Compute the implicit form of the plane through the point

$$\mathbf{p} = \begin{bmatrix} 4 \\ 0 \\ 0 \end{bmatrix}$$

which is perpendicular to the vector

$$\mathbf{n} = \begin{bmatrix} 1 \\ 1 \\ 1 \end{bmatrix}.$$

All we need to compute is D:

$$D = -(1 \times 4 + 1 \times 0 + 1 \times 0) = -4.$$

Thus, the plane equation is

$$x_1 + x_2 + x_3 - 4 = 0.$$

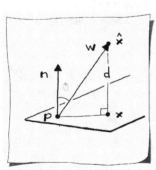

Sketch 10.9.

Origin to plane distance D.

Similar to a 2D implicit line, if the coefficients A, B, C correspond to the normal to the plane, then $|D|$ describes the distance of the plane to the origin. Notice in Sketch 10.8 that this is the *perpendicular distance*. A 2D cross section of the geometry is illustrated in Sketch 10.9. We equate two definitions of the cosine of an angle,

$$\cos(\theta) = \frac{D}{\|\mathbf{p}\|} \quad \text{and} \quad \cos(\theta) = \frac{\mathbf{n} \cdot \mathbf{p}}{\|\mathbf{n}\|\|\mathbf{p}\|},$$

and remember that the normal is unit length, to find that $D = \mathbf{n} \cdot \mathbf{p}$.

The point normal form reflects the (perpendicular) distance of a point from a plane. This situation is illustrated in Sketch 10.10. The distance d of an arbitrary point $\hat{\mathbf{x}}$ from the point normal form of the plane is

$$d = A\hat{x}_1 + B\hat{x}_2 + C\hat{x}_3 + D.$$

The reason for this follows precisely as it did for the implicit line in Section 3.3. See Section 11.1 for more on this topic.

Suppose we would like to find the distance of many points to a given plane. Then it is computationally more efficient to have the plane in (10.13) corresponding to the point normal form: the new coefficients

Sketch 10.10.

Point to plane distance.

will be A', B', C', D'. In order to do this we need to know one point $\mathbf{p}$ in the plane. This can be found by setting two x_i coordinates to zero and solving for the third, e.g., $x_1 = x_2 = 0$ and solving for x_3. With this point, we normalize the vector of coefficients and define:

$$\begin{bmatrix} A' \\ B' \\ C' \end{bmatrix} = \frac{\mathbf{n}}{\|\mathbf{n}\|}.$$

Now solve for D':

$$D' = -(A'p_1 + B'p_2 + C'p_3).$$

Example 10.6

Let's continue with the plane from the previous example,

$$x_1 + x_2 + x_3 - 4 = 0.$$

Clearly it is not in point normal form because the length of the vector $\|\mathbf{n}\| \neq 1$.

In order to convert it to point normal form, we need one point $\mathbf{p}$ in the plane. Set $x_2 = x_3 = 0$ and solve for $x_1 = 4$, thus

$$\mathbf{p} = \begin{bmatrix} 4 \\ 0 \\ 0 \end{bmatrix}.$$

Normalize $\mathbf{n}$ above, thus forming

$$\begin{bmatrix} A' \\ B' \\ C' \end{bmatrix} = \frac{\mathbf{n}}{\|\mathbf{n}\|} = \begin{bmatrix} \frac{1}{\sqrt{3}} \\ \frac{1}{\sqrt{3}} \\ \frac{1}{\sqrt{3}} \end{bmatrix}.$$

Now solve for the plane coefficient

$$D' = -\left(\frac{1}{\sqrt{3}} \times 4 + \frac{1}{\sqrt{3}} \times 0 + \frac{1}{\sqrt{3}} \times 0 \right) = \frac{-4}{\sqrt{3}},$$

making the point normal plane equation

$$\frac{1}{\sqrt{3}}x_1 + \frac{1}{\sqrt{3}}x_2 + \frac{1}{\sqrt{3}}x_3 - \frac{4}{\sqrt{3}} = 0.$$

Determine the distance d of the point

$$\mathbf{q} = \begin{bmatrix} 4 \\ 4 \\ 4 \end{bmatrix}$$

from the plane:

$$d = \frac{1}{\sqrt{3}} \times 4 + \frac{1}{\sqrt{3}} \times 4 + \frac{1}{\sqrt{3}} \times 4 - \frac{4}{\sqrt{3}} = \frac{8}{\sqrt{3}} \approx 4.6.$$

Notice that $d > 0$; this is because the point $\mathbf{q}$ is on the same side of the plane as the normal direction. The distance of the origin to the plane is $d = D' = -4/\sqrt{3}$, which is negative because it is on the opposite side of the plane to which the normal points. This is analogous to the 2D implicit line.

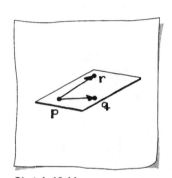

Sketch 10.11.

Parametric plane.

The implicit plane equation is wonderful for determining if a point is in a plane; however, it is not so useful for creating points in a plane. For this we have the *parametric form* of a plane.

The *given* information for defining a parametric representation of a plane usually comes in one of two ways:

- three points, or

- a point and two vectors.

If we start with the first scenario, we choose three points $\mathbf{p}, \mathbf{q}, \mathbf{r}$, then choose one of these points and form two vectors $\mathbf{v}$ and $\mathbf{w}$ bound to that point as shown in Sketch 10.11:

$$\mathbf{v} = \mathbf{q} - \mathbf{p} \quad \text{and} \quad \mathbf{w} = \mathbf{r} - \mathbf{p}. \tag{10.14}$$

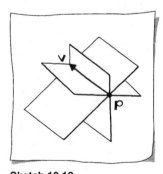

Sketch 10.12.

Family of planes through a point and vector.

Why not just specify one point and a vector in the plane, analogous to the implicit form of a plane? Sketch 10.12 illustrates that this is not enough information to uniquely define a plane. Many planes fit that data.

Two vectors bound to a point are the data we'll use to define a plane $\mathbf{P}$ in parametric form as

$$\mathbf{P}(s,t) = \mathbf{p} + s\mathbf{v} + t\mathbf{w}. \tag{10.15}$$

The two independent parameters, s and t, determine a point $\mathbf{P}(s,t)$ in the plane.[1] Notice that (10.15) can be rewritten as

$$\begin{aligned} \mathbf{P}(s,t) &= \mathbf{p} + s(\mathbf{q}-\mathbf{p}) + t(\mathbf{r}-\mathbf{p}) \\ &= (1-s-t)\mathbf{p} + s\mathbf{q} + t\mathbf{r}. \end{aligned} \qquad (10.16)$$

As described in Section 8.1, $(1-s-t, s, t)$ are the *barycentric coordinates* of a point $\mathbf{P}(s,t)$ with respect to the triangle with vertices $\mathbf{p}, \mathbf{q}$, and $\mathbf{r}$.

Another method for specifying a plane, as illustrated in Sketch 10.13, is as the *bisector of two points*. This is how a plane is defined in Euclidean geometry—the locus of points equidistant from two points. The line between two given points defines the normal to the plane, and the midpoint of this line segment defines a point in the plane. With this information it is most natural to express the plane in implicit form.

Sketch 10.13.
A plane defined as the bisector of two points.

10.5 Application: Lighting and Shading

Let's look at an application of a handful of the tools that we have developed so far: lighting and shading for computer graphics. One of the most basic elements needed to calculate the lighting of a 3D object (model) is the *normal*, which was introduced in Section 10.4. Although a lighted model might look smooth, it is represented simply with vertices and planar facets, most often triangles. The normal of each planar facet is used in conjunction with the light source location and our eye location to calculate the lighting (color) of each vertex, and one such method is called the *Phong illumination model*; details of the method may be found in graphics texts such as [11, 19]. Figure 10.2 illustrates normals drawn emanating from the centroid of the facets. Determining the color of a facet is called *shading*. A nice example is illustrated in Figure 10.3.

We calculate the normal by using the *cross product* from Section 10.2. Suppose we have a triangle defined by points $\mathbf{p}, \mathbf{q}, \mathbf{r}$. Then we form two vectors, $\mathbf{v}$ and $\mathbf{w}$ as defined in (10.14) from these points. The normal $\mathbf{n}$ is

$$\mathbf{n} = \frac{\mathbf{v} \wedge \mathbf{w}}{\|\mathbf{v} \wedge \mathbf{w}\|}.$$

The normal is by convention considered to be of unit length. Why use $\mathbf{v} \wedge \mathbf{w}$ instead of $\mathbf{w} \wedge \mathbf{v}$? Triangle vertices imply an *orientation*.

---

[1]This is a slight deviation in notation: an uppercase boldface letter rather than a lowercase one denoting a point.

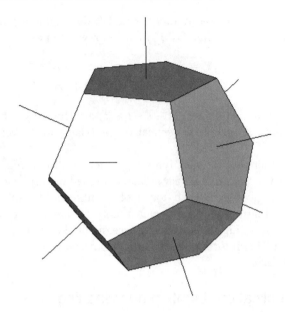

Figure 10.2.
Hedgehog plot: the normal of each facet is drawn at the centroid.

From this we follow the right-hand rule to determine the normal direction. It is important to have a rule, as just described, so the facets for a model are consistently defined. In turn, the lighting will be consistently calculated at all points on the model.

Figure 10.3 illustrates *flat shading*, which is the fastest and most rough-looking shading method. Each facet is given one color, so a lighting calculation is done at one point, say the centroid, based on the facet's normal. Figure 10.4 illustrates *Gouraud or smooth shading*, which produces smoother looking models but involves more calculation than flat shading. At each vertex of a triangle, the lighting is calculated. Each of these lighting vectors, $\mathbf{i_p}, \mathbf{i_q}, \mathbf{i_r}$, each vector indicating red, green, and blue components of light, are then interpolated across the triangle. For instance, a point $\mathbf{x}$ in the triangle with barycentric coordinates (u, v, w),

$$\mathbf{x} = u\mathbf{p} + v\mathbf{q} + w\mathbf{r},$$

will be assigned color

$$\mathbf{i_x} = u\mathbf{i_p} + v\mathbf{i_q} + w\mathbf{i_r},$$

which is a handy application of barycentric coordinates!

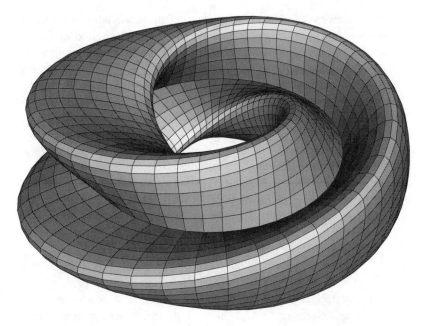

Figure 10.3.
Flat shading: the normal to each planar facets is used to calculate the illumination of
each facet.

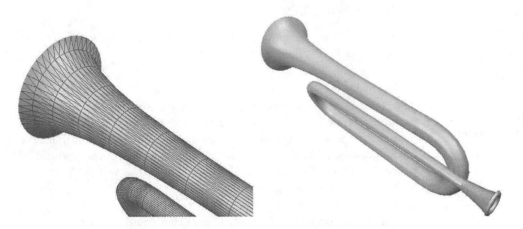

Figure 10.4.
Smooth shading: a normal at each vertex is used to calculate the illumination over
each facet. Left: zoomed-in and displayed with triangles. Right: the smooth shaded
bugle.

Still unanswered in the smooth shading method: what normals do we assign the vertices in order to achieve a smooth shaded model? If we used the same normal at each vertex, for small enough triangles, we would wind up with (expensive) flat shading. The answer: a *vertex normal* is calculated as the average of the triangle normals for the triangles in the *star* of the vertex. Better normals can be generated by weighting the contribution of each triangle normal based on the area of the triangle.

The direction of the normal $\mathbf{n}$ relative to our eye's position can be used to eliminate facets from the rendering pipeline. For a closed surface, such as that in Figure 10.3, the "back-facing" facets will be obscured by the "front-facing" facets. If the centroid of a triangle is $\mathbf{c}$ and the eye's position is $\mathbf{e}$, then form the vector

$$\mathbf{v} = (\mathbf{e} - \mathbf{c})/\|\mathbf{e} - \mathbf{c}\|.$$

If

$$\mathbf{n} \cdot \mathbf{v} < 0$$

then the triangle is back-facing, and we need not render it. This process is called *culling*. A great savings in rendering time can be achieved with culling.

Planar facet normals play an important role computer graphics, as demonstrated in this section. For more advanced applications, consult a graphics text such as [19].

10.6   Scalar Triple Product

In Section 10.2 we encountered the area P of the parallelogram formed by vectors $\mathbf{v}$ and $\mathbf{w}$ measured as

$$P = \|\mathbf{v} \wedge \mathbf{w}\|.$$

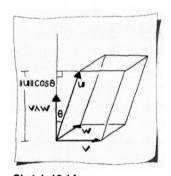

Sketch 10.14.
Scalar triple product for the volume.

The next natural question is how do we measure the *volume* of the *parallelepiped*, or skew box, formed by three vectors. See Sketch 10.14. The volume is a product of a face area and the corresponding height of the skew box. As illustrated in the sketch, after choosing $\mathbf{v}$ and $\mathbf{w}$ to form the face, the height is $\|\mathbf{u}\| \cos \theta$. Thus, the volume V is

$$V = \|\mathbf{u}\|\|\mathbf{v} \wedge \mathbf{w}\| \cos \theta.$$

Bringing together the dot product and cross product, then

$$V = \mathbf{u} \cdot (\mathbf{v} \wedge \mathbf{w}). \qquad (10.17)$$

This is called the *scalar triple product*, and it is a number representing a signed volume.

The sign reveals something about the orientation of the three vectors. If $\cos\theta > 0$, resulting in a positive volume, then $\mathbf{u}$ is on the same side of the plane formed by $\mathbf{v}$ and $\mathbf{w}$ as $\mathbf{v} \wedge \mathbf{w}$. If $\cos\theta < 0$, resulting in a negative volume, then $\mathbf{u}$ is on the opposite side of the plane as $\mathbf{v} \wedge \mathbf{w}$. If $\cos\theta = 0$ resulting in zero volume, then $\mathbf{u}$ lies in this plane—the vectors are *coplanar*.

From the discussion above and the anti-symmetry of the cross product, we see that the scalar triple product is invariant under *cyclic permutations*. This means that the we get the same volume for the following:

$$\begin{aligned} V &= \mathbf{u} \cdot (\mathbf{v} \wedge \mathbf{w}) \\ &= \mathbf{w} \cdot (\mathbf{u} \wedge \mathbf{v}) \\ &= \mathbf{v} \cdot (\mathbf{w} \wedge \mathbf{u}). \end{aligned} \tag{10.18}$$

In Section 4.10 we introduced the 2×2 determinant as a tool to calculate area. The scalar triple product is really just a fancy name for a 3×3 determinant, but we'll get to that in Section 12.7.

10.7   Linear Spaces

We shall now concentrate on the space involving 3D vectors. The set of all 3D vectors is referred to as a 3D *linear space* if we associate with it the operation of forming linear combinations.[2] This means that if $\mathbf{v}$ and $\mathbf{w}$ are two vectors in a linear space, then any vector $\mathbf{u} = r\mathbf{u} + s\mathbf{v}$ is also in that space. The vector $\mathbf{u}$ is then said to be a *linear combination* of $\mathbf{v}$ and $\mathbf{w}$. Generalizing, we say that if $\mathbf{v}_1, \mathbf{v}_2, \ldots, \mathbf{v}_k$ are vectors in a linear space, then the linear combination

$$\mathbf{u} = r_1\mathbf{v}_1 + r_2\mathbf{v}_2 + \ldots + r_k\mathbf{v}_k$$

is also in that space.

Now let us return to just two vectors $\mathbf{v}$ and $\mathbf{w}$. Consider all vectors $\mathbf{u}$ which may be expressed as

$$\mathbf{u} = s\mathbf{v} + t\mathbf{w} \tag{10.19}$$

with arbitrary scalars s, t. Clearly, all vectors $\mathbf{u}$ of the form (10.19) form a subset of all 3D vectors. But beyond that, they form a linear space themselves—a 2D space, to be more precise. For if two vectors $\mathbf{u}_1$ and $\mathbf{u}_2$ are in this space, then they can be written as

---

[2]The term *vector space* is also used.

$$\mathbf{u}_1 = s_1\mathbf{v} + t_1\mathbf{w} \quad \text{and} \quad \mathbf{u}_2 = s_2\mathbf{v} + t_2\mathbf{w} \qquad (10.20)$$

and thus any linear combination of them can be written as

$$\alpha\mathbf{u}_1 + \beta\mathbf{u}_2 = (\alpha s_1 + \beta s_2)\mathbf{v} + (\alpha t_1 + \beta t_2)\mathbf{w}$$

which is again in the same space. We call the set of all vectors of the form (10.19) a *subspace* of the linear space of all 3D vectors.

The term subspace is justified since not all 3D vectors are in it. Take for instance the vector $\mathbf{n} = \mathbf{v} \wedge \mathbf{w}$, which is perpendicular to both $\mathbf{v}$ and $\mathbf{w}$. There is no way to write this vector as a linear combination of $\mathbf{v}$ and $\mathbf{w}$!

We say our subspace has dimension 2 since it is generated, or *spanned* by two vectors. These vectors have to be non-collinear; otherwise, they just define a line, or a 1D subspace. If two vectors are collinear, then they are also called *linearly dependent*. Conversely, if they are not collinear, they are called *linearly independent*.

Given two linearly independent vectors $\mathbf{v}$ and $\mathbf{w}$, how do we decide if another vector $\mathbf{u}$ is in the subspace spanned by $\mathbf{v}$ and $\mathbf{w}$? We would have to find numbers s and t such that (10.19) holds. Finding these numbers amounts to solving a *linear system*, as discussed in Chapter 14. If s and t can be determined, then $\mathbf{u}$ is in the subspace, otherwise, it is not.

We'll revisit this topic in a more abstract setting for n-dimensional vectors in Chapter 15.

- 3D vector
- 3D point
- vector length
- unit vector
- dot product
- subspace
- cross product
- orthonormal
- area
- Lagrange's identity
- 3D line
- plane
- point-plane distance
- normal

- plane-origin distance
- barycentric coordinates
- scalar triple product
- volume
- linear space
- linear combination
- linearly independent
- linearly dependent
- triangle normal
- back-facing triangle
- lighting model
- flat and Gouraud shading
- vertex normal
- culling

10.8 Exercises

For the following exercises, use the following points and vectors:

$$\mathbf{p} = \begin{bmatrix} 0 \\ 0 \\ 1 \end{bmatrix}, \mathbf{q} = \begin{bmatrix} 1 \\ 1 \\ 1 \end{bmatrix}, \mathbf{r} = \begin{bmatrix} 4 \\ 2 \\ 4 \end{bmatrix}, \mathbf{v} = \begin{bmatrix} 1 \\ 0 \\ 0 \end{bmatrix}, \mathbf{w} = \begin{bmatrix} 1 \\ 1 \\ 1 \end{bmatrix}, \mathbf{u} = \begin{bmatrix} 0 \\ 0 \\ 1 \end{bmatrix}.$$

1. Normalize the vector $\mathbf{r}$. What is the length of the vector $2\mathbf{r}$?

2. Find the angle between the vectors $\mathbf{v}$ and $\mathbf{w}$.

3. Compute $\mathbf{v} \wedge \mathbf{w}$.

4. Compute the area of the parallelogram formed by vectors $\mathbf{v}$ and $\mathbf{w}$.

5. What is the sine of the angle between $\mathbf{v}$ and $\mathbf{w}$?

6. Find three vectors so that their cross product is associative.

7. Form the point normal plane equation for a plane through point $\mathbf{p}$ and with normal direction $\mathbf{r}$.

8. Form the point normal plane equation for the plane defined by points $\mathbf{p}$, $\mathbf{q}$, and $\mathbf{r}$.

9. Form a parametric plane equation for the plane defined by points $\mathbf{p}$, $\mathbf{q}$, and $\mathbf{r}$.

10. Form an equation of the plane that bisects the points $\mathbf{p}$ and $\mathbf{q}$.

11. Find the volume of the parallelepiped defined by vectors $\mathbf{v}$, $\mathbf{w}$, and $\mathbf{u}$.

12. Given the line l defined by point $\mathbf{q}$ and vector $\mathbf{v}$, what is the length of the projection of vector $\mathbf{w}$ bound to $\mathbf{q}$ onto l?

13. Given the line l defined by point $\mathbf{q}$ and vector $\mathbf{v}$, what is the (perpendicular) distance of the point $\mathbf{q} + \mathbf{w}$ (where $\mathbf{w}$ is a vector) to the line l?

14. What is $\mathbf{w} \wedge 6\mathbf{w}$?

15. For the plane in Exercise 7, what is the distance of this plane to the origin?

16. For the plane in Exercise 7, what is the distance of the point $\mathbf{q}$ to this plane?

17. Given the triangle formed by points $\mathbf{p}, \mathbf{q}, \mathbf{r}$, and colors

$$\mathbf{i_p} = \begin{bmatrix} 1 \\ 0 \\ 0 \end{bmatrix}, \quad \mathbf{i_q} = \begin{bmatrix} 0 \\ 1 \\ 0 \end{bmatrix}, \quad \mathbf{i_r} = \begin{bmatrix} 1 \\ 0 \\ 0 \end{bmatrix},$$

what color $\mathbf{i_c}$ would be assigned to the centroid of this triangle using Gouraud shading? Note: the color vectors are scaled between $[0,1]$. Black would be represented by

$$\begin{bmatrix} 0 \\ 0 \\ 0 \end{bmatrix},$$

white by

$$\begin{bmatrix} 1 \\ 1 \\ 1 \end{bmatrix},$$

and red by

$$\begin{bmatrix} 1 \\ 0 \\ 0 \end{bmatrix}.$$

18. Given a 2D linear subspace formed by vectors $\mathbf{w}$ and $\mathbf{v}$, is $\mathbf{u}$ an element of that subspace?

19. Decompose $\mathbf{w}$ into

$$\mathbf{w} = \mathbf{u}_1 + \mathbf{u}_2$$

where $\mathbf{u}_1$ and $\mathbf{u}_2$ are perpendicular. Additionally, find $\mathbf{u}_3$ to complete an orthogonal frame. Hint: orthogonal projections are the topic of Section 2.8.

11

Interactions in 3D

Figure 11.1.
Ray tracing: 3D intersections are a key element of rendering a raytraced image.
(Courtesy of Ben Steinberg, Arizona State University.)

The tools of points, lines, and planes are our most basic 3D geometry building blocks. But in order to build real objects, we must be able to compute with these building blocks. For example, a cube is defined by its six bounding planes; but in most cases, one also needs to know its eight vertices. These are found by intersecting appropriate planes. A similar problem is to determine whether a given point is inside or outside the cube. This chapter outlines the basic algorithms for these types of problems. The ray traced image of Figure 11.1 was generated by using the tools developed in this chapter.[1]

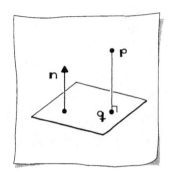

Sketch 11.1.
A point and a plane.

11.1 Distance Between a Point and a Plane

Let a plane be given by its implicit form $\mathbf{n} \cdot \mathbf{x} + c = 0$. If we also have a point $\mathbf{p}$, what is its distance d to the plane, and what is its closest point $\mathbf{q}$ on the plane? See Sketch 11.1 for the geometry. Notice how close this problem is to the foot of a point from Section 3.7.

Clearly, the vector $\mathbf{p} - \mathbf{q}$ must be perpendicular to the plane, i.e., parallel to the plane's normal $\mathbf{n}$. Thus, $\mathbf{p}$ can be written as

$$\mathbf{p} = \mathbf{q} + t\mathbf{n};$$

if we find t, our problem is solved. This is easy, since $\mathbf{q}$ must also satisfy the plane equation:

$$\mathbf{n} \cdot [\mathbf{p} - t\mathbf{n}] + c = 0.$$

Thus,

$$t = \frac{c + \mathbf{n} \cdot \mathbf{p}}{\mathbf{n} \cdot \mathbf{n}}. \tag{11.1}$$

It is good practice to assure that $\mathbf{n}$ is normalized, i.e., $\mathbf{n} \cdot \mathbf{n} = 1$, and then

$$t = c + \mathbf{n} \cdot \mathbf{p}. \tag{11.2}$$

Note that $t = 0$ is equivalent to $\mathbf{n} \cdot \mathbf{p} + c = 0$; in that case, $\mathbf{p}$ is on the plane to begin with!

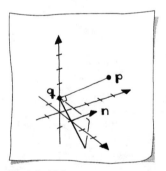

Sketch 11.2.
Example point and a plane.

Example 11.1

Consider the plane

$$x_1 + x_2 + x_3 - 1 = 0$$

and the point

$$\mathbf{p} = \begin{bmatrix} 2 \\ 2 \\ 3 \end{bmatrix},$$

as shown in Sketch 11.2.

---
[1]See Section 11.3 for a description of the technique used to generate this image.

According to (11.1), we find $t = 2$. Thus,

$$\mathbf{q} = \begin{bmatrix} 2 \\ 2 \\ 3 \end{bmatrix} - 2 \times \begin{bmatrix} 1 \\ 1 \\ 1 \end{bmatrix} = \begin{bmatrix} 0 \\ 0 \\ 1 \end{bmatrix}.$$

The vector $\mathbf{p} - \mathbf{q}$ is given by

$$\mathbf{p} - \mathbf{q} = t\mathbf{n}.$$

Thus, the length of $\mathbf{p} - \mathbf{q}$, or the *distance* of $\mathbf{p}$ to the plane, is given by $t\|\mathbf{n}\|$. If $\mathbf{n}$ is normalized, then $\|\mathbf{p} - \mathbf{q}\| = t$; this means that we simply insert $\mathbf{p}$ into the plane equation and obtain for the distance d:

$$d = c + \mathbf{n} \cdot \mathbf{p}.$$

It is also clear that if $t > 0$, then $\mathbf{n}$ points towards $\mathbf{p}$, and away from it if $t < 0$ (see Sketch 11.3) where the plane is drawn "edge on." Compare with the almost identical Sketch 3.3!

Again, a numerical caveat: If a point is very close to a plane, it becomes very hard numerically to decide which side it is on!

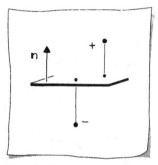

Sketch 11.3.
Points around a plane.

11.2 Distance Between Two Lines

Two 3D lines typically do not meet—such lines are called *skew*. It might be of interest to know how close they are to meeting; in other words, what is the *distance* between the lines? See Sketch 11.4 for an illustration.

Let the two lines $\mathbf{l}_1$ and $\mathbf{l}_2$ be given by

$$\mathbf{x}_1(s_1) = \mathbf{p}_1 + s_1\mathbf{v}_1, \quad \text{and}$$
$$\mathbf{x}_2(s_2) = \mathbf{p}_2 + s_2\mathbf{v}_2,$$

respectively. Let $\mathbf{x}_1$ be the point on $\mathbf{l}_1$ closest to $\mathbf{l}_2$, also let $\mathbf{x}_2$ be the point on $\mathbf{l}_2$ closest to $\mathbf{l}_1$. It should be clear that the vector $\mathbf{x}_2 - \mathbf{x}_1$ is perpendicular to both $\mathbf{l}_1$ and $\mathbf{l}_2$. Thus

$$[\mathbf{x}_2 - \mathbf{x}_1]\mathbf{v}_1 = 0,$$
$$[\mathbf{x}_2 - \mathbf{x}_1]\mathbf{v}_2 = 0,$$

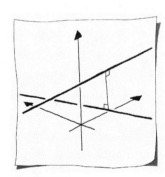

Sketch 11.4.
Skew lines in 3D.

or

$$[\mathbf{p}_2 - \mathbf{p}_1]\mathbf{v}_1 = s_1\mathbf{v}_1 \cdot \mathbf{v}_1 - s_2\mathbf{v}_1 \cdot \mathbf{v}_2,$$
$$[\mathbf{p}_2 - \mathbf{p}_1]\mathbf{v}_2 = s_1\mathbf{v}_1 \cdot \mathbf{v}_2 - s_2\mathbf{v}_2 \cdot \mathbf{v}_2.$$

These are two equations in the two unknowns s_1 and s_2, and are thus readily solved using the methods from Chapter 5.

Example 11.2

Let $\mathbf{l}_1$ be given by

$$\mathbf{x}_1(s_1) = \begin{bmatrix} 0 \\ 0 \\ 0 \end{bmatrix} + s_1 \begin{bmatrix} 1 \\ 0 \\ 0 \end{bmatrix}.$$

This means, of course, that $\mathbf{l}_1$ is the $\mathbf{e}_1$-axis. For $\mathbf{l}_2$, we assume

$$\mathbf{x}_2(s_2) = \begin{bmatrix} 0 \\ 1 \\ 1 \end{bmatrix} + s_2 \begin{bmatrix} 0 \\ 1 \\ 0 \end{bmatrix}.$$

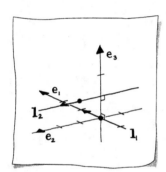

Sketch 11.5.

Skew lines closest point example.

This line is parallel to the $\mathbf{e}_2$-axis; both lines are shown in Sketch 11.5.

Our linear system becomes

$$0 = s_1,$$
$$1 = -s_2.$$

Inserting these values, we have

$$\mathbf{x}_1(0) = \begin{bmatrix} 0 \\ 0 \\ 0 \end{bmatrix} \quad \text{and} \quad \mathbf{x}_2(-1) = \begin{bmatrix} 0 \\ 0 \\ 1 \end{bmatrix}.$$

These are the two points of closest proximity.

Two 3D lines intersect if the two points $\mathbf{x}_1$ and $\mathbf{x}_2$ are identical.[2]

---

[2] In "real-life," that means within a tolerance!

A condition for two 3D lines to intersect is found from the observation that the three vectors $\mathbf{v}_1, \mathbf{v}_2, \mathbf{p}_2 - \mathbf{p}_1$ must be coplanar, or linearly dependent. This would lead to the condition

$$\det[\mathbf{v}_1, \mathbf{v}_2, \mathbf{p}_2 - \mathbf{p}_1] = 0.$$

From a numerical viewpoint, it is safer to compare the distance between the points $\mathbf{x}_1$ and $\mathbf{x}_2$; in the field of *Computer-Aided Design* (*CAD*), one usually has known tolerances (e.g., 0.001") for distances. It is much harder to come up with a meaningful tolerance for a determinant.

11.3   Lines and Planes: Intersections

One of the basic techniques in computer graphics is called *ray tracing*. Figure 11.1 illustrates this technique. A scene is given as an assembly of planes (usually restricted to triangles). A computer-generated image needs to compute proper lighting; this is done by tracing light rays through the scene. The ray intersects a plane, it is reflected, then it intersects the next plane, etc. Sketch 11.6 gives an example.

The basic problem to be solved is this: Given a plane $\mathbf{P}$ and a line $\mathbf{l}$, what is their *intersection point* $\mathbf{x}$? It is most convenient to represent the plane by assuming that we know a point $\mathbf{q}$ on it as well as its normal vector $\mathbf{n}$ (see Sketch 11.7). Then the unknown point $\mathbf{x}$, being on $\mathbf{P}$, must satisfy

$$[\mathbf{x} - \mathbf{q}] \cdot \mathbf{n} = 0. \tag{11.3}$$

By definition, the intersection point is also on the line (the ray, in computer graphics jargon), given by a point $\mathbf{p}$ and a vector $\mathbf{v}$:

$$\mathbf{x} = \mathbf{p} + t\mathbf{v}. \tag{11.4}$$

At this point, we do not know the correct value for t; once we have it, our problem is solved.

The solution is obtained by substituting the expression for $\mathbf{x}$ from (11.4) into (11.3):

$$[\mathbf{p} + t\mathbf{v} - \mathbf{q}] \cdot \mathbf{n} = 0.$$

Thus,

$$[\mathbf{p} - \mathbf{q}] \cdot \mathbf{n} + t\mathbf{v} \cdot \mathbf{n} = 0$$

and

$$t = \frac{[\mathbf{q} - \mathbf{p}] \cdot \mathbf{n}}{\mathbf{v} \cdot \mathbf{n}}. \tag{11.5}$$

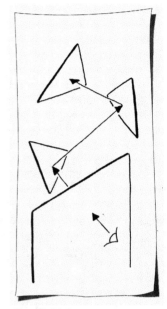

Sketch 11.6.
A ray is traced through a scene.

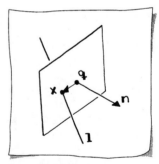

Sketch 11.7.
A line and a plane.

The intersection point $\mathbf{x}$ is now computed as

$$\mathbf{x} = \mathbf{p} + \frac{[\mathbf{q} - \mathbf{p}] \cdot \mathbf{n}}{\mathbf{v} \cdot \mathbf{n}} \mathbf{v}. \tag{11.6}$$

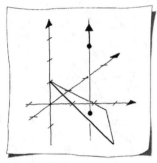

Sketch 11.8.

Intersecting a line and a plane.

Example 11.3

Take the plane

$$x_1 + x_2 + x_3 - 1 = 0$$

and the line

$$\mathbf{p}(t) = \begin{bmatrix} 1 \\ 1 \\ 2 \end{bmatrix} + t \begin{bmatrix} 0 \\ 0 \\ 1 \end{bmatrix}$$

as shown in Sketch 11.8.

We need a point $\mathbf{q}$ on the plane; set $x_1 = x_2 = 0$ and solve for x_3, resulting in $x_3 = 1$. This amounts to intersecting the plane with the $\mathbf{e}_3$-axis. From (11.5), we find $t = -3$ and then (11.6) gives the intersection point as

$$\mathbf{x} = \begin{bmatrix} 1 \\ 1 \\ 2 \end{bmatrix} - 3 \begin{bmatrix} 0 \\ 0 \\ 1 \end{bmatrix} = \begin{bmatrix} 1 \\ 1 \\ -1 \end{bmatrix}.$$

It never hurts to carry out a sanity check: Verify that this $\mathbf{x}$ does indeed satisfy the plane equation!

A word of caution: In (11.5), we happily divide by the dot product $\mathbf{v} \cdot \mathbf{n}$—but that better not be zero![3] If it is, then the ray "grazes" the plane, i.e., it is parallel to it. Then no intersection exists.

The same problem—intersecting a line with a plane—may be solved if the plane is given in *parametric form*. Then the unknown intersection point $\mathbf{x}$ must satisfy

$$\mathbf{x} = \mathbf{q} + u_1 \mathbf{r}_1 + u_2 \mathbf{r}_2.$$

Since we know that $\mathbf{x}$ is also on the line $\mathbf{l}$, we may set

$$\mathbf{p} + t\mathbf{v} = \mathbf{q} + u_1 \mathbf{r}_1 + u_2 \mathbf{r}_2.$$

---

[3]Keep in mind that real numbers are rarely *equal* to zero. A tolerance needs to be used; 0.001 should work if both $\mathbf{n}$ and $\mathbf{v}$ are normalized.

This equation is short for three individual equations, one for each coordinate. We thus have three equations in three unknowns t, u_1, u_2, and solve them according to methods of Chapter 14.

11.4 Intersecting a Triangle and a Line

A plane is, by definition, an unbounded object. In many applications, planes are parts of objects; one is only interested in a small part of a plane. For example, the six faces of a cube are bounded planes, so are the four faces of a tetrahedron.

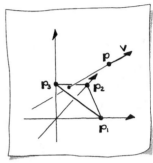

We will now examine the case of a 3D *triangle* as an example of a bounded plane. If we intersect a 3D triangle with a line (a ray), then we are not interested in an intersection point *outside* the triangle— only an interior one will count.

Let the triangle be given by three points $\mathbf{p}_1, \mathbf{p}_2, \mathbf{p}_3$ and the line by a point $\mathbf{p}$ and a direction $\mathbf{v}$ (see Sketch 11.9).

The plane may be written in parametric form as

$$\mathbf{x}(u_1, u_2) = \mathbf{p}_1 + u_1(\mathbf{p}_2 - \mathbf{p}_1) + u_2(\mathbf{p}_3 - \mathbf{p}_1).$$

Sketch 11.9.
Intersecting a triangle and a line.

We thus arrive at

$$\mathbf{p} + t\mathbf{v} = \mathbf{p}_1 + u_1(\mathbf{p}_2 - \mathbf{p}_1) + u_2(\mathbf{p}_3 - \mathbf{p}_1),$$

a linear system in the unknowns t, u_1, u_2. The solution is inside the triangle if both u_1 and u_2 are between zero and one, and their sum is less than or equal to one. This is so since we may view $(u_1, u_2, 1 - u_1 - u_2)$ as *barycentric coordinates* of the triangle. These are positive exactly for points inside the triangle. See Section 8.1 for a review of barycentric coordinates in a triangle.

11.5 Lines and Planes: Reflections

The next problem is that of line or plane *reflection*. Given a point $\mathbf{x}$ on a plane $\mathbf{P}$ and an "incoming" direction $\mathbf{v}$, what is the reflected, or "outgoing" direction $\mathbf{v}'$? See Sketch 11.10, where we look at the plane $\mathbf{P}$ "edge on." We assume that $\mathbf{v}$, $\mathbf{v}'$ and $\mathbf{n}'$ are of unit length.

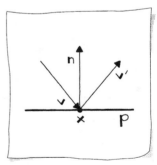

From physics, you might recall that the angle between $\mathbf{v}$ and the plane normal $\mathbf{n}$ must equal that of $\mathbf{v}'$ and $\mathbf{n}$, except for a sign change. We conveniently record this fact using a dot product:

$$-\mathbf{v} \cdot \mathbf{n} = \mathbf{v}' \cdot \mathbf{n}. \tag{11.7}$$

Sketch 11.10.
A reflection.

The normal vector $\mathbf{n}$ is thus the *angle bisector* of $\mathbf{v}$ and $\mathbf{v}'$. From inspection of Sketch 11.10, we also infer the symmetry property

$$c\mathbf{n} = \mathbf{v}' - \mathbf{v} \tag{11.8}$$

for some real number c. This means that some multiple of the normal vector may be written as the sum $\mathbf{v}' + (-\mathbf{v})$.

We now solve (11.8) for $\mathbf{v}'$ and insert into (11.7):

$$-\mathbf{v} \cdot \mathbf{n} = [c\mathbf{n} + \mathbf{v}] \cdot \mathbf{n},$$

and solve for c:

$$c = -2\mathbf{v} \cdot \mathbf{n}. \tag{11.9}$$

Here, we made use of the fact that $\mathbf{n}$ is a unit vector and thus $\mathbf{n} \cdot \mathbf{n} = 1$.

The reflected vector $\mathbf{v}'$ is now given by using our value for c in (11.9):

$$\mathbf{v}' = \mathbf{v} - [2\mathbf{v} \cdot \mathbf{n}]\mathbf{n}. \tag{11.10}$$

In the special case of $\mathbf{v}$ being perpendicular to the plane, i.e., $\mathbf{v} = -\mathbf{n}$, we obtain $\mathbf{v}' = -\mathbf{v}$ as expected. Also note that the point of reflection does not enter the equations at all.

We may rewrite (11.10) as

$$\mathbf{v}' = \mathbf{v} - 2[\mathbf{v}^{\mathrm{T}}\mathbf{n}]\mathbf{n}$$

which in turn may be reformulated to

$$\mathbf{v}' = \mathbf{v} - 2[\mathbf{n}\mathbf{n}^{\mathrm{T}}]\mathbf{v}.$$

You see this after multiplying out all products involved. Note that $\mathbf{n}\mathbf{n}^{\mathrm{T}}$ is a symmetric matrix, not a dot product! This is an example of a *dyadic matrix*; this type of matrix has rank one.

Now we are in a position to formulate a reflection as a linear map. It is of the form $\mathbf{v}' = H\mathbf{v}$ with

$$H = I - 2\mathbf{n}\mathbf{n}^{\mathrm{T}}. \tag{11.11}$$

The matrix H is known as a *Householder matrix*. We'll look at this matrix again in Section 16.1 in the context of solving a linear system.

In computer graphics, this reflection problem is an integral part of calculating the lighting (color) at each vertex. A brief introduction to the lighting model may be found in Section 10.5. The light is positioned somewhere on $\mathbf{v}$, and after the light hits the point $\mathbf{x}$, the reflected light travels in the direction of $\mathbf{v}'$. We use a dot product to measure the cosine of the angle between the direction of our eye $(\mathbf{e} - \mathbf{x})$ and the reflection vector. The smaller the angle, the more reflected light hits our eye. This type of lighting is called *specular reflection*; it is the element of light that is dependent on our position and it produces a highlight—a shiny spot.

11.6 Intersecting Three Planes

Suppose we are given three planes with implicit equations

$$\mathbf{n}_1 \cdot \mathbf{x} + c_1 = 0,$$
$$\mathbf{n}_2 \cdot \mathbf{x} + c_2 = 0,$$
$$\mathbf{n}_3 \cdot \mathbf{x} + c_3 = 0.$$

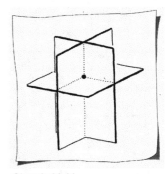

Where do they intersect? The answer is some point $\mathbf{x}$, which lies on each of the planes. See Sketch 11.11 for an illustration.

Sketch 11.11.
Intersecting three planes.

The solution is surprisingly simple; just condense the three plane equations into matrix form:

$$\begin{bmatrix} \mathbf{n}_1^{\mathrm{T}} \\ \mathbf{n}_2^{\mathrm{T}} \\ \mathbf{n}_3^{\mathrm{T}} \end{bmatrix} \begin{bmatrix} x_1 \\ x_2 \\ x_3 \end{bmatrix} = \begin{bmatrix} -c_1 \\ -c_2 \\ -c_3 \end{bmatrix}. \tag{11.12}$$

We have three equations in the three unknowns x_1, x_2, x_3!

Example 11.4

The following example is shown in Sketch 11.12. The equations of the planes in that sketch are

$$x_1 + x_3 = 1, \quad x_3 = 1, \quad x_2 = 2.$$

The linear system is

$$\begin{bmatrix} 1 & 0 & 1 \\ 0 & 0 & 1 \\ 0 & 1 & 0 \end{bmatrix} \begin{bmatrix} x_1 \\ x_2 \\ x_3 \end{bmatrix} = \begin{bmatrix} 1 \\ 1 \\ 2 \end{bmatrix}.$$

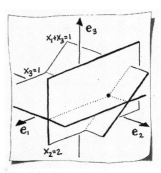

Sketch 11.12.
Intersecting three planes example.

Solving it by Gauss elimination (see Chapter 14), we obtain

$$\begin{bmatrix} x_1 \\ x_2 \\ x_3 \end{bmatrix} = \begin{bmatrix} 0 \\ 2 \\ 1 \end{bmatrix}.$$

While simple to solve, the three-planes problem does not always have a solution. Two lines in 2D do not intersect if they are parallel; in this case, their normal vectors are also parallel, or linearly dependent.

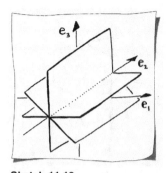

Sketch 11.13.

Three planes intersecting in one line.

The situation is analogous in 3D. If the normal vectors $\mathbf{n}_1, \mathbf{n}_2, \mathbf{n}_3$ are linearly dependent, then there is no solution to the intersection problem.

Example 11.5

The normal vectors are

$$\mathbf{n}_1 = \begin{bmatrix} 1 \\ 0 \\ 0 \end{bmatrix}, \quad \mathbf{n}_2 = \begin{bmatrix} 1 \\ 0 \\ 1 \end{bmatrix}, \quad \mathbf{n}_3 = \begin{bmatrix} 0 \\ 0 \\ 1 \end{bmatrix}.$$

Since $\mathbf{n}_2 = \mathbf{n}_1 + \mathbf{n}_3$, they are indeed linearly dependent, and thus the planes defined by them do not intersect in one point (see Sketch 11.13).

11.7   Intersecting Two Planes

Odd as it may seem, intersecting two planes is harder than intersecting three of them. The problem is this: Two planes are given in their implicit form

$$\mathbf{n} \cdot \mathbf{x} + c = 0, \tag{11.13}$$

$$\mathbf{m} \cdot \mathbf{x} + d = 0. \tag{11.14}$$

Find their intersection, which is a line. We would like the solution to be of the form

$$\mathbf{x}(t) = \mathbf{p} + t\mathbf{v}. \tag{11.15}$$

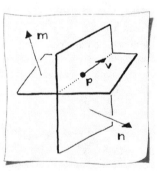

Sketch 11.14.

Intersecting two planes.

This situation is depicted in Sketch 11.14

The direction vector $\mathbf{v}$ of this line is easily found; since it lies in each of the planes, it must be perpendicular to both their normal vectors:

$$\mathbf{v} = \mathbf{n} \wedge \mathbf{m}.$$

We still need a point $\mathbf{p}$ on the line.

To this end, we come up with an auxiliary plane that intersects both given planes. The intersection point is clearly on the desired line. Let us assume for now that c and d are not both zero. Define the third plane by

$$\mathbf{v} \cdot \mathbf{x} = 0.$$

This plane passes through the origin and has normal vector $\mathbf{v}$, i.e., it is perpendicular to the desired line (see Sketch 11.15).

We now solve the three-plane intersection problem for the two given planes and the auxiliary plane for the missing point $\mathbf{p}$, and our line is determined.

In the case $c = d = 0$, both given planes pass through the origin, and it can serve as the point $\mathbf{p}$.

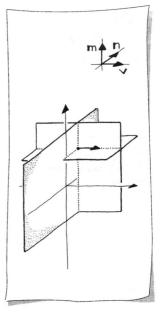

Sketch 11.15.
The auxiliary plane (shaded).

11.8   Creating Orthonormal Coordinate Systems

Often times when working in 3D, life is made easier by creating a local coordinate system. We have seen one example of this already: digitizing. In Section 1.5 we eluded to a coordinate frame as a means to "capture" a cat. Let's look at that example as a motivation for creating an *orthonormal coordinate system* with the Gram-Schmidt method.

The digitizer needs a coordinate frame in order to store coordinates for the cat. The cat sits on a table; suppose we record three points on the table: $\mathbf{p}, \mathbf{q}, \mathbf{r}$. From these three points, we form two vectors,

$$\mathbf{v}_1 = \mathbf{q} - \mathbf{p} \quad \text{and} \quad \mathbf{v}_2 = \mathbf{r} - \mathbf{p}.$$

This will establish $\mathbf{p}$ as the origin. A simple cross product will supply us with a vector normal to the table: $\mathbf{v}_3 = \mathbf{v}_1 \wedge \mathbf{v}_2$. Now we can state the *3D Gram-Schmidt problem*: given three linearly independent vectors, $\mathbf{v}_1, \mathbf{v}_2, \mathbf{v}_3$, find an orthonormal set of vectors $\mathbf{b}_1, \mathbf{b}_2, \mathbf{b}_3$.

Orthogonal projections, as described in Section 2.8, are the foundation of this method. After we have an orthogonal frame, we normalize all the vectors. Creation of an orthogonal set of vectors proceeds as follows:

$$\hat{\mathbf{b}}_1 = \mathbf{v}_1, \tag{11.16}$$

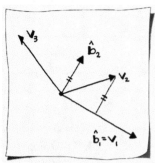

Sketch 11.16.

The first step of Gram-Schmidt orthonormalization.

$$\hat{\mathbf{b}}_2 = \mathbf{v}_2 - \left[\frac{\mathbf{v}_2\hat{\mathbf{b}}_1}{\left\|\hat{\mathbf{b}}_1\right\|^2}\right]\hat{\mathbf{b}}_1, \qquad (11.17)$$

$$\hat{\mathbf{b}}_3 = \mathbf{v}_3 - \left[\frac{\mathbf{v}_3\hat{\mathbf{b}}_1}{\left\|\hat{\mathbf{b}}_1\right\|^2}\right]\hat{\mathbf{b}}_1 - \left[\frac{\mathbf{v}_3\hat{\mathbf{b}}_2}{\left\|\hat{\mathbf{b}}_2\right\|^2}\right]\hat{\mathbf{b}}_2. \qquad (11.18)$$

Next, normalize each vector to form the orthonormal set of vectors:

$$\mathbf{b}_1 = \frac{\hat{\mathbf{b}}_1}{\left\|\hat{\mathbf{b}}_1\right\|}, \quad \mathbf{b}_2 = \frac{\hat{\mathbf{b}}_2}{\left\|\hat{\mathbf{b}}_2\right\|}, \quad \mathbf{b}_3 = \frac{\hat{\mathbf{b}}_3}{\left\|\hat{\mathbf{b}}_3\right\|}.$$

As illustrated in Sketch 11.16, (11.17) simply forms a vector which is the difference between $\mathbf{v}_2$ and the projection of $\mathbf{v}_2$ on $\hat{\mathbf{b}}_1$. Likewise, (11.18) projects $\mathbf{v}_3$ into the plane defined by $\hat{\mathbf{b}}_1$ and $\hat{\mathbf{b}}_2$, resulting in the vector

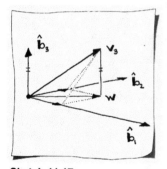

Sketch 11.17.

The second step of Gram-Schmidt orthonormalization.

$$\mathbf{w} = \left[\frac{\mathbf{v}_3\hat{\mathbf{b}}_1}{\left\|\hat{\mathbf{b}}_1\right\|^2}\right]\hat{\mathbf{b}}_1 + \left[\frac{\mathbf{v}_3\hat{\mathbf{b}}_2}{\left\|\hat{\mathbf{b}}_2\right\|^2}\right]\hat{\mathbf{b}}_2$$

which is illustrated in Sketch 11.17. The difference of this vector from $\mathbf{v}_3$ produces $\hat{\mathbf{b}}_3$.

As you might have observed, in 3D the Gram-Schmidt method is more work than simply applying the cross product repeatedly. The real advantage of the Gram-Schmidt method is for dimensions higher than three, where we don't have the cross product. However, understanding the process in 3D makes the n-dimensional formulas easier to follow. A much more general version is discussed in Section 15.4.

- distance of point and plane
- distance between two lines
- plane and line intersection
- triangle and line intersection
- reflection vector
- Householder matrix
- intersection of three planes
- intersection of two planes
- Gram-Schmidt orthonormalization

11.9 Exercises

For Exercises 1 and 2, we will deal with two planes. $\mathbf{P}_1$ goes through a point $\mathbf{p}$ and has normal vector $\mathbf{n}$:

$$\mathbf{p} = \begin{bmatrix} 1 \\ 2 \\ 0 \end{bmatrix}, \quad \mathbf{n} = \begin{bmatrix} -1 \\ 0 \\ 0 \end{bmatrix}.$$

The plane $\mathbf{P}_2$ is given by its implicit form

$$x_1 + 2x_2 - 2x_3 - 1 = 0.$$

Also, let a line $\mathbf{l}$ go through the point $\mathbf{q}$ and have direction $\mathbf{v}$:

$$\mathbf{q} = \begin{bmatrix} -1 \\ 2 \\ 0 \end{bmatrix}, \quad \mathbf{v} = \begin{bmatrix} 0 \\ 1 \\ 0 \end{bmatrix}.$$

1. Find the intersection of $\mathbf{P}_1$ with the line $\mathbf{l}$.

2. Find the intersection of $\mathbf{P}_2$ with the line $\mathbf{l}$.

3. Does the ray $\mathbf{p} + t\mathbf{v}$ with

$$\mathbf{p} = \begin{bmatrix} -1 \\ -1 \\ 0 \end{bmatrix} \quad \text{and} \quad \mathbf{v} = \begin{bmatrix} 1 \\ 1 \\ 1 \end{bmatrix}$$

   intersect the triangle with vertices

$$\begin{bmatrix} 3 \\ 0 \\ 0 \end{bmatrix}, \quad \begin{bmatrix} 0 \\ 2 \\ 1 \end{bmatrix}, \quad \begin{bmatrix} 2 \\ 2 \\ 3 \end{bmatrix}?$$

4. Revisit Example 11.2, but set the point defining the line $\mathbf{l}_2$ to be

$$\mathbf{p}_2 = \begin{bmatrix} 0 \\ 0 \\ 1 \end{bmatrix}.$$

   The lines have not changed; how do you obtain the (unchanged) solutions $\mathbf{x}_1$ and $\mathbf{x}_2$?

5. Let $\mathbf{a}$ be an arbitrary vector. It may be projected along a direction $\mathbf{v}$ onto the plane $\mathbf{P}$ with normal vector $\mathbf{n}$. What is its image $\mathbf{a}'$?

6. Given the point $\mathbf{p}$ in the plane P_1 what is the reflected direction of the vector

$$\mathbf{v} = \begin{bmatrix} 1/3 \\ 2/3 \\ -2/3 \end{bmatrix}?$$

7. Find the intersection of the three planes:

$$x_1 + x_2 = 1, \quad x_1 = 1, \quad x_3 = 4.$$

8. Find the intersection of the planes:

$$x_1 = 1 \quad \text{and} \quad x_3 = 4.$$

9. Given vectors

$$\mathbf{v}_1 = \begin{bmatrix} 1 \\ 0 \\ 0 \end{bmatrix}, \quad \mathbf{v}_2 = \begin{bmatrix} 1 \\ 1 \\ 0 \end{bmatrix}, \quad \mathbf{v}_3 = \begin{bmatrix} -1 \\ -1 \\ 1 \end{bmatrix},$$

carry out the Gram-Schmidt orthonormalization.

10. Given the vectors $\mathbf{v}_i$ in Exercise 9, what are the cross products that will produce the orthonormal vectors $\mathbf{b}_i$?

12

Linear Maps in 3D

Figure 12.1.
Flight simulator: 3D linear maps are necessary to create the twists and turns in a flight simulator. (Image is from the NASA website http://www.nasa.gov).

The flight simulator is an important part in the training of airplane pilots. It has a real cockpit, but what you see outside the windows is computer imagery. As you take a right turn, the terrain below changes

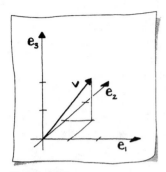

Sketch 12.1.

A vector in the $[\mathbf{e}_1, \mathbf{e}_2, \mathbf{e}_3]$-coordinate system.

accordingly; as you dive downwards, it comes closer to you. When you change the (simulated) position of your plane, the simulation software must recompute a new view of the terrain, clouds, or other aircraft. This is done through the application of 3D affine and linear maps.[1] Figure 12.1 shows an image that was generated by an actual flight simulator. For each frame of the simulated scene, complex 3D computations are necessary, most of them consisting of the types of maps discussed in this section.

12.1 Matrices and Linear Maps

The general concept of a linear map in 3D is the same as that for a 2D map. Let $\mathbf{v}$ be a vector in the standard $[\mathbf{e}_1, \mathbf{e}_2, \mathbf{e}_3]$-coordinate system, i.e.,

$$\mathbf{v} = v_1 \mathbf{e}_1 + v_2 \mathbf{e}_2 + v_3 \mathbf{e}_3.$$

(See Sketch 12.1 for an illustration.)

Let another coordinate system, the $[\mathbf{a}_1, \mathbf{a}_2, \mathbf{a}_3]$-coordinate system, be given by the origin $\mathbf{0}$ and three vectors $\mathbf{a}_1, \mathbf{a}_2, \mathbf{a}_3$. What vector $\mathbf{v}'$ in the $[\mathbf{a}_1, \mathbf{a}_2, \mathbf{a}_3]$-system corresponds to $\mathbf{v}$ in the $[\mathbf{e}_1, \mathbf{e}_2, \mathbf{e}_3]$-system? Simply the vector with the same coordinates relative to the $[\mathbf{a}_1, \mathbf{a}_2, \mathbf{a}_3]$-system! Thus,

$$\mathbf{v}' = v_1 \mathbf{a}_1 + v_2 \mathbf{a}_2 + v_3 \mathbf{a}_3. \tag{12.1}$$

This is illustrated by Sketch 12.2 and the following example.

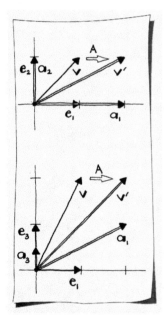

Sketch 12.2.

The matrix A maps $\mathbf{v}$ in the $[\mathbf{e}_1, \mathbf{e}_2, \mathbf{e}_3]$-coordinate system to the vector $\mathbf{v}'$ in the $[\mathbf{a}_1, \mathbf{a}_2, \mathbf{a}_3]$-coordinate system.

Example 12.1

Let

$$\mathbf{v} = \begin{bmatrix} 1 \\ 1 \\ 2 \end{bmatrix}, \quad \mathbf{a}_1 = \begin{bmatrix} 2 \\ 0 \\ 1 \end{bmatrix}, \quad \mathbf{a}_2 = \begin{bmatrix} 0 \\ 1 \\ 0 \end{bmatrix}, \quad \mathbf{a}_3 = \begin{bmatrix} 0 \\ 0 \\ 1/2 \end{bmatrix}.$$

Then

$$\mathbf{v}' = 1 \cdot \begin{bmatrix} 2 \\ 0 \\ 1 \end{bmatrix} + 1 \cdot \begin{bmatrix} 0 \\ 1 \\ 0 \end{bmatrix} + 2 \cdot \begin{bmatrix} 0 \\ 0 \\ 1/2 \end{bmatrix} = \begin{bmatrix} 2 \\ 1 \\ 2 \end{bmatrix}.$$

---

[1] Actually, perspective maps are also needed here. They will be discussed in Section 13.5.

You should recall that we had the same configuration earlier for the 2D case—(12.1) corresponds directly to (4.2) of Section 4.1. In Section 4.2, we then introduced the matrix form. That is now an easy project for this chapter—nothing changes except the matrices will be 3×3 instead of 2×2. In 3D, a matrix equation looks like this:

$$\mathbf{v}' = A\mathbf{v}, \qquad (12.2)$$

i.e., just the same as for the 2D case. Written out in detail, there is a difference:

$$\begin{bmatrix} v_1' \\ v_2' \\ v_3' \end{bmatrix} = \begin{bmatrix} a_{1,1} & a_{1,2} & a_{1,3} \\ a_{2,1} & a_{2,2} & a_{2,3} \\ a_{3,1} & a_{3,2} & a_{3,3} \end{bmatrix} \begin{bmatrix} v_1 \\ v_2 \\ v_3 \end{bmatrix}. \qquad (12.3)$$

All matrix properties from Sections 4.2 and 4.3 carry over almost verbatim.

Example 12.2

Returning to our example, it is quite easy to condense it into a matrix equation:

$$\begin{bmatrix} 2 & 0 & 0 \\ 0 & 1 & 0 \\ 1 & 0 & 1/2 \end{bmatrix} \begin{bmatrix} 1 \\ 1 \\ 2 \end{bmatrix} = \begin{bmatrix} 2 \\ 1 \\ 2 \end{bmatrix}.$$

Again, if we multiply a matrix A by a vector $\mathbf{v}$, the ith component of the result vector is obtained as the dot product of the ith row of A and $\mathbf{v}$.

The matrix A represents a *linear map*. Given the vector $\mathbf{v}$ in the $[\mathbf{e}_1, \mathbf{e}_2, \mathbf{e}_3]$-system, there is a vector $\mathbf{v}'$ in the $[\mathbf{a}_1, \mathbf{a}_2, \mathbf{a}_3]$-system such that $\mathbf{v}'$ has the same components in the $[\mathbf{a}_1, \mathbf{a}_2, \mathbf{a}_3]$-system as did $\mathbf{v}$ in the $[\mathbf{e}_1, \mathbf{e}_2, \mathbf{e}_3]$-system. The matrix A finds the components of $\mathbf{v}'$ relative to the $[\mathbf{e}_1, \mathbf{e}_2, \mathbf{e}_3]$-system.

With the 2×2 matrices of Section 4.3, we introduced the *transpose* A^{T} of a matrix A. We will need this for 3×3 matrices, and it is obtained by interchanging rows and columns, i.e.,

$$\begin{bmatrix} 2 & 3 & -4 \\ 3 & 9 & -4 \\ -1 & -9 & 4 \end{bmatrix}^{T} = \begin{bmatrix} 2 & 3 & -1 \\ 3 & 9 & -9 \\ -4 & -4 & 4 \end{bmatrix}.$$

The boldface row of A has become the boldface column of A^{T}. As a concise formula,

$$a_{i,j}^{\mathrm{T}} = a_{j,i}.$$

12.2   Scalings

A scaling is a linear map which enlarges or reduces vectors:

$$\mathbf{v}' = \begin{bmatrix} s_{1,1} & 0 & 0 \\ 0 & s_{2,2} & 0 \\ 0 & 0 & s_{3,3} \end{bmatrix} \mathbf{v} \qquad (12.4)$$

If all scale factors $s_{i,i}$ are larger than one, then all vectors are enlarged, see Figure 12.2. If all $s_{i,i}$ are positive yet less than one, all vectors are shrunk.

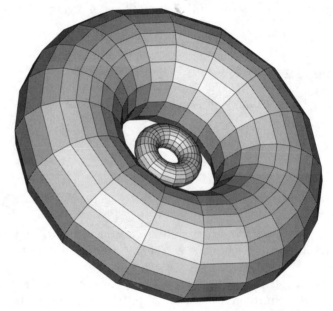

Figure 12.2.
Scalings in 3D: the small torus is scaled to form the large torus.

Example 12.3

In this example,

$$\begin{bmatrix} s_{1,1} & 0 & 0 \\ 0 & s_{2,2} & 0 \\ 0 & 0 & s_{3,3} \end{bmatrix} = \begin{bmatrix} 1/2 & 0 & 0 \\ 0 & 1 & 0 \\ 0 & 0 & 2 \end{bmatrix},$$

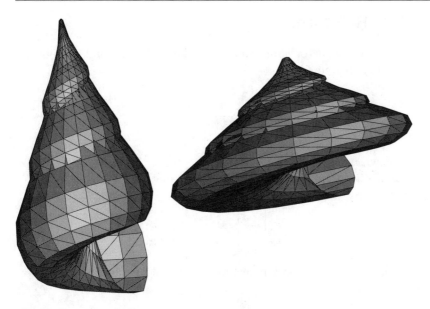

Figure 12.3.

Non-uniform scalings in 3D: the shell on the left is scaled by 1/2, 1, 2 in the $\mathbf{e}_1$, $\mathbf{e}_2$, $\mathbf{e}_3$-directions, respectively, and then translated resulting in the shell on the right.

we stretch in the $\mathbf{e}_1$-direction, shrink in the $\mathbf{e}_2$-direction, and leave the $\mathbf{e}_3$-direction unchanged. See Figure 12.3.

Negative numbers for the $s_{i,i}$ will cause a flip in addition to a scale. So, for instance

$$\begin{bmatrix} -2 & 0 & 0 \\ 0 & 1 & 0 \\ 0 & 0 & -1 \end{bmatrix}$$

will stretch and reverse the $\mathbf{e}_1$-direction, leave the $\mathbf{e}_2$-direction unchanged, and will reverse the $\mathbf{e}_3$-direction.

How do scalings affect volumes? If we map the *unit cube*, given by the three vectors $\mathbf{e}_1, \mathbf{e}_2, \mathbf{e}_3$ with a scaling, we get a rectangular box. Its side lengths are $s_{1,1}$ in the $\mathbf{e}_1$-direction, $s_{2,2}$ in the $\mathbf{e}_2$-direction, and $s_{3,3}$ in the $\mathbf{e}_3$-direction. Hence, its volume is given by $s_{1,1}s_{2,2}s_{3,3}$. A scaling, thus changes the volume of an object by a factor that equals the product of its diagonal elements.[2]

---

[2]We have only shown this for the unit cube. But it is true for any other object as well.

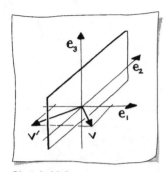

Sketch 12.3.

Reflection of a vector about
the e_2, e_3-plane.

12.3   Reflections

If we reflect a vector about the e_2, e_3-plane, then its first component
should change in sign:

$$\begin{bmatrix} v_1 \\ v_2 \\ v_3 \end{bmatrix} \longrightarrow \begin{bmatrix} -v_1 \\ v_2 \\ v_3 \end{bmatrix},$$

as shown in Sketch 12.3.

This reflection is achieved by a scaling matrix:

$$\begin{bmatrix} -v_1 \\ v_2 \\ v_3 \end{bmatrix} = \begin{bmatrix} -1 & 0 & 0 \\ 0 & 1 & 0 \\ 0 & 0 & 1 \end{bmatrix} \begin{bmatrix} v_1 \\ v_2 \\ v_3 \end{bmatrix}.$$

The following is also a reflection, as Sketch 12.4[3] shows:

$$\begin{bmatrix} v_1 \\ v_2 \\ v_3 \end{bmatrix} \longrightarrow \begin{bmatrix} v_3 \\ v_2 \\ v_1 \end{bmatrix}.$$

It interchanges the first and third component of a vector, and is thus a
reflection about the plane $x_1 = x_3$. This is an implicit plane equation,
as discussed in Section 10.4.

This map is achieved by the following matrix equation:

$$\begin{bmatrix} v_1 \\ v_2 \\ v_3 \end{bmatrix} = \begin{bmatrix} 0 & 0 & 1 \\ 0 & 1 & 0 \\ 1 & 0 & 0 \end{bmatrix} \begin{bmatrix} v_1 \\ v_2 \\ v_3 \end{bmatrix}.$$

By their very nature, reflections do not change volumes—but they
do change their signs. See Section 12.7 for more details.

Sketch 12.4.

Reflection of a vector about
the $x_1 = x_3$ plane.

12.4   Shears

What map takes a cube to the slanted cube of Sketch 12.5? That
slanted cube, by the way, is called a *parallelepiped*, but *skew box*
will do here. The answer: a shear. Shears in 3D are more complicated than the 2D shears from Section 4.7 because there are so
many more directions to shear. Let's look at some of the shears more
commonly used.

---

[3]In that sketch, the plane $x_1 = x_3$ is shown "edge on."

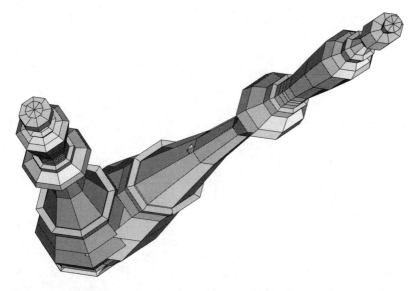

Figure 12.4.
Shears in 3D: a king chess piece sheared in the $\mathbf{e}_1$- and $\mathbf{e}_2$-directions. The $\mathbf{e}_3$-direction is the king's axis.

Consider the shear that maps $\mathbf{e}_1$ and $\mathbf{e}_2$ to themselves, and that also maps $\mathbf{e}_3$ to

$$\begin{bmatrix} a \\ b \\ 1 \end{bmatrix}.$$

The shear matrix S_1 that accomplishes the desired task is easily found:

$$S_1 = \begin{bmatrix} 1 & 0 & a \\ 0 & 1 & b \\ 0 & 0 & 1 \end{bmatrix}.$$

It is illustrated in Sketch 12.5 with $a = 1$ and $b = 1$, and in Figure 12.4.

What shear maps $\mathbf{e}_2$ and $\mathbf{e}_3$ to themselves, and also maps

$$\begin{bmatrix} a \\ b \\ c \end{bmatrix} \quad \text{to} \quad \begin{bmatrix} a \\ 0 \\ 0 \end{bmatrix}.$$

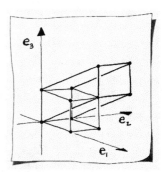

Sketch 12.5.
A 3D shear parallel to the $\mathbf{e}_1$, $\mathbf{e}_2$-plane.

This shear is given by the matrix

$$S_2 = \begin{bmatrix} 1 & 0 & 0 \\ \frac{-b}{a} & 1 & 0 \\ \frac{-c}{a} & 0 & 1 \end{bmatrix}.$$

One quick check gives:

$$\begin{bmatrix} 1 & 0 & 0 \\ \frac{-b}{a} & 1 & 0 \\ \frac{-c}{a} & 0 & 1 \end{bmatrix} \begin{bmatrix} a \\ b \\ c \end{bmatrix} = \begin{bmatrix} a \\ 0 \\ 0 \end{bmatrix};$$

thus our map does what it was meant to do. (This is the shear of the Gauss elimination step that we will encounter in Section 14.2.)

Let's look at the matrix for the shear a little more generally. Although it is possible to shear in any direction, it is more common to shear parallel to a coordinate axis or coordinate plane. With the help of rotations, the following shear matrices should be sufficient for any need.

Write the shear matrix S as

$$S = \begin{bmatrix} 1 & s_{1,2} & s_{1,3} \\ s_{2,1} & 1 & s_{2,3} \\ s_{3,1} & s_{3,2} & 1 \end{bmatrix}. \tag{12.5}$$

Suppose we apply this shear to a vector $\mathbf{v}$ resulting in

$$\mathbf{v}' = S\mathbf{v}.$$

An $s_{i,j}$ element is a factor by which the jth component of $\mathbf{v}$ affects the ith component of $\mathbf{v}'$. However, not all $s_{i,j}$ entries can be nonzero. Here are three scenarios for this matrix:

$$\begin{bmatrix} 1 & 0 & 0 \\ \star & 1 & \diamond \\ \star & \diamond & 1 \end{bmatrix} \quad \begin{bmatrix} 1 & \star & \diamond \\ 0 & 1 & 0 \\ \diamond & \star & 1 \end{bmatrix} \quad \begin{bmatrix} 1 & \diamond & \star \\ \diamond & 1 & \star \\ 0 & 0 & 1 \end{bmatrix}$$

with $\star$ denoting possible nonzero entries, and of the two $\diamond$ entries in each matrix, only one can be nonzero. Notice that one row must come from the identity matrix. The corresponding column is where the $\star$ entries lie.

If we take the $\diamond$ entries to be zero in the three matrices above, then we have created shears parallel to the $\mathbf{e}_2, \mathbf{e}_3$-, $\mathbf{e}_1, \mathbf{e}_3$-, and $\mathbf{e}_1, \mathbf{e}_2$-planes, respectively.

The shear matrix

$$\begin{bmatrix} 1 & \star & \star \\ 0 & 1 & 0 \\ 0 & 0 & 1 \end{bmatrix}$$

shears parallel to the $\mathbf{e}_1$-axis. Matrices for the other axes follow similarly.

How does a shear affect volume? For a geometric feeling, notice the simple shear S_1 from above. It maps the unit cube to a skew box with the same base and the same height—thus it does not change volume! All shears are volume preserving, although for some of the shears above it's difficult to see this from a geometric perspective. After reading Section 12.7, revisit these shear matrices and check the volumes for yourself.

12.5   Projections

Recall from 2D that a projection reduces dimensionality; it "flattens" geometry. In 3D this means that a vector is projected into a (2D) plane. Two examples are illustrated in Figure 12.5. Projections that are linear maps are *parallel projections*. There are two categories. If the projection direction is perpendicular to the projection plane then it is an orthographic projection, otherwise it is an oblique projection.

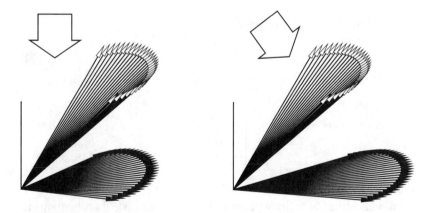

Figure 12.5.

Projections in 3D: on the left is an orthographic projection, and on the right is an oblique projection.

Projections are essential in computer graphics to view 3D geometry on a 2D screen. A parallel projection is a linear map, as opposed to a perspective projection which is not. A parallel projection preserves relative dimensions of an object, thus it is used in drafting to produce accurate views of a design.

Example 12.4

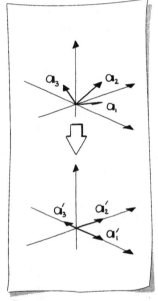

Sketch 12.6.
Projection example.

Let's construct an orthographic projection. Take the three vectors

$$\mathbf{a}_1 = \begin{bmatrix} 2 \\ 0 \\ 1 \end{bmatrix}, \quad \mathbf{a}_2 = \begin{bmatrix} 0 \\ 2 \\ 1 \end{bmatrix}, \quad \mathbf{a}_3 = \begin{bmatrix} -1 \\ 0 \\ 1 \end{bmatrix}.$$

If we flatten them out into the $\mathbf{e}_1, \mathbf{e}_2$-plane, they become

$$\mathbf{a}_1' = \begin{bmatrix} 2 \\ 0 \\ 0 \end{bmatrix}, \quad \mathbf{a}_2' = \begin{bmatrix} 0 \\ 2 \\ 0 \end{bmatrix}, \quad \mathbf{a}_3' = \begin{bmatrix} -1 \\ 0 \\ 0 \end{bmatrix}$$

(see Sketch 12.6). This action is achieved by the linear map

$$\begin{bmatrix} v_1 \\ v_2 \\ 0 \end{bmatrix} = \begin{bmatrix} 1 & 0 & 0 \\ 0 & 1 & 0 \\ 0 & 0 & 0 \end{bmatrix} \begin{bmatrix} v_1 \\ v_2 \\ v_3 \end{bmatrix},$$

as you should convince yourself!

Similarly, the matrix

$$\begin{bmatrix} 1 & 0 & 0 \\ 0 & 0 & 0 \\ 0 & 0 & 1 \end{bmatrix}$$

will flatten any vector into the $\mathbf{e}_1, \mathbf{e}_3$-plane.

We will examine oblique projections in the context of affine maps in Section 13.4. Finally, we note that projections have a significant effect on the volume of objects. Since everything is flat after a projection, it has zero 3D volume.

12.6   Rotations

Suppose you want to rotate a vector $\mathbf{v}$ around the $\mathbf{e}_3$-axis by $90°$ to a vector $\mathbf{v}'$. Sketch 12.7 illustrates such a rotation:

$$\mathbf{v} = \begin{bmatrix} 2 \\ 0 \\ 1 \end{bmatrix} \quad \rightarrow \quad \mathbf{v}' = \begin{bmatrix} 0 \\ 2 \\ 1 \end{bmatrix}.$$

A rotation around $\mathbf{e}_3$ by different angles would result in different vectors, but they all will have one thing in common: their third components will not be changed by the rotation. Thus, if we rotate a vector around $\mathbf{e}_3$, the rotation action will only change its first and second components. This suggests another look at the 2D rotation matrices from Section 4.6! Our desired rotation matrix R_3 looks much like the one from (4.16):

$$R_3 = \begin{bmatrix} \cos\alpha & -\sin\alpha & 0 \\ \sin\alpha & \cos\alpha & 0 \\ 0 & 0 & 1 \end{bmatrix}. \qquad (12.6)$$

Figure 12.6 illustrates the letter "L" rotated through several angles about the $\mathbf{e}_3$-axis.

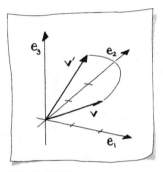

Sketch 12.7.
Rotation example.

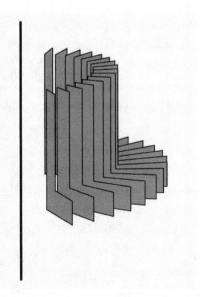

Figure 12.6.
Rotations in 3D: the letter "L" rotated about the $\mathbf{e}_3$-axis.

Example 12.5

Let us verify that R_3 performs as promised with $\alpha = 90°$:

$$\begin{bmatrix} 0 & -1 & 0 \\ 1 & 0 & 0 \\ 0 & 0 & 1 \end{bmatrix} \begin{bmatrix} 2 \\ 0 \\ 1 \end{bmatrix} = \begin{bmatrix} 0 \\ 2 \\ 1 \end{bmatrix},$$

so it works!

Similarly, we may rotate around the $\mathbf{e}_2$-axis; the corresponding matrix is

$$R_2 = \begin{bmatrix} \cos\alpha & 0 & \sin\alpha \\ 0 & 1 & 0 \\ -\sin\alpha & 0 & \cos\alpha \end{bmatrix}. \qquad (12.7)$$

Notice the pattern here. The rotation matrix for a rotation about the $\mathbf{e}_i$-axis is characterized by the ith row being $\mathbf{e}_i^{\text{T}}$ and the ith column being $\mathbf{e}_i$. For completeness, the last rotation matrix about the $\mathbf{e}_1$-axis:

$$R_1 = \begin{bmatrix} 1 & 0 & 0 \\ 0 & \cos\alpha & -\sin\alpha \\ 0 & \sin\alpha & \cos\alpha \end{bmatrix}. \qquad (12.8)$$

The direction of rotation by a positive angle follows the right-hand rule: curl your fingers with the rotation, and your thumb points in the direction of the rotation axis.

If you examine the column vectors of a rotation matrix, you will see that each one is a unit length vector, and they are orthogonal to each other. Thus, the column vectors form an orthonormal set of vectors, and a rotation matrix is an orthogonal matrix. These properties hold for the row space of the matrix too. As a result, we have that

$$RR^{\text{T}} = I$$
$$R^{-1} = R^{\text{T}}$$

Additionally, if R rotates by θ, then R^{-1} rotates by $-\theta$.

How about a rotation by α degrees around an arbitrary vector $\mathbf{a}$? The principle is illustrated in Sketch 12.8. The derivation of the following matrix is more tedious than called for here, so we just give the result:

$$R = \begin{bmatrix} a_1^2 + C(1-a_1^2) & a_1a_2(1-C) - a_3S & a_1a_3(1-C) + a_2S \\ a_1a_2(1-C) + a_3S & a_2^2 + C(1-a_2^2) & a_2a_3(1-C) - a_1S \\ a_1a_3(1-C) - a_2S & a_2a_3(1-C) + a_1S & a_3^2 + C(1-a_3^2) \end{bmatrix}$$

$$(12.9)$$

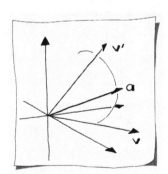

Sketch 12.8.

Rotation about an arbitrary vector.

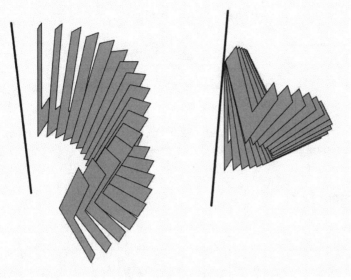

Figure 12.7.
Rotations in 3D: the letter "L" is rotated about axes which are not the coordinate axes. On the right the point on the "L" that touches the rotation axes does not move.

where we have set $C = \cos\alpha$ and $S = \sin\alpha$. It is necessary that $\|\mathbf{a}\| = 1$ in order for the rotation to take place without scaling.[4] Figure 12.7 illustrates two examples of rotations about an arbitrary axis.

Example 12.6

With a complicated result as (12.9), a sanity check is not a bad idea. So let $\alpha = 90°$,

$$\mathbf{a} = \begin{bmatrix} 0 \\ 0 \\ 1 \end{bmatrix} \quad \text{and} \quad \mathbf{v} = \begin{bmatrix} 1 \\ 0 \\ 0 \end{bmatrix}.$$

This means that we want to rotate $\mathbf{v}$ around $\mathbf{a}$, or the $\mathbf{e}_3$-axis, by $90°$ as shown in Sketch 12.9. In advance, we know what R should be. In (12.9), $C = 0$ and $S = 1$, and we calculate

$$R = \begin{bmatrix} 0 & -1 & 0 \\ 1 & 0 & 0 \\ 0 & 0 & 1 \end{bmatrix},$$

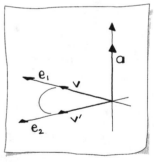

Sketch 12.9.
A simple example of a rotation about a vector.

---

[4] As an exercise, you can construct the transformations to do this using 3D affine maps from Chapter 13 and orthonormal coordinate frames from Section 11.8.

which is the expected matrix! We obtain

$$\mathbf{v}' = \begin{bmatrix} 0 \\ 1 \\ 0 \end{bmatrix}.$$

With some confidence that (12.9) works, let's try a more complicated example.

Example 12.7

Let $\alpha = 90°$,

$$\mathbf{a} = \begin{bmatrix} \frac{1}{\sqrt{3}} \\ \frac{1}{\sqrt{3}} \\ \frac{1}{\sqrt{3}} \end{bmatrix} \quad \text{and} \quad \mathbf{v} = \begin{bmatrix} 1 \\ 0 \\ 0 \end{bmatrix}.$$

With $C = 0$ and $S = 1$ in (12.9), we calculate

$$R = \begin{bmatrix} \frac{1}{3} & \frac{1}{3} - \frac{1}{\sqrt{3}} & \frac{1}{3} + \frac{1}{\sqrt{3}} \\ \frac{1}{3} + \frac{1}{\sqrt{3}} & \frac{1}{3} & \frac{1}{3} - \frac{1}{\sqrt{3}} \\ \frac{1}{3} - \frac{1}{\sqrt{3}} & \frac{1}{3} + \frac{1}{\sqrt{3}} & \frac{1}{3} \end{bmatrix},$$

We obtain

$$\mathbf{v}' = \begin{bmatrix} \frac{1}{3} \\ \frac{1}{3} + \frac{1}{\sqrt{3}} \\ \frac{1}{3} - \frac{1}{\sqrt{3}} \end{bmatrix}.$$

Convince yourself that $\|\mathbf{v}'\| = \|\mathbf{v}\|$.
Continue this example with the vector

$$\mathbf{v} = \begin{bmatrix} 1 \\ 1 \\ 1 \end{bmatrix}.$$

Surprised by the result?

It should be intuitively clear that rotations do not change volumes. Recall from 2D that rotations are *rigid body motions*.

12.7 Volumes and Linear Maps: Determinants

Most linear maps change volumes; some don't. Since this is an important aspect of the action of a map, this section will discuss the effect of a linear map on volume. The unit cube in the $[\mathbf{e}_1, \mathbf{e}_2, \mathbf{e}_3]$-system has volume one. A linear map A will change that volume to that of the skew box spanned by the images of $\mathbf{e}_1, \mathbf{e}_2, \mathbf{e}_3$, i.e., by the volume spanned by the vectors $\mathbf{a}_1, \mathbf{a}_2, \mathbf{a}_3$—the column vectors of A. What is the volume spanned by $\mathbf{a}_1, \mathbf{a}_2, \mathbf{a}_3$?

First, let's look at what we have done so far with areas and volumes. Recall the 2×2 determinant from Section 4.10. Through Sketch 4.8, the area of a 2D parallelogram was shown to be equivalent to a determinant. In fact, in Section 10.2 it was shown that the cross product can be used to calculate this area for a parallelogram embedded in 3D. With a very geometric approach, the *scalar triple product* of triple Section 10.6 gives us the means to calculate the volume of a parallelepiped by simply using a "base area times height" calculation. Let's revisit that formula and look at it from the perspective of linear maps.

So, using linear maps, we want to illustrate that the volume of the parallelepiped, or skew box, simply reduces to a 3D determinant calculation. Proceeding directly with a sketch in the 3D case would be difficult to follow. For 3D, let's augment the determinant idea with the tools from Section 5.4. There we demonstrated how shears—area-preserving linear maps—can be used to transform a matrix to upper triangular. These are the Gauss elimination steps.

First, let's introduce a 3×3 determinant of a matrix A. It is easily remembered as an alternating sum of 2×2 determinants:

$$|A| = a_{1,1} \begin{vmatrix} a_{2,2} & a_{2,3} \\ a_{3,2} & a_{3,3} \end{vmatrix} - a_{2,1} \begin{vmatrix} a_{1,2} & a_{1,3} \\ a_{3,2} & a_{3,3} \end{vmatrix} + a_{3,1} \begin{vmatrix} a_{1,2} & a_{1,3} \\ a_{2,2} & a_{2,3} \end{vmatrix}. \quad (12.10)$$

The representation in (12.10) is called the *cofactor expansion*. Each (signed) 2×2 determinant is the *cofactor* of the $a_{i,j}$ it is paired with in the sum. The sign comes from factor $(-1)^{i+j}$. For example, the cofactor of $a_{2,1}$ is

$$(-1)^{2+1} \begin{vmatrix} a_{1,2} & a_{1,3} \\ a_{3,2} & a_{3,3} \end{vmatrix}.$$

The cofactor is also written as $(-1)^{i+j} M_{i,j}$ where $M_{i,j}$ is called the *minor* of $a_{i,j}$. As a result, (12.10) is also known as *expansion by minors*. We'll look into this method more in Section 14.3.

If (12.10) is expanded, then an interesting form for writing the determinant arises. The formula is nearly impossible to remember, but the following trick is not. Copy the first two columns after the last column. Next, form the product of the three "diagonals" and add them. Then, form the product of the three "anti-diagonals" and subtract them. The three "plus" products may be written as:

$$
\begin{array}{ccccc}
a_{1,1} & a_{1,2} & a_{1,3} & \square & \square \\
\square & a_{2,2} & a_{2,3} & a_{2,1} & \square \\
\square & \square & a_{3,3} & a_{3,1} & a_{3,2}
\end{array}
$$

and the three "minus" products as:

$$
\begin{array}{ccccc}
\square & \square & a_{1,3} & a_{1,1} & a_{1,2} \\
\square & a_{2,2} & a_{2,3} & a_{2,1} & \square \\
a_{3,1} & a_{3,2} & a_{3,3} & \square & \square
\end{array} .
$$

The complete formula for the 3×3 determinant is

$$
\begin{aligned}
\det A \;=\; & a_{1,1}a_{2,2}a_{3,3} + a_{1,2}a_{2,3}a_{3,1} + a_{1,3}a_{2,1}a_{3,2} \\
& - a_{3,1}a_{2,2}a_{1,3} - a_{3,2}a_{2,3}a_{1,1} - a_{3,3}a_{2,1}a_{1,2}.
\end{aligned}
$$

Example 12.8

What is the volume spanned by the three vectors

$$
\mathbf{a}_1 = \begin{bmatrix} 4 \\ 0 \\ 0 \end{bmatrix}, \quad
\mathbf{a}_2 = \begin{bmatrix} -1 \\ 4 \\ 4 \end{bmatrix}, \quad
\mathbf{a}_3 = \begin{bmatrix} 0.1 \\ -0.1 \\ 0.1 \end{bmatrix} ?
$$

All we have to do is to compute

$$
\det[\mathbf{a}_1, \mathbf{a}_2, \mathbf{a}_3] = 4 \times 4 \times 0.1 - (-0.1) \times 4 \times 4 = 3.2.
$$

In this computation, we did not write down zero terms.

As we have seen in Section 12.4, a 3D shear preserves volume. Therefore, we can apply a series of shears to the matrix A, resulting in a new matrix

$$
\tilde{A} = \begin{bmatrix}
\tilde{a}_{1,1} & \tilde{a}_{1,2} & \tilde{a}_{1,3} \\
0 & \tilde{a}_{2,2} & \tilde{a}_{2,3} \\
0 & 0 & \tilde{a}_{3,3}
\end{bmatrix} .
$$

The determinant of $\tilde{A}$ is

$$
|\tilde{A}| = \tilde{a}_{1,1}\tilde{a}_{2,2}\tilde{a}_{3,3},
$$

with of course, $|A| = |\tilde{A}|$. For 3×3 matrices, we don't actually calculate the volume of three vectors by proceeding with the Gauss elimination steps, or shears.[5] We would just directly calculate the 3×3 determinant from (12.10). What is interesting about this development is now we can illustrate, as in Sketch 12.10, how the determinant defines the volume of the skew box. The first two column vectors of $\tilde{A}$ lie in the $[\mathbf{e}_1, \mathbf{e}_2]$-plane. Their determinant defines the area of the parallelogram that they span; this determinant is $\tilde{a}_{1,1}\tilde{a}_{2,2}$. The height of the skew box is simply the $\mathbf{e}_3$ component of $\tilde{\mathbf{a}}_3$. Thus, we have an easy to visualize interpretation of the 3×3 determinant. And, from a slightly different perspective, we have revisited the geometric development of the determinant of the scalar triple product in Section 10.6.

Let's conclude this section with some *rules for determinants*. Suppose we have two 3×3 matrices, A and B. The column vectors of A are $\begin{bmatrix} \mathbf{a}_1 & \mathbf{a}_2 & \mathbf{a}_3 \end{bmatrix}$.

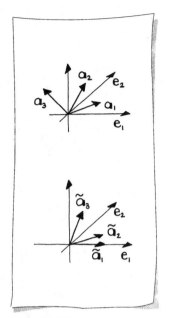

Sketch 12.10.

Determinant and volume in 3D.

- The determinant of the transpose matrix equals that of the matrix: $|A| = |A^{\mathrm{T}}|$. This property allows us to exchange the words "row" or "column" when working with determinants.

- Exchanging two columns changes the sign of the determinant: $\begin{vmatrix} \mathbf{a}_2 & \mathbf{a}_1 & \mathbf{a}_3 \end{vmatrix} = -|A|$.

- Exchanging two columns, or a *cyclic permutations*, does not change the sign of the determinant: $\begin{vmatrix} \mathbf{a}_2 & \mathbf{a}_3 & \mathbf{a}_1 \end{vmatrix} = |A|$. (See the scalar triple product in Section 10.6 for more permutations.)

- Multiplying one column by a scalar c results in the determinant being multiplied by c: $\begin{vmatrix} c\mathbf{a}_1 & \mathbf{a}_2 & \mathbf{a}_3 \end{vmatrix} = c|A|$.

- As an extension of the previous item: $|cA| = c^3|A|$.

- Multiples of rows (or columns) can be added together without changing the determinant. For example, the shears of Gauss elimination do not change the determinant.

- If A has a row (or column) of zeros then $|A| = 0$.

- If A has two identical rows (columns) then $|A| = 0$.

---

[5]However, we will use Gauss elimination for $n \times n$ systems. The equation given in this section needs to be adjusted if pivoting is included in Gauss elimination. See Section 14.3 for details.

- The sum of determinants is not the determinant of the sum, $|A| + |B| \neq |A + B|$, in general.

- The product of determinants is the determinant of the product: $|AB| = |A||B|$.

- A is invertible if and only if $|A| \neq 0$.

- If A is invertible then

$$|A^{-1}| = \frac{1}{|A|}.$$

12.8   Combining Linear Maps

If we apply a linear map A to a vector $\mathbf{v}$ and then apply a map B to the result, we may write this as

$$\mathbf{v}' = BA\mathbf{v}.$$

Matrix multiplication is defined just as in the 2D case; the element $c_{i,j}$ of the product matrix $C = BA$ is obtained as the dot product of the ith row of B with the jth column of A. A handy way to write the matrices so as to keep the dot products in order:

$$\begin{array}{c|c} & A \\ \hline B & C \end{array}.$$

Instead of a complicated formula, an example should suffice.

$$\begin{array}{ccc|ccc}
 & & & 1 & 5 & -4 \\
 & & & -1 & -2 & 0 \\
 & & & 2 & 3 & -4 \\
\hline
0 & 0 & -1 & -2 & -3 & 4 \\
1 & -2 & 0 & 3 & 9 & -4 \\
-2 & 1 & 1 & -1 & -9 & 4
\end{array}$$

We have computed the dot product of the boldface row of B and column of A to produce the boldface entry of C. In this example, B and A are 3×3 matrices, and thus the result is another 3×3 matrix. In the example in Section 12.1, a 3×3 matrix A is multiplied by a 3×1 matrix (vector) $\mathbf{v}$ resulting in a 3×1 matrix or vector. Thus two matrices need not be the same size in order to multiply them.

There is a rule, however! Suppose we are to multiply two matrices A and B together as AB. The sizes of A and B are

$$m \times n \quad \text{and} \quad n \times p, \qquad (12.11)$$

respectively. The resulting matrix will be of size $m \times p$—the "outside" dimensions in 12.11. In order to form AB, it is necessary that the "inside" dimensions, n, be equal. The matrix multiplication scheme from Section 4.2 simplifies hand-calculations by illustrating the resulting dimensions.

As in the 2D case, matrix multiplication does not commute! That is, $AB \neq BA$ in most cases. An interesting difference between 2D and 3D is the fact that in 2D, rotations *did* commute; however, in 3D they do not. For example, in 2D, rotating first by α and then by β is no different from doing it the other way around. In 3D, that is not the case. Let's look at an example to illustrate this point.

Example 12.9

Let's look at a rotation by $-90°$ around the $\mathbf{e}_1$-axis with matrix R_1 and a rotation by $-90°$ around the $\mathbf{e}_3$-axis with matrix R_3:

$$R_1 = \begin{bmatrix} 1 & 0 & 0 \\ 0 & 0 & 1 \\ 0 & -1 & 0 \end{bmatrix} \quad \text{and} \quad R_3 = \begin{bmatrix} 0 & 1 & 0 \\ -1 & 0 & 0 \\ 0 & 0 & 1 \end{bmatrix}.$$

Figure 12.8 illustrates what the algebra tells us:

$$R_3 R_1 = \begin{bmatrix} 0 & 0 & 1 \\ -1 & 0 & 0 \\ 0 & -1 & 0 \end{bmatrix} \quad \text{is not equal to} \quad R_1 R_3 = \begin{bmatrix} 0 & 1 & 0 \\ 0 & 0 & 1 \\ 1 & 0 & 0 \end{bmatrix}.$$

Also helpful for understanding what is happening, is to track the transformation of a point $\mathbf{p}$ on "L." Form the vector $\mathbf{v} = \mathbf{p} - \mathbf{o}$, and let's track

$$\mathbf{v} = \begin{bmatrix} 0 \\ 0 \\ 1 \end{bmatrix}.$$

In Figure 12.8 on the left, observe the transformation of $\mathbf{v}$:

$$R_1 \mathbf{v} = \begin{bmatrix} 0 \\ 1 \\ 0 \end{bmatrix} \quad \text{and} \quad R_3 R_1 \mathbf{v} = \begin{bmatrix} 1 \\ 0 \\ 0 \end{bmatrix}.$$

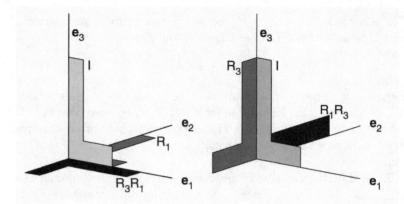

Figure 12.8.
Combining 3D rotations: left and right, the original "L" is labeled I for identity matrix. On the left, R_1 is applied and then R_3, the result is labeled R_3R_1. On the right, R_3 is applied and then R_1, the result is labeled R_1R_3. This shows that 3D rotations do not commute.

Now, on the right, observe the transformation of $\mathbf{v}$:

$$R_3\mathbf{v} = \begin{bmatrix} 0 \\ 0 \\ 1 \end{bmatrix} \quad \text{and} \quad R_1R_3\mathbf{v} = \begin{bmatrix} 0 \\ 1 \\ 0 \end{bmatrix}.$$

So it does matter which rotation we perform first!

12.9   More on Matrices

A handful of matrix properties are explained and illustrated in Chapter 4. Here we restate them so they are conveniently together. These properties hold for $n \times n$ matrices (the topic of Chapter 14):

- preserve scalings: $A(c\mathbf{v}) = cA\mathbf{v}$

- preserve summations: $A(\mathbf{u} + \mathbf{v}) = A\mathbf{u} + A\mathbf{v}$

- preserve linear combinations: $A(a\mathbf{u} + b\mathbf{v}) = aA\mathbf{u} + bA\mathbf{v}$

- distributive law: $A\mathbf{v} + B\mathbf{v} = (A + B)\mathbf{v}$

- commutative law for addition: $A + B = B + A$

- associative law for addition: $A + (B + C) = (A + B) + C$

- associative law for multiplication: $A(BC) = (AB)C$

- distributive law: $A(B + C) = AB + AC$
$$(B + C)A = BA + CA$$

- scalar laws:

  - $a(B + C) = aB + aC$
  - $(a + b)C = aC + bC$
  - $(ab)C = a(bC)$
  - $a(BC) = (aB)C = B(aC)$

- determinants: $|AB| = |A| \cdot |B|$

- laws involving exponents:

  - $A^r = \underbrace{A \cdot \ldots \cdots A}_{r \text{ times}}$
  - $A^{r+s} = A^r A^s$
  - $A^{rs} = (A^r)^s$
  - $A^0 = I$

- laws involving the transpose:

  - $[A + B]^{\mathrm{T}} = A^{\mathrm{T}} + B^{\mathrm{T}}$
  - $A^{\mathrm{T}^{\mathrm{T}}} = A$
  - $[cA]^{\mathrm{T}} = cA^{\mathrm{T}}$
  - $[AB]^{\mathrm{T}} = B^{\mathrm{T}} A^{\mathrm{T}}$

12.10   Inverse Matrices

In Section 5.5, we saw how inverse matrices undo linear maps. A linear map A takes a vector $\mathbf{v}$ to its image $\mathbf{v}'$. The inverse map, A^{-1}, will take $\mathbf{v}'$ back to $\mathbf{v}$, i.e., $A^{-1}\mathbf{v}' = \mathbf{v}$ or $A^{-1}A\mathbf{v} = \mathbf{v}$. Thus, the combined action of $A^{-1}A$ has no effect on any vector $\mathbf{v}$, which we can write as

$$A^{-1}A = I, \tag{12.12}$$

where I is the 3×3 identity matrix. If we applied A^{-1} to $\mathbf{v}$ first, and then applied A, there would not be any action either; in other words,

$$AA^{-1} = I, \tag{12.13}$$

too.

A matrix is not always invertible. For example, the projections from Section 12.5 are rank deficient, and therefore not invertible. This is apparent from Sketch 12.6: once we flatten the vectors $\mathbf{a}_i$ to $\mathbf{a}_i'$ in 2D, there isn't enough information available in the $\mathbf{a}_i'$ to return them to 3D.

As we discovered in Section 12.6 on rotations, orthogonal matrices, which are constructed from a set of orthonormal vectors, possess the nice property $R^{\mathrm{T}} = R^{-1}$. Undoing a rotation is simple and requires no computation; this provides for a huge savings in computer graphics where rotating objects is a common operation.

Scaling also has an inverse which is simple to compute. If

$$S = \begin{bmatrix} s_{1,1} & 0 & 0 \\ 0 & s_{2,2} & 0 \\ 0 & 0 & s_{3,3} \end{bmatrix},$$

then

$$S^{-1} = \begin{bmatrix} 1/s_{1,1} & 0 & 0 \\ 0 & 1/s_{2,2} & 0 \\ 0 & 0 & 1/s_{3,3} \end{bmatrix}.$$

So here are more rules for matrices! These involve calculating with inverse matrices.

$$A^{-n} = (A^{-1})^n = \underbrace{A^{-1} \cdot \; \cdots \; \cdots A^{-1}}_{n \text{ times}}$$
$$(A^{-1})^{-1} = A$$
$$(kA)^{-1} = \frac{1}{k}A^{-1}$$
$$(AB)^{-1} = B^{-1}A^{-1}$$

See Section 14.5 for details on calculating A^{-1}.

- 3D linear map
- scale
- rotation
- shear
- reflection
- projection
- orthographic projection
- oblique projection
- determinant
- volume

- cofactor expansion
- expansion by minors
- multiply matrices
- linear combinations
- transpose matrix
- rigid body motions
- non-commutative property
  of matrix multiplication
- rules of matrix arithmetic
- inverse matrix

12.11 Exercises

1. Describe the linear map given by the matrix

$$\begin{bmatrix} 0 & 1 & 0 \\ 0 & 0 & -1 \\ 1 & 0 & 0 \end{bmatrix}.$$

   Hint: you might want to try a few simple examples to get a feeling for what is going on.

2. What matrix scales by 2 in the e_1-direction, scales by 1/4 in the e_2-direction, and reverses direction and scales by 4 in the e_3-direction? Map the unit cube with this matrix. What is the volume of the resulting parallelepiped?

3. What is the matrix which reflects a vector about the plane $x_1 = x_2$? Map the unit cube with this matrix. What is the volume of the resulting parallelepiped?

4. What is the shear matrix which maps

$$\begin{bmatrix} a \\ b \\ c \end{bmatrix} \quad \text{to} \quad \begin{bmatrix} 0 \\ 0 \\ c \end{bmatrix}?$$

   Map the unit cube with this matrix. What is the volume of the resulting parallelepiped?

5. What matrix rotates around the e_1-axis by α degrees?

6. What matrix rotates by $45°$ around the vector $\begin{bmatrix} -1 \\ 0 \\ -1 \end{bmatrix}$?

7. Compute

$$\begin{bmatrix} 0 & 0 & 1 \\ 1 & -2 & 0 \\ -2 & 1 & 1 \end{bmatrix} \cdot \begin{bmatrix} 1 & 5 & -4 \\ -1 & -2 & 0 \\ 2 & 3 & -4 \end{bmatrix}.$$

8. What is the matrix for a shear parallel to an arbitrary plane in a specified direction. Assume the given information is a plane with normal $\mathbf{n}$ and a vector $\mathbf{s}$ in this plane. *Hint: First write this as a vector equation, then build the matrix.*

9. What is the transpose of the matrix

$$A = \begin{bmatrix} 1 & 5 & -4 \\ -1 & -2 & 0 \\ 2 & 3 & -4 \end{bmatrix} ?$$

10. For the matrix A in Exercise 9, what is $(A^{\mathrm{T}})^{\mathrm{T}}$?

11. Given the projection matrix

$$A = \begin{bmatrix} 1 & 0 & -1 \\ 0 & 1 & 0 \\ 0 & 0 & 0 \end{bmatrix}$$

what is the projection direction? If the projection plane is $x_3 = 0$, then what type of projection is it? *Hint: See Section 4.9.*

12. What is the kernel of the matrix in Exercise 11?

13. Find the inverse for each of the following matrices.

rotation:
$$\begin{bmatrix} 1/\sqrt{2} & 0 & 1/\sqrt{2} \\ 0 & 1 & 0 \\ -1/\sqrt{2} & 0 & 1/\sqrt{2} \end{bmatrix}$$

scale:
$$\begin{bmatrix} 1/2 & 0 & 0 \\ 0 & 1/4 & 0 \\ 0 & 0 & 2 \end{bmatrix}$$

projection:
$$\begin{bmatrix} 1 & 0 & -1 \\ 0 & 1 & 0 \\ 0 & 0 & 0 \end{bmatrix}$$

13

Affine Maps in 3D

Figure 13.1.
Affine maps in 3D: fighter jets twisting and turning through 3D space.

This chapter wraps up the basic geometry tools. Affine maps in 3D are a primary tool for modeling and computer graphics. Figure 13.1 illustrates the use of various affine maps. This chapter goes a little farther than just affine maps by introducing projective maps—the maps used to create realistic 3D images.

13.1 Affine Maps

Linear maps relate vectors to vectors. Affine maps relate points to points. A 3D affine map is written just as a 2D one, namely as

$$\mathbf{x}' = \mathbf{p} + A(\mathbf{x} - \mathbf{o}). \tag{13.1}$$

In general, we will assume that the origin of $\mathbf{x}$'s coordinate system has three zero coordinates, and drop the $\mathbf{o}$ term:

$$\mathbf{x}' = \mathbf{p} + A\mathbf{x}. \tag{13.2}$$

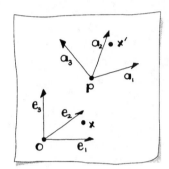

Sketch 13.1.

An affine map in 3D.

Sketch 13.1 gives an example. Recall, the column vectors of A are the vectors $\mathbf{a}_1, \mathbf{a}_2, \mathbf{a}_3$. The point $\mathbf{p}$ tells us where to move the origin of the $[\mathbf{e}_1, \mathbf{e}_2, \mathbf{e}_3]$-system; again, the real action of an affine map is captured by the matrix. Thus, by studying matrix actions, or linear maps, we will learn more about affine maps.

We now list some of the important properties of 3D affine maps. They are straightforward generalizations of the 2D cases, and so we just give a brief listing.

1. Affine maps leave *ratios* invariant (see Sketch 13.2).

2. Affine maps take *parallel planes* to parallel planes (see Figure 13.2).

3. Affine maps take *intersecting planes* to intersecting planes. In particular, the intersection line of the mapped planes is the map of the original intersection line.

4. Affine maps leave *barycentric combinations* invariant. If

$$\mathbf{x} = c_1\mathbf{p}_1 + c_2\mathbf{p}_2 + c_3\mathbf{p}_3 + c_4\mathbf{p}_4,$$

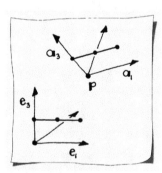

Sketch 13.2.

Affine maps leave ratios invariant. This map is a rigid body motion.

where $c_1 + c_2 + c_3 + c_4 = 1$, then after an affine map we have

$$\mathbf{x}' = c_1\mathbf{p}_1' + c_2\mathbf{p}_2' + c_3\mathbf{p}_3' + c_4\mathbf{p}_4'.$$

For example, the *centroid* of a tetrahedron will be mapped to the centroid of the mapped tetrahedron (see Sketch 13.3).

Most 3D maps do not offer much over their 2D counterparts—but some do. We will go through all of them in detail now.

Figure 13.2.
Affine map property: parallel planes get mapped to parallel planes via an affine map.

13.2   Translations

A translation is simply (13.2) with $A = I$, the 3×3 identity matrix:

$$I = \begin{bmatrix} 1 & 0 & 0 \\ 0 & 1 & 0 \\ 0 & 0 & 1 \end{bmatrix},$$

that is

$$\mathbf{x}' = \mathbf{p} + I\mathbf{x}.$$

Thus, the new $[\mathbf{a}_1, \mathbf{a}_2, \mathbf{a}_3]$-system has its coordinate axes parallel to the $[\mathbf{e}_1, \mathbf{e}_2, \mathbf{e}_3]$-system. The term $I\mathbf{x} = \mathbf{x}$ needs to be interpreted as a *vector* in the $[\mathbf{e}_1, \mathbf{e}_2, \mathbf{e}_3]$-system for this to make sense! Figure 13.3 shows an example of repeated 3D translations.

Just as in 2D, a translation is a rigid body motion. The volume of an object is not changed.

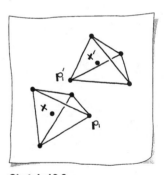

Sketch 13.3.
The centroid is mapped to the centroid.

13.3   Mapping Tetrahedra

A 3D affine map is determined by four point pairs $\mathbf{p}_i \rightarrow \mathbf{p}'_i$ for $i = 1, 2, 3, 4$. In other words, an affine map is determined by a tetrahedron

Figure 13.3.
Translations in 3D: three translated teapots.

and its image. What is the image of an arbitrary point $\mathbf{x}$ under this affine map?

Affine maps leave *barycentric combinations* unchanged. This will be the key to finding $\mathbf{x}'$, the image of $\mathbf{x}$. If we can write $\mathbf{x}$ in the form

$$\mathbf{x} = u_1\mathbf{p}_1 + u_2\mathbf{p}_2 + u_3\mathbf{p}_3 + u_4\mathbf{p}_4, \qquad (13.3)$$

then we know that the image has the same relationship with the $\mathbf{p}_i'$:

$$\mathbf{x}' = u_1\mathbf{p}_1' + u_2\mathbf{p}_2' + u_3\mathbf{p}_3' + u_4\mathbf{p}_4'. \qquad (13.4)$$

So all we need to do is find the u_i! These are called the *barycentric coordinates* of $\mathbf{x}$ with respect to the $\mathbf{p}_i$, quite in analogy to the triangle case (see Section 6.5).

We observe that (13.3) is short for three individual coordinate equations. Together with the barycentric combination condition

$$u_1 + u_2 + u_3 + u_4 = 1,$$

we have four equations for the four unknowns $u_1, \ldots, u_4$, which we can solve by consulting Chapter 14.

Example 13.1

Let the original tetrahedron be given by the four points $\mathbf{p}_i$

$$\begin{bmatrix} 0 \\ 0 \\ 0 \end{bmatrix}, \quad \begin{bmatrix} 1 \\ 0 \\ 0 \end{bmatrix}, \quad \begin{bmatrix} 0 \\ 1 \\ 0 \end{bmatrix}, \quad \begin{bmatrix} 0 \\ 0 \\ 1 \end{bmatrix}.$$

Let's assume we want to map this tetrahedron to the four points $\mathbf{p}_i'$

$$\begin{bmatrix} 0 \\ 0 \\ 0 \end{bmatrix}, \quad \begin{bmatrix} -1 \\ 0 \\ 0 \end{bmatrix}, \quad \begin{bmatrix} 0 \\ -1 \\ 0 \end{bmatrix}, \quad \begin{bmatrix} 0 \\ 0 \\ -1 \end{bmatrix}.$$

This is a pretty straightforward map if you consult Sketch 13.4.
Let's see where the point

$$\mathbf{x} = \begin{bmatrix} 1 \\ 1 \\ 1 \end{bmatrix}$$

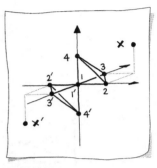

Sketch 13.4.
An example tetrahedron map.

ends up! First, we find that

$$\begin{bmatrix} 1 \\ 1 \\ 1 \end{bmatrix} = -2 \begin{bmatrix} 0 \\ 0 \\ 0 \end{bmatrix} + \begin{bmatrix} 1 \\ 0 \\ 0 \end{bmatrix} + \begin{bmatrix} 0 \\ 1 \\ 0 \end{bmatrix} + \begin{bmatrix} 0 \\ 0 \\ 1 \end{bmatrix},$$

i.e., the barycentric coordinates of $\mathbf{x}$ with respect to the original $\mathbf{p}_i$ are $(-2, 1, 1, 1)$. Note how they sum to one! Now it is simple to compute the image of $\mathbf{x}$; compute $\mathbf{x}'$ using the same barycentric coordinates with respect to the $\mathbf{p}_i'$:

$$\mathbf{x}' = -2 \begin{bmatrix} 0 \\ 0 \\ 0 \end{bmatrix} + \begin{bmatrix} -1 \\ 0 \\ 0 \end{bmatrix} + \begin{bmatrix} 0 \\ -1 \\ 0 \end{bmatrix} + \begin{bmatrix} 0 \\ 0 \\ -1 \end{bmatrix} = \begin{bmatrix} -1 \\ -1 \\ -1 \end{bmatrix}.$$

A different approach would be to find the 3×3 matrix A and point $\mathbf{p}$ which describe the affine map. Construct a coordinate system from the $\mathbf{p}_i$ tetrahedron. One way to do this is to choose $\mathbf{p}_1$ as the origin[1] and the three axes are defined as $\mathbf{p}_i - \mathbf{p}_1$ for $i = 2, 3, 4$. The coordinate system of the $\mathbf{p}_i'$ tetrahedron must be based on the same indices. Once we have defined A and $\mathbf{p}$ then we will be able to map $\mathbf{x}$ by this map:

$$\mathbf{x}' = A[\mathbf{x} - \mathbf{p}_1] + \mathbf{p}_1'.$$

[1] Any of the four $\mathbf{p}_i$ would do, so for the sake of concreteness, we chose the first one.

Thus, the point $\mathbf{p} = \mathbf{p}'_1$. In order to determine A, let's write down some known relationships. Referring to Sketch 13.5, we know

$$A[\mathbf{p}_2 - \mathbf{p}_1] = \mathbf{p}'_2 - \mathbf{p}'_1,$$
$$A[\mathbf{p}_3 - \mathbf{p}_1] = \mathbf{p}'_3 - \mathbf{p}'_1,$$
$$A[\mathbf{p}_4 - \mathbf{p}_1] = \mathbf{p}'_4 - \mathbf{p}'_1,$$

which may be written matrix form as

$$A \begin{bmatrix} \mathbf{p}_2 - \mathbf{p}_1 & \mathbf{p}_3 - \mathbf{p}_1 & \mathbf{p}_4 - \mathbf{p}_1 \end{bmatrix} = \begin{bmatrix} \mathbf{p}'_2 - \mathbf{p}'_1 & \mathbf{p}'_3 - \mathbf{p}'_1 & \mathbf{p}'_4 - \mathbf{p}'_1 \end{bmatrix}.$$
$$(13.5)$$

Thus,

$$A = \begin{bmatrix} \mathbf{p}'_2 - \mathbf{p}'_1 & \mathbf{p}'_3 - \mathbf{p}'_1 & \mathbf{p}'_4 - \mathbf{p}'_1 \end{bmatrix} \begin{bmatrix} \mathbf{p}_2 - \mathbf{p}_1 & \mathbf{p}_3 - \mathbf{p}_1 & \mathbf{p}_4 - \mathbf{p}_1 \end{bmatrix}^{-1},$$
$$(13.6)$$

and A is defined.

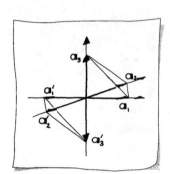

Sketch 13.5.

The relationship between tetrahedra.

Example 13.2

Revisiting the previous example, we now want to construct the matrix A. By selecting $\mathbf{p}_1$ as the origin for the $\mathbf{p}_i$ tetrahedron coordinate system there is no translation; $\mathbf{p}_1$ is the origin in the $[\mathbf{e}_1, \mathbf{e}_2, \mathbf{e}_3]$-system and $\mathbf{p}'_1 = \mathbf{p}_1$. We now compute A. (A is the product matrix in the bottom right position):

$$
\begin{array}{ccc|ccc}
 & & & 1 & 0 & 0 \\
 & & & 0 & 1 & 0 \\
 & & & 0 & 0 & 1 \\
\hline
-1 & 0 & 0 & -1 & 0 & 0 \\
0 & -1 & 0 & 0 & -1 & 0 \\
0 & 0 & -1 & 0 & 0 & -1
\end{array}
$$

In order to compute $\mathbf{x}'$, we have

$$\mathbf{x}' = \begin{bmatrix} -1 & 0 & 0 \\ 0 & -1 & 0 \\ 0 & 0 & -1 \end{bmatrix} \begin{bmatrix} 1 \\ 1 \\ 1 \end{bmatrix} = \begin{bmatrix} -1 \\ -1 \\ -1 \end{bmatrix}.$$

This is the same result as in the previous example.

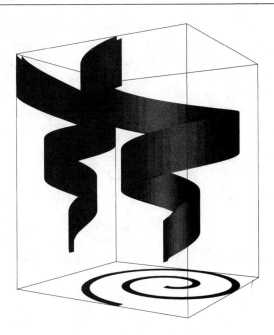

Figure 13.4.
Projections in 3D: a 3D helix is projected into two different 2D planes.

13.4   Projections

Take any object outside; let the sun shine on it, and you can observe a shadow. This shadow is the *parallel projection* of your object onto a plane. Everything we draw is a projection of necessity—paper is 2D, after all, whereas most interesting objects are 3D. Figure 13.4 gives an example.

Projections reduce dimensionality; as basic linear maps, we encountered them in Sections 4.8 and 12.5. As affine maps, they map 3D points onto a plane. As illustrated in Sketch 13.6, a parallel projection is defined by a *direction* of projection $\mathbf{d}$ and a *projection plane* P. A point $\mathbf{x}$ is projected into P, and is represented as $\mathbf{x}_p$ in the sketch. This information in turn defines a *projection angle* θ between $\mathbf{d}$ and the line joining the perpendicular projection point $\mathbf{x}_o$ in P. This angle is used to categorize parallel projections as *orthographic* or *oblique*.

Orthographic projections are special; the direction is perpendicular to the plane. There are special names for many particular projection angles; see a computer graphics text such as [11] for more details.

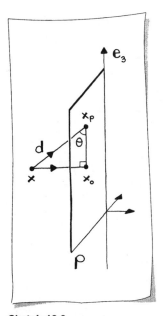

Sketch 13.6.
Oblique and orthographic parallel projections.

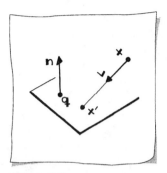

Sketch 13.7.

Projecting a point on a plane.

Construction of a projection requires defining a projection plane and a projection direction. Let $\mathbf{x}$ be the 3D point to be projected, let $\mathbf{n}\cdot[\mathbf{q}-\mathbf{x}] = 0$ be the projection plane, and let $\mathbf{v}$ indicate the projection direction; this geometry is illustrated in Sketch 13.7. We have already encountered this problem in Section 11.3 where it is called line/plane intersection. There, we established that $\mathbf{x}'$, the image of $\mathbf{x}$ under the projection is given by (11.6), which we repeat here:

$$\mathbf{x}' = \mathbf{x} + \frac{[\mathbf{q} - \mathbf{x}] \cdot \mathbf{n}}{\mathbf{v} \cdot \mathbf{n}}\mathbf{v}. \tag{13.7}$$

How do we write (13.7) as an affine map? Without much effort, we find

$$\mathbf{x}' = \mathbf{x} - \frac{\mathbf{n} \cdot \mathbf{x}}{\mathbf{v} \cdot \mathbf{n}}\mathbf{v} + \frac{\mathbf{q} \cdot \mathbf{n}}{\mathbf{v} \cdot \mathbf{n}}\mathbf{v}.$$

We know that we may write dot products in matrix form (see Section 4.12):

$$\mathbf{x}' = \mathbf{x} - \frac{\mathbf{n}^{\mathrm{T}}\mathbf{x}}{\mathbf{v} \cdot \mathbf{n}}\mathbf{v} + \frac{\mathbf{q} \cdot \mathbf{n}}{\mathbf{v} \cdot \mathbf{n}}\mathbf{v}.$$

Next, we observe that

$$\left[\mathbf{n}^{\mathrm{T}} \cdot \mathbf{x}\right] \mathbf{v} = \mathbf{v} \left[\mathbf{n}^{\mathrm{T}}\mathbf{x}\right].$$

Since matrix multiplication is associative (see Section 4.13), we also have

$$\mathbf{v} \left[\mathbf{n}^{\mathrm{T}}\mathbf{x}\right] = \left[\mathbf{v}\mathbf{n}^{\mathrm{T}}\right] \mathbf{x},$$

and thus

$$\mathbf{x}' = \left[I - \frac{\mathbf{v}\mathbf{n}^{\mathrm{T}}}{\mathbf{v} \cdot \mathbf{n}}\right] \mathbf{x} + \frac{\mathbf{q} \cdot \mathbf{n}}{\mathbf{v} \cdot \mathbf{n}}\mathbf{v}. \tag{13.8}$$

This is of the form $\mathbf{x}' = A\mathbf{x}+\mathbf{p}$ and hence is an affine map![2] The term $\mathbf{v}\mathbf{n}^{\mathrm{T}}$ might appear odd, yet it is well-defined. It is a 3×3 matrix, as in the following example.

Example 13.3

Given vectors

$$\mathbf{v} = \begin{bmatrix} 1 \\ 2 \\ 0 \end{bmatrix} \text{ and } \mathbf{n} = \begin{bmatrix} 1 \\ 3 \\ 3 \end{bmatrix},$$

---

[2]Technically, we should incorporate the origin by replacing $\mathbf{v}$ with $\mathbf{v} + \mathbf{o}$ to have a point, and replacing $\mathbf{x}$ with $\mathbf{x} - \mathbf{o}$ to have a vector.

calculate $\mathbf{vn}^{\mathrm{T}}$. The result is the matrix:

	1	3	3
1	1	3	3
2	2	6	6
0	0	0	0

All rows of the matrix in Example 13.3 are multiples of each other, and so are all columns. Matrices which are generated like this are called *dyadic*; their rank is one.

Example 13.4

Suppose we are given the projection plane $x_1 + x_2 + x_3 - 1 = 0$, and a point $\mathbf{x}$ and a direction $\mathbf{v}$ are given by

$$\mathbf{x} = \begin{bmatrix} 3 \\ 2 \\ 4 \end{bmatrix} \quad \text{and} \quad \mathbf{v} = \begin{bmatrix} 0 \\ 0 \\ -1 \end{bmatrix}.$$

If we project $\mathbf{x}$ along $\mathbf{v}$ onto the plane, what is $\mathbf{x}'$? First, we need the plane's normal direction. Calling it $\mathbf{n}$, we have

$$\mathbf{n} = \begin{bmatrix} 1 \\ 1 \\ 1 \end{bmatrix}.$$

Now, choose a point $\mathbf{q}$ in the plane. Let's choose

$$\mathbf{q} = \begin{bmatrix} 1 \\ 0 \\ 0 \end{bmatrix}$$

for simplicity. Now we are ready to calculate the quantities in (13.8):

$$\mathbf{v} \cdot \mathbf{n} = -1,$$

		1	1	1
$\mathbf{vn}^{\mathrm{T}} =$	0	0	0	0
	0	0	0	0
	-1	-1	-1	-1

$$\frac{\mathbf{q} \cdot \mathbf{n}}{\mathbf{v} \cdot \mathbf{n}} \mathbf{v} = \begin{bmatrix} 0 \\ 0 \\ 1 \end{bmatrix}.$$

Putting all the pieces together:

$$\mathbf{x}' = \left[I - \begin{bmatrix} 0 & 0 & 0 \\ 0 & 0 & 0 \\ 1 & 1 & 1 \end{bmatrix} \right] \begin{bmatrix} 3 \\ 2 \\ 4 \end{bmatrix} + \begin{bmatrix} 0 \\ 0 \\ 1 \end{bmatrix} = \begin{bmatrix} 3 \\ 2 \\ -4 \end{bmatrix}.$$

Just to double check, enter $\mathbf{x}'$ into the plane equation

$$3 + 2 - 4 - 1 = 0,$$

and we see that

$$\begin{bmatrix} 3 \\ 2 \\ -4 \end{bmatrix} = \begin{bmatrix} 3 \\ 2 \\ 4 \end{bmatrix} + 8 \begin{bmatrix} 0 \\ 0 \\ -1 \end{bmatrix},$$

which together verify that this is the correct point. Sketch 13.8 should convince you that this is indeed the correct answer.

Sketch 13.8.
A projection example.

Which of the two possibilities, (13.7) or the affine map (13.8) should you use? Clearly, (13.7) is more straightforward and less involved. Yet in some computer graphics or CAD system environments, it may be desirable to have all maps in a unified format, i.e., $A\mathbf{x} + \mathbf{p}$. We'll revisit this unified format idea in Section 13.5.

13.5 Homogeneous Coordinates and Perspective Maps

There is a way to condense the form $\mathbf{x}' = A\mathbf{x} + \mathbf{p}$ of an affine map into just one matrix multiplication

$$\underline{\mathbf{x}}' = M\underline{\mathbf{x}}. \tag{13.9}$$

This is achieved by setting

$$M = \begin{bmatrix} a_{1,1} & a_{1,2} & a_{1,3} & p_1 \\ a_{2,1} & a_{2,2} & a_{2,3} & p_2 \\ a_{3,1} & a_{3,2} & a_{3,3} & p_3 \\ 0 & 0 & 0 & 1 \end{bmatrix}$$

and

$$\underline{\mathbf{x}} = \begin{bmatrix} x_1 \\ x_2 \\ x_3 \\ 1 \end{bmatrix}, \quad \underline{\mathbf{x}}' = \begin{bmatrix} x_1' \\ x_2' \\ x_3' \\ 1 \end{bmatrix}.$$

The 4D point $\underline{\mathbf{x}}$ is called the *homogeneous form* of the affine point $\mathbf{x}$. You should verify for yourself that (13.9) is indeed the same affine map as before!

The homogeneous representation of a vector $\underline{\mathbf{v}}$ must have the form,

$$\underline{\mathbf{v}} = \begin{bmatrix} v_1 \\ v_2 \\ v_3 \\ 0 \end{bmatrix}.$$

This form allows us to apply the linear map to the vector,

$$\underline{\mathbf{v}}' = M\underline{\mathbf{v}}.$$

By having a zero fourth component, we disregard the translation, which we know has no effect on vectors. Recall that a vector is defined as the difference of two points.

This method of condensing transfomation information into one matrix is implemented in the popular computer graphics *Application Programmer's Interface (API)*, OpenGL [20]. It is very convenient and efficient to have all this information (plus more, as we will see), in one data structure.

The homogeneous form is more general than just adding a fourth coordinate $x_4 = 1$ to a point. If, perhaps as the result of some computation, the fourth coordinate does not equal one, one gets from the homogeneous point $\underline{\mathbf{x}}$ to its affine counterpart $\mathbf{x}$ by dividing through by x_4. Thus, one affine point has infinitely many homogeneous representations!

Example 13.5

This example shows two homogeneous representations of one affine point. (The symbol $\approx$ should be read "corresponds to.")

$$\begin{bmatrix} 1 \\ -1 \\ 3 \end{bmatrix} \approx \begin{bmatrix} 10 \\ -10 \\ 30 \\ 10 \end{bmatrix} \approx \begin{bmatrix} -2 \\ 2 \\ -6 \\ -2 \end{bmatrix}.$$

Using the homogeneous matrix form of (13.9), the matrix M for the point into a plane projection from (13.8) becomes

$$\left[\begin{array}{c|c}\begin{array}{ccc} \mathbf{v}\cdot\mathbf{n} & 0 & 0 \\ 0 & \mathbf{v}\cdot\mathbf{n} & 0 \\ 0 & 0 & \mathbf{v}\cdot\mathbf{n} \end{array} - \mathbf{v}\mathbf{n}^{\mathrm{T}} & (\mathbf{q}\cdot\mathbf{n})\mathbf{v} \\ \hline \begin{array}{ccc} 0 & 0 & 0 \end{array} & \mathbf{v}\cdot\mathbf{n} \end{array}\right].$$

Here, the element $m_{4,4} = \mathbf{v}\cdot\mathbf{n}$. Thus, $\underline{x}_4 = \mathbf{v}\cdot\mathbf{n}$, and we will have to divide $\underline{\mathbf{x}}$'s coordinates by $\underline{x}_4$ in order to obtain the corresponding affine point.

A simple change in our equations will lead us from parallel projections onto a plane to *perspective projections*. Instead of using a constant direction $\mathbf{v}$ for all projections, now the direction depends on the point $\mathbf{x}$. More precisely, let it be the line from $\mathbf{x}$ to the origin of our coordinate system. Then, as shown in Sketch 13.9, $\mathbf{v} = -\mathbf{x}$, and (13.7) becomes

$$\mathbf{x}' = \mathbf{x} + \frac{[\mathbf{q}-\mathbf{x}]\cdot\mathbf{n}}{\mathbf{x}\cdot\mathbf{n}}\mathbf{x},$$

which quickly simplifies to

$$\mathbf{x}' = \frac{\mathbf{q}\cdot\mathbf{n}}{\mathbf{x}\cdot\mathbf{n}}\mathbf{x}. \qquad (13.10)$$

In homogeneous form, this is described by the following matrix

$$M : \left[\begin{array}{ccc|c} & I[\mathbf{q}\cdot\mathbf{n}] & & \mathbf{o} \\ \hline 0 & 0 & 0 & \mathbf{x}\cdot\mathbf{n} \end{array}\right].$$

Perspective projections are not affine maps anymore! To see this, a simple example will suffice.

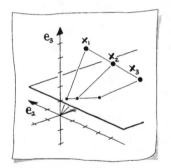

Sketch 13.9.
Perspective projection.

Example 13.6

Take the plane $x_3 = 1$; let

$$\mathbf{q} = \begin{bmatrix} 0 \\ 0 \\ 1 \end{bmatrix}$$

be a point on the plane. Now $\mathbf{q}\cdot\mathbf{n} = 1$ and $\mathbf{x}\cdot\mathbf{n} = x_3$, resulting in the map

$$\mathbf{x}' = \frac{1}{x_3}\mathbf{x}.$$

Take the three points

$$\mathbf{x}_1 = \begin{bmatrix} 2 \\ 0 \\ 4 \end{bmatrix}, \quad \mathbf{x}_2 = \begin{bmatrix} 3 \\ -1 \\ 3 \end{bmatrix}, \quad \mathbf{x}_3 = \begin{bmatrix} 4 \\ -2 \\ 2 \end{bmatrix}.$$

This example is illustrated in Sketch 13.9. Note that $\mathbf{x}_2 = \frac{1}{2}\mathbf{x}_1 + \frac{1}{2}\mathbf{x}_3$, i.e., $\mathbf{x}_2$ is the midpoint of $\mathbf{x}_1$ and $\mathbf{x}_3$.

Their images are

$$\mathbf{x}_1' = \begin{bmatrix} 1/2 \\ 0 \\ 1 \end{bmatrix}, \quad \mathbf{x}_2' = \begin{bmatrix} 1 \\ -1/3 \\ 1 \end{bmatrix}, \quad \mathbf{x}_3' = \begin{bmatrix} 2 \\ -1 \\ 1 \end{bmatrix}.$$

The perspective map destroyed the midpoint relation! Now,

$$\mathbf{x}_2' = \frac{2}{3}\mathbf{x}_1' + \frac{1}{3}\mathbf{x}_3'.$$

Thus, the ratio of three points is changed by perspective maps. As a consequence, two parallel lines will not be mapped to parallel lines. Because of this effect, perspective maps are a good model for how we perceive 3D space around us. Parallel lines do seemingly intersect in a distance, and are thus not *perceived* as being parallel! Figure 13.5 is a parallel projection and Figure 13.6 illustrates the same geometry with a perspective projection. Notice in the perspective image, the sides of the bounding cube that move into the page are no longer parallel.

As we saw above, $m_{4,4}$ allows us to specify perspective projections. The other elements of the bottom row of M are used for *projective maps*, a more general mapping than a perspective projection. The topic of this chapter is affine maps, so we'll leave a detailed discussion of these elements to another source: A mathematical treatment of this map is supplied by [6] and a computer graphics treatment is supplied by [19]. In short, these entries are used in computer graphics for mapping a *viewing volume*[3] in the shape of a frustum to one in the shape of a cube, while preserving the perspective projection effect. Algorithms in the graphics pipeline are very much simplified by only dealing with geometry known to be in a cube.

---

[3]The dimension and shape of the viewing volume defines *what* will be display and *how* it will be displayed (orthographic or perspective).

Figure 13.5.
Parallel projection: a 3D helix and two orthographic projections on the left and bottom walls of the bounding cube—not visible due to the orthographic projection used for the whole scene.

Figure 13.6.
Perspective projection: a 3D helix and two orthographic projections on the left and bottom walls of the bounding cube—visible due to the perspective projection used for the whole scene.

Figure 13.7.
Perspective maps: an experiment by A. Dürer.

The study of perspective goes back to the fourteenth century—before that, artists simply could not draw realistic 3D images. One of the foremost researchers in the area of perspective maps was A. Dürer.[4] See Figure 13.7 for one of his experiments.

---

[4]From *The Complete Woodcuts of Albrecht Dürer*, edited by W. Durth, Dover Publications Inc., New York, 1963.

- affine map
- affine map properties
- barycentric combination
- centroid
- mapping four points to four points

- parallel projection
- orthographic projection
- oblique projection
- homogeneous coordinates
- perspective projection
- dyadic matrix
- rank

13.6 Exercises

We'll use four points

$$\mathbf{x}_1 = \begin{bmatrix} 1 \\ 0 \\ 0 \end{bmatrix}, \quad \mathbf{x}_2 = \begin{bmatrix} 0 \\ 1 \\ 0 \end{bmatrix}, \quad \mathbf{x}_3 = \begin{bmatrix} 0 \\ 0 \\ -1 \end{bmatrix}, \quad \mathbf{x}_4 = \begin{bmatrix} 0 \\ 0 \\ 1 \end{bmatrix},$$

and four points

$$\mathbf{y}_1 = \begin{bmatrix} -1 \\ 0 \\ 0 \end{bmatrix}, \quad \mathbf{y}_2 = \begin{bmatrix} 0 \\ -1 \\ 0 \end{bmatrix}, \quad \mathbf{y}_3 = \begin{bmatrix} 0 \\ 0 \\ -1 \end{bmatrix}, \quad \mathbf{y}_4 = \begin{bmatrix} 0 \\ 0 \\ 1 \end{bmatrix},$$

and also the plane through $\mathbf{q}$ with normal $\mathbf{n}$:

$$\mathbf{q} = \begin{bmatrix} 1 \\ 0 \\ 0 \end{bmatrix}, \quad \mathbf{n} = \frac{1}{5} \begin{bmatrix} 3 \\ 0 \\ 4 \end{bmatrix}.$$

1. Using a direction

$$\mathbf{v} = \frac{1}{4} \begin{bmatrix} 2 \\ 0 \\ 2 \end{bmatrix},$$

   what are the images of the $\mathbf{x}_i$ when projected onto the plane with this direction?

2. Using the same $\mathbf{v}$ as in the previous problem, what are the images of the $\mathbf{y}_i$?

3. What are the images of the $\mathbf{x}_i$ when projected onto the plane by a perspective projection through the origin?

4. What are the images of the $\mathbf{y}_i$ when projected onto the plane by a perspective projection through the origin?

5. Compute the centroid $\mathbf{c}$ of the $\mathbf{x}_i$ and then the centroid $\mathbf{c}'$ of their perspective images (from Exercise 3). Is $\mathbf{c}'$ the image of $\mathbf{c}$ under the perspective map?

6. An affine map $\mathbf{x}_i \to \mathbf{y}_i; i = 1, 2, 3, 4$ is uniquely defined. What is it?

7. What is the image of

$$\mathbf{p} = \begin{bmatrix} 1 \\ 1 \\ 1 \end{bmatrix}$$

under the map from the previous problem? Use two ways to compute it.

8. What are the geometric properties of the affine map from the last two problems?

9. We claimed that (13.8) reduces to (13.10). This necessitates that

$$\left[I - \frac{\mathbf{v}\mathbf{n}^{\mathrm{T}}}{\mathbf{n} \cdot \mathbf{v}} \right] \mathbf{x} = \mathbf{0}.$$

Show that this is indeed true.

10. What is the affine map that rotates the point $\mathbf{q}$ defined above, $90°$ about the line defined as

$$\mathbf{l}(t) = \begin{bmatrix} 1 \\ 1 \\ 0 \end{bmatrix} + t \begin{bmatrix} 1 \\ 0 \\ 0 \end{bmatrix}?$$

Hint: This is a simple construction, and does not require (12.9).

14

General Linear Systems

Figure 14.1.
Linear systems: a shoe last is constructed that fits data points extracted by a CMM from a physical model.

In Chapter 5, we studied linear systems of two equations in two unknowns. A whole chapter for such a humble task seems like a bit of overkill—its main purpose was really to lay the groundwork for this chapter.

Linear systems arise in virtually every area of science and engineering—some are as big as 1,000,000 equations in as many unknowns. Such huge systems require more sophisticated treatment than the methods introduced here. They *will* allow you to solve systems with several thousand equations without a problem. Figure 14.1 illustrates a surface fitting problem that necessitated solving a system with approximately 300 equations.

This chapter explains the basic ideas underlying linear systems. Readers eager for hands-on experience should download linear system solvers from the web. The most prominent collection of routines is LINPACK, which was written in FORTRAN. The new generation of LINPACK is CLAPACK, and is written in C and can be found at http://www.netlib.org/clapack/.[1]

14.1 The Problem

A linear system is a set of equations like this:

$$3u_1 - 2u_2 - 10u_3 + u_4 = 0$$
$$u_1 - u_3 = 4$$
$$u_1 + u_2 - 2u_3 + 3u_4 = 1$$
$$u_2 + 2u_4 = -4.$$

The unknowns are the numbers $u_1, \ldots, u_4$. There are as many equations as there are unknowns, four in this example. We rewrite this system in matrix form:

$$\begin{bmatrix} 3 & -2 & -10 & 1 \\ 1 & 0 & -1 & 0 \\ 1 & 1 & -2 & 3 \\ 0 & 1 & 0 & 2 \end{bmatrix} \begin{bmatrix} u_1 \\ u_2 \\ u_3 \\ u_4 \end{bmatrix} = \begin{bmatrix} 0 \\ 4 \\ 1 \\ -4 \end{bmatrix}.$$

Our example was a 4×4 linear system. A general, $n \times n$ linear system looks like this:

$$a_{1,1}u_1 + a_{1,2}u_2 + \ldots + a_{1,n}u_n = b_1$$
$$a_{2,1}u_1 + a_{2,2}u_2 + \ldots + a_{2,n}u_n = b_2$$
$$\vdots$$
$$a_{n,1}u_1 + a_{n,2}u_2 + \ldots + a_{n,n}u_n = b_n.$$

---

[1]For more scientific computing software, go to http://www.netlib.org/.

In matrix form, it becomes

$$\begin{bmatrix} a_{1,1} & a_{1,2} & \cdots & a_{1,n} \\ a_{2,1} & a_{2,2} & \cdots & a_{2,n} \\ & & \vdots & \\ a_{n,1} & a_{n,2} & \cdots & a_{n,n} \end{bmatrix} \begin{bmatrix} u_1 \\ u_2 \\ \vdots \\ u_n \end{bmatrix} = \begin{bmatrix} b_1 \\ b_2 \\ \vdots \\ b_n \end{bmatrix}, \tag{14.1}$$

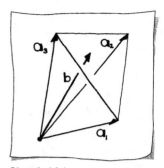

Sketch 14.1.

A solvable 3×3 system.

or even shorter

$$A\mathbf{u} = \mathbf{b}.$$

The *coefficient matrix* A has n rows and n columns. For example, the first row is

$$a_{1,1}, a_{1,2}, \ldots, a_{1,n},$$

and the second column is

$$a_{1,2}$$
$$a_{2,2}$$
$$\vdots$$
$$a_{n,2}.$$

Equation (14.1) is a compact way of writing n equations for the n unknowns $u_1, \ldots, u_n$. In the 2×2 case, such systems had nice geometric interpretations; in the general case, that interpretation needs n-dimensional linear spaces, and is not very intuitive. Still the methods that we developed for the 2×2 case can be gainfully employed here!

General linear systems defy geometric intuition, yet some underlying principles are of a geometric nature. It is best explained for the example $n = 3$. We are given a vector $\mathbf{b}$ and we try to write it as a linear combination of vectors $\mathbf{a}_1, \mathbf{a}_2, \mathbf{a}_3$. If the $\mathbf{a}_i$ are truly 3D, i.e., if they form a tetrahedron, then a *unique solution* may be found (see Sketch 14.1). But if the three $\mathbf{a}_i$ all lie in a plane (i.e., if the volume formed by them is zero), then you cannot write $\mathbf{b}$ as a linear combination of them, unless it is itself in that 2D plane. In this case, you cannot expect uniqueness for your answer. Sketch 14.2 covers these cases. In general, a linear system is uniquely solvable if the $\mathbf{a}_i$ have a nonzero n-dimensional volume. If they do not, they span a k-dimensional *subspace* (with $k < n$)—non-unique solutions only exist if $\mathbf{b}$ is itself in that subspace. A linear system is called *consistent* if at least one solution exits.

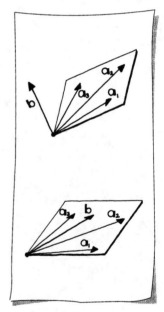

Sketch 14.2.

Top: no solution, bottom: non-unique solution.

14.2   The Solution via Gauss Elimination

The key to success in the 2×2 case was the application of a shear so that the matrix A was transformed to *upper triangular*; All entries below the diagonal are zero. Then it was possible to apply *back substitution* to solve for the unknowns. The shear was constructed to map the first column vector of the matrix onto the $\mathbf{e}_1$-axis. Revisiting an example from Chapter 5, we have

$$\begin{bmatrix} 2 & 4 \\ 1 & 6 \end{bmatrix} \begin{bmatrix} u_1 \\ u_2 \end{bmatrix} = \begin{bmatrix} 4 \\ 4 \end{bmatrix}.$$

The shear used was

$$S_1 = \begin{bmatrix} 1 & 0 \\ -1/2 & 1 \end{bmatrix},$$

which when applied to the system as

$$S_1 A \mathbf{u} = S_1 \mathbf{b}$$

produced the system

$$\begin{bmatrix} 2 & 4 \\ 0 & 4 \end{bmatrix} \begin{bmatrix} u_1 \\ u_2 \end{bmatrix} = \begin{bmatrix} 4 \\ 2 \end{bmatrix}.$$

Algebraically, what this shear did was to change the *rows* of the system in the following manner:

$$\text{row}_1 \leftarrow \text{row}_1 \quad \text{and} \quad \text{row}_2 \leftarrow \text{row}_2 - \frac{1}{2}\text{row}_1.$$

Each of these are an *elementary row operation*. Back substitution came next, with

$$u_2 = \frac{1}{4} \times 2 = \frac{1}{2}$$

and then

$$u_1 = \frac{1}{2}(4 - 4u_2) = 1.$$

The divisions in the back substitution equations are actually scalings, thus they could be rewritten in terms of a scale matrix:

$$S_2 = \begin{bmatrix} 1/2 & 0 \\ 0 & 1/4 \end{bmatrix},$$

and then the system would be transformed via

$$S_2 S_1 A \mathbf{u} = S_2 S_1 \mathbf{b}.$$

The corresponding upper triangular matrix and system is

$$\begin{bmatrix} 1 & 2 \\ 0 & 1 \end{bmatrix} \begin{bmatrix} u_1 \\ u_2 \end{bmatrix} = \begin{bmatrix} 2 \\ 1/2 \end{bmatrix}.$$

Check for yourself that we get the same result.

Thus, we see the geometric steps for solving a linear system have methodical algebraic interpretations. This algebraic approach is what will be followed for the rest of the chapter. For general linear systems the matrices, such as S_1 and S_2 above, are not actually constructed due to the speed and storage expense. Notice that the shear to zero one element in the matrix, only changed the elements in one row, thus it is unnecessary to manipulate the other row. This is an important observation for large systems.

In the general case, as in the 2×2 case, *pivoting* will be used. Recall for the 2×2 case this meant that the equations were reordered such that the matrix element $a_{1,1}$ is the largest one in the first column.

Example 14.1

Let's step through the necessary shears for a 3×3 linear system. The goal is to get it in upper triangular form so we may use back substitution to solve for the unknowns. The system is

$$\begin{bmatrix} 2 & -2 & 0 \\ 4 & 0 & -2 \\ 4 & 2 & -4 \end{bmatrix} \begin{bmatrix} u_1 \\ u_2 \\ u_3 \end{bmatrix} = \begin{bmatrix} 4 \\ -2 \\ 0 \end{bmatrix}.$$

The matrix element $a_{1,1}$ is not the largest in the first column, so we reorder:

$$\begin{bmatrix} 4 & 0 & -2 \\ 2 & -2 & 0 \\ 4 & 2 & -4 \end{bmatrix} \begin{bmatrix} u_1 \\ u_2 \\ u_3 \end{bmatrix} = \begin{bmatrix} -2 \\ 4 \\ 0 \end{bmatrix}.$$

To zero entries in the first column apply:

$$\text{row}_2 \leftarrow \text{row}_2 - \frac{1}{2}\text{row}_1$$

$$\text{row}_3 \leftarrow \text{row}_3 - \text{row}_1,$$

and the system becomes

$$\begin{bmatrix} 4 & 0 & -2 \\ 0 & -2 & 1 \\ 0 & 2 & -2 \end{bmatrix} \begin{bmatrix} u_1 \\ u_2 \\ u_3 \end{bmatrix} = \begin{bmatrix} -2 \\ 5 \\ 2 \end{bmatrix}.$$

Now the first column consists of only zeroes except for $a_{1,1}$, meaning that it is lined up with the $\mathbf{e}_1$-axis.

Now work on the second column vector. First, check if pivoting is necessary; this means checking that $a_{2,2}$ is the largest in absolute value of all values in the second column that are below the diagonal. No pivoting is necessary. To zero the last element in this vector apply

$$\text{row}_3 \leftarrow \text{row}_3 + \text{row}_2,$$

which produces

$$\begin{bmatrix} 4 & 0 & -2 \\ 0 & -2 & 1 \\ 0 & 0 & -1 \end{bmatrix} \begin{bmatrix} u_1 \\ u_2 \\ u_3 \end{bmatrix} = \begin{bmatrix} -2 \\ 5 \\ 7 \end{bmatrix}.$$

By chance, the second column is aligned with $\mathbf{e}_2$ because $a_{1,2} = 0$. If this extra zero had not appeared, then we would have mapped this 3D vector into the $[\mathbf{e}_1, \mathbf{e}_2]$-plane.

Now we are ready for back substitution:

$$u_3 = \frac{1}{-1}(7)$$

$$u_2 = \frac{1}{-2}(5 - u_3)$$

$$u_1 = \frac{1}{4}(-2 + 2u_3).$$

This implicitly incorporates a scaling matrix. We obtain the solution

$$\begin{bmatrix} u_1 \\ u_2 \\ u_3 \end{bmatrix} = \begin{bmatrix} -4 \\ -6 \\ -7 \end{bmatrix}.$$

It is usually a good idea to insert the solution into the original equations:

$$\begin{bmatrix} 2 & -2 & 0 \\ 4 & 0 & -2 \\ 4 & 2 & -4 \end{bmatrix} \begin{bmatrix} -4 \\ -6 \\ -7 \end{bmatrix} = \begin{bmatrix} 4 \\ -2 \\ 0 \end{bmatrix}.$$

It works!

This example illustrates each of the *elementary row operations* that take place during Gauss elimination:

- Shears result in adding a multiple of one row to another.

- Scaling results in multiplying a row by a scalar.

- Pivoting results in the exchange of two rows.

Here is the algorithm for solving a general $n \times n$ system of linear equations.

Gauss Elimination with Pivoting

Given: A coefficient matrix A and a right-hand side $\mathbf{b}$ describing a linear system
$$A\mathbf{u} = \mathbf{b},$$
which is short for the more detailed (14.1).

Find: The unknowns $u_1, \ldots, u_n$.

Algorithm:

For $j = 1, \ldots, n-1$: (note: j counts columns)

Pivoting Step:

Find element in largest absolute value in column j
from $a_{j,j}$ to $a_{n,j}$; this is element $a_{r,j}$.
 If $r > j$, exchange equations r and j.

If $a_{j,j} = 0$, the system is not solvable.

Elimination Step for column j:

For $i = j+1, \ldots, n$: (elements below diagonal of column j)
 Construct the *multiplier* $g_{i,j} = a_{i,j}/a_{j,j}$
 $a_{i,j} = 0$
 For $k = j+1, \ldots, n$ (each element in row i after column j)
  $a_{i,k} = a_{i,k} - g_{i,j}a_{j,k}$
 $b_i = b_i - g_{i,j}b_j$

After this loop, all elements below the diagonal have been set to zero.
The matrix is now in *upper triangular* form.
We call this transformed matrix $A^\triangle$.

Back substitution:
$u_n = b_n/a_{n,n}$
For $j = n - 1, \dots, 1$
$\qquad u_j = \frac{1}{a_{j,j}}[b_j - a_{j,j+1}u_{j+1} - \dots - a_{j,n}u_n].$

In a programming environment, it can be convenient to form an *augmented matrix* which is the matrix A augmented with the vector **b**. Here is the idea for a 3×3 linear system:

$$\begin{bmatrix} a_{1,1} & a_{1,2} & a_{1,3} & b_1 \\ a_{2,1} & a_{2,2} & a_{2,3} & b_2 \\ a_{3,1} & a_{3,2} & a_{3,3} & b_3 \end{bmatrix}.$$

Then the k steps would run to $n + 1$, and there would be no need for the extra line for the b_i element.

Notice that it is possible to store the $g_{i,j}$ in the zero elements of A rather than explicitly setting the $a_{i,j}$ element equal to zero. The operations in the elimination step above may also be written in matrix form. If A is the current matrix, then at step j, to produce zeroes under $a_{j,j}$ the matrix product $G_j A$ is formed, where

$$G_j = \begin{bmatrix} 1 & & & & & & \\ & \ddots & & & & & \\ & & 1 & & & & \\ & & & 1 & & & \\ & & & -g_{j+1,j} & 1 & & \\ & & & \vdots & & \ddots & \\ & & & -g_{n,j} & & & 1 \end{bmatrix}. \qquad (14.2)$$

The elements $g_{i,j}$ of G_j are the *multipliers*. The matrix G_j is called a *Gauss matrix*. All entries except for the diagonal and the entries $g_{i,j}$ are zero. As we see with the algorithm, it is not necessary to explicitly form the Gauss matrix. In fact, it is more efficient with regard to speed and storage not to. Using the Gauss matrix blindly would result in many unnecessary calculations.

Example 14.2

We look at one more example, taken from [2]. Let the system be given by

$$\begin{bmatrix} 2 & 2 & 0 \\ 1 & 1 & 2 \\ 2 & 1 & 1 \end{bmatrix} \begin{bmatrix} u_1 \\ u_2 \\ u_3 \end{bmatrix} = \begin{bmatrix} 6 \\ 9 \\ 7 \end{bmatrix}.$$

We start the algorithm with $j = 1$, and observe that no element in column 1 exceeds $a_{1,1}$ in absolute value, so no pivoting is necessary at this step. Proceed with the elimination step for row 2 by constructing the multiplier

$$g_{2,1} = a_{2,1}/a_{1,1} = 1/2.$$

Change row 2 as follows:

$$\text{row}_2 \leftarrow \text{row}_2 - 1/2\text{row}_1.$$

Remember, this includes changing the element b_2. Similarly for row 3,

$$g_{3,1} = a_{3,1}/a_{1,1} = 2/2 = 1$$

then

$$\text{row}_3 \leftarrow \text{row}_3 - \text{row}_1.$$

Step $j = 1$ is complete and the linear system is now

$$\begin{bmatrix} 2 & 2 & 0 \\ 0 & 0 & 2 \\ 0 & -1 & 1 \end{bmatrix} \begin{bmatrix} u_1 \\ u_2 \\ u_3 \end{bmatrix} = \begin{bmatrix} 6 \\ 6 \\ 1 \end{bmatrix}.$$

Next is column 2, so $j = 2$. Observe that $a_{2,2} = 0$, whereas $a_{3,2} = -1$. We exchange equations 2 and 3 and the system becomes

$$\begin{bmatrix} 2 & 2 & 0 \\ 0 & -1 & 1 \\ 0 & 0 & 2 \end{bmatrix} \begin{bmatrix} u_1 \\ u_2 \\ u_3 \end{bmatrix} = \begin{bmatrix} 6 \\ 1 \\ 6 \end{bmatrix}. \tag{14.3}$$

If blindly following the algorithm above, we would proceed with the elimination for row 3 by forming the multiplier

$$g_{3,2} = a_{3,2}/a_{2,2} = 0/-1 = 0.$$

Then operate on the third row

$$\text{row}_3 \leftarrow \text{row}_3 - 0 \times \text{row}_2,$$

which doesn't change the row at all—ignoring numerical instabilities. Without putting a special check for a zero multiplier, this unnecessary work takes place. Tolerances are very important here.

Apply back substitution by first solving for the last unknown:

$$u_3 = 3.$$

Start the back substitution loop with $j = 2$:

$$u_2 = \frac{1}{-1}[1 - u_3] = 2,$$

and finally

$$u_1 = \frac{1}{2}[6 - 2u_2] = 1.$$

It's a good idea to check the solution:

$$\begin{bmatrix} 2 & 2 & 0 \\ 1 & 1 & 2 \\ 2 & 1 & 1 \end{bmatrix} \begin{bmatrix} 1 \\ 2 \\ 3 \end{bmatrix} = \begin{bmatrix} 6 \\ 9 \\ 7 \end{bmatrix} !$$

Just before back substitution, we could scale to achieve ones along the diagonal of the matrix. Let's do precisely that to the linear system in Example 14.2. Multiply both sides of (14.3) by

$$\begin{bmatrix} 1/2 & 0 & 0 \\ 0 & -1 & 0 \\ 0 & 0 & 1/2 \end{bmatrix}.$$

This transforms the linear system to

$$\begin{bmatrix} 1 & 1 & 0 \\ 0 & 1 & -1 \\ 0 & 0 & 1 \end{bmatrix} \begin{bmatrix} u_1 \\ u_2 \\ u_3 \end{bmatrix} = \begin{bmatrix} 3 \\ -1 \\ 63 \end{bmatrix}.$$

This matrix with $rank = n$ is said to be in *row echelon form*, which is upper triangular with ones along the diagonal. If the matrix is *rank deficient*, $rank < n$, then the rows with all zeros should be the last rows. Of course it is more efficient to do the scaling as part of back substitution.

14.3 Determinants

With the introduction of the scalar triple product, Section 10.6 provided a geometric derivation of 3×3 determinants. And then in Section 12.7 we learned more about determinants from the perspective of linear maps. Let's revisit that approach for $n \times n$ determinants.

When we take A to upper triangular form $A^\triangle$ we apply a sequence of shears to its initial column vectors. Shears do not change volumes; thus the column vectors of $A^\triangle$ span the same volume as did those of A. This volume is given by the product of the diagonal entries and is called the *determinant* of A:

$$\det A = a_{1,1}^\triangle \times \ldots \times a_{n,n}^\triangle.$$

However this equivalence is not entirely correct from an implementation viewpoint: often it is necessary during Gauss elimination to exchange rows. As we learned in Section 12.7, this action will change the sign of the determinant. Therefore, if we proceed with k row exchanges, then the determinant becomes

$$\det A = (-1)^k [a_{1,1}^\triangle \times \ldots \times a_{n,n}^\triangle]. \qquad (14.4)$$

In general, this is the best (and most stable) method for finding the determinant.

Example 14.3

Let's revisit Example 14.2 to illustrate how to calculate the determinant with the upper triangular form, and how row exchanges influence the sign of the determinant.

Use the technique of cofactor expansion, as defined by (12.10) to find the determinant of the given 3×3 matrix A from the example:

$$\det A = 2 \begin{vmatrix} 1 & 2 \\ 1 & 1 \end{vmatrix} - 2 \begin{vmatrix} 1 & 2 \\ 2 & 1 \end{vmatrix} = 4.$$

Now, apply (14.4) to the upper triangular form $A^\triangle$ from the example, and notice that we did one row exchange, $k = 1$:

$$\det A^\triangle = (-1)^1 [2 \times -1 \times 2] = 4.$$

So the shears of Gauss elimination have not changed the determinant.

The technique of *cofactor expansion* that was used for the 3×3 matrix in Example 14.3 may be generalized to $n \times n$ matrices. Choose any column or row of the matrix, for example entries $a_{1,j}$ as above, and then

$$\det A = a_{1,1} C_{1,1} + a_{1,2} C_{1,2} + \ldots + a_{1,n} C_{1,n}$$

where each cofactor is defined as

$$C_{i,j} = (-1)^{i+j} M_{i,j},$$

and the $M_{i,j}$ are called the *minors*; each is the determinant of the matrix with the ith row and j^{th} column removed. The $M_{i,j}$ are $(n-1) \times (n-1)$ determinants, and they are computed by yet another cofactor expansion. This process is repeated until we have 2×2 determinants. This technique is also known as *expansion by minors*.

Example 14.4

Let's look at repeated application of cofactor expansion to find the determinant. Suppose we are given the following matrix,

$$A = \begin{bmatrix} 2 & 2 & 0 & 4 \\ 0 & -1 & 1 & 3 \\ 0 & 0 & 2 & 0 \\ 0 & 0 & 0 & 5 \end{bmatrix}.$$

We may choose any row or column from which to form the cofactors, so in this example, we will have less work to do if we choose the first column. The cofactor expansion is

$$\det A = 2 \begin{vmatrix} -1 & 1 & 3 \\ 0 & 2 & 0 \\ 0 & 0 & 5 \end{vmatrix} = 2(-1) \begin{vmatrix} 2 & 0 \\ 0 & 5 \end{vmatrix} = 2(-1)(10) = -20.$$

Since the matrix is in upper triangular form, we could use (14.4) and immediately see that this is in fact the correct determinant.

Cofactor expansion is more a theoretical tool than a computational one. This method of calculating the determinant plays an important theoretical role in the analysis of linear systems, and there are advanced theorems involving cofactor expansion and the inverse of a matrix. Computationally, Gauss elimination and the calculation of $\det A^{\triangle}$ is superior.

In our first encounter with solving linear systems via Cramer's rule in Section 5.3, we learned that the solution to a linear system may be found by simply forming quotients of areas. Now with our knowledge of $n \times n$ determinants, let's revisit *Cramer's rule*. If $A\mathbf{u} = \mathbf{b}$ is an

$n \times n$ linear system such that $\det A \neq 0$, then the system has the following unique solution:

$$u_1 = \frac{\det A_1}{\det A}, \quad u_2 = \frac{\det A_2}{\det A}, \quad \ldots, \quad u_n = \frac{\det A_n}{\det A}, \qquad (14.5)$$

where A_i is the matrix obtained by replacing the entries in the ith column by the entries of **b**. Cramer's rule is an important theoretical tool; however, it should *not* be used for linear systems greater than 3×3! It does prove to be a handy method to solve such systems by hand.

Example 14.5

Let's solve the linear system from Example 14.2 using Cramer's rule. Following (14.5), we have

$$u_1 = \frac{\begin{vmatrix} 6 & 2 & 0 \\ 9 & 1 & 2 \\ 7 & 1 & 1 \end{vmatrix}}{\begin{vmatrix} 2 & 2 & 0 \\ 1 & 1 & 2 \\ 2 & 1 & 1 \end{vmatrix}}, \quad u_2 = \frac{\begin{vmatrix} 2 & 6 & 0 \\ 1 & 9 & 2 \\ 2 & 7 & 1 \end{vmatrix}}{\begin{vmatrix} 2 & 2 & 0 \\ 1 & 1 & 2 \\ 2 & 1 & 1 \end{vmatrix}}, \quad u_3 = \frac{\begin{vmatrix} 2 & 2 & 6 \\ 1 & 1 & 9 \\ 2 & 1 & 7 \end{vmatrix}}{\begin{vmatrix} 2 & 2 & 0 \\ 1 & 1 & 2 \\ 2 & 1 & 1 \end{vmatrix}}.$$

We have computed the determinant of the coefficient matrix A in Exercise 14.3, $\det A = 4$. With the application of cofactor expansion for each numerator, we find that

$$u_1 = \frac{4}{4} = 1, \quad u_2 = \frac{8}{4} = 2, \quad u_3 = \frac{12}{4} = 3,$$

which is identical to the solution found with Gauss elimination.

Rules for working with determinants are given in Section 12.7.

14.4   Overdetermined Systems

Sometimes systems arise that have more equations than unknowns, such as

$$4u_1 + u_2 = 1$$
$$-u_1 + 4u_2 + 2u_3 = -1$$
$$4u_3 = 0$$
$$u_1 + u_2 + u_3 = 2.$$

This is a system of four equations in three unknowns. In matrix form:

$$\begin{bmatrix} 4 & 1 & 0 \\ -1 & 4 & 2 \\ 0 & 0 & 4 \\ 1 & 1 & 1 \end{bmatrix} \begin{bmatrix} u_1 \\ u_2 \\ u_3 \end{bmatrix} = \begin{bmatrix} 1 \\ -1 \\ 0 \\ 2 \end{bmatrix}. \tag{14.6}$$

Systems like this will in general not have solutions. But there is a recipe for finding an *approximate solution*. As usual, write the system as

$$A\mathbf{u} = \mathbf{b},$$

with a matrix that has more rows than columns. Simply multiply both sides by A^T:

$$A^\mathrm{T} A\mathbf{u} = A^\mathrm{T}\mathbf{b}. \tag{14.7}$$

This is a linear system with a square matrix $A^\mathrm{T}A$! Even more, that matrix is symmetric. The equations of this matrix are called the *normal equations*. The solution to the new system (14.7) (when it has one) is the one that minimizes the *error*

$$\|A\mathbf{u} - \mathbf{b}\|^2.$$

It is called the *least squares solution* of the original system.

Example 14.6

Returning to the system in (14.6), the least squares solution is the solution of the linear system

$$\begin{bmatrix} 18 & 1 & -1 \\ 1 & 18 & 9 \\ -1 & 9 & 21 \end{bmatrix} \begin{bmatrix} u_1 \\ u_2 \\ u_3 \end{bmatrix} = \begin{bmatrix} 7 \\ -2 \\ 0 \end{bmatrix}.$$

Example 14.7

As a second example, consider the problem of fitting a straight line to a set of 2D data points. Let the points be given by

$$\mathbf{p}_2 = \begin{bmatrix} -2 \\ -2 \end{bmatrix}, \quad \mathbf{p}_1 = \begin{bmatrix} 0 \\ 0 \end{bmatrix}, \quad \mathbf{p}_3 = \begin{bmatrix} 3 \\ 0 \end{bmatrix}, \quad \mathbf{p}_4 = \begin{bmatrix} 0 \\ 1 \end{bmatrix}.$$

We want to find a line of the form

$$x_2 = ax_1 + b,$$

that is close to the given points.[2] For the solution, see Figure 14.2. Ideally, we would like all of our data points to be on the line. This would require each of the following equations to be satisfied for one pair of a and b:

$$-2a + b = -2,$$
$$b = 0,$$
$$3a + b = 0,$$
$$b = 1.$$

This is an overdetermined linear system for the unknowns a and b—obvious from looking at the second and fourth equations. Now our overdetermined linear system is of the form

$$\begin{bmatrix} -2 & 1 \\ 0 & 1 \\ 3 & 1 \\ 0 & 1 \end{bmatrix} \begin{bmatrix} a \\ b \end{bmatrix} = \begin{bmatrix} -2 \\ 0 \\ 0 \\ 1 \end{bmatrix}.$$

For the least squares system, we obtain

$$\begin{bmatrix} 13 & 1 \\ 1 & 4 \end{bmatrix} \begin{bmatrix} a \\ b \end{bmatrix} = \begin{bmatrix} 4 \\ -1 \end{bmatrix}.$$

It has the solution

$$\begin{bmatrix} a \\ b \end{bmatrix} = \begin{bmatrix} 0.33 \\ -0.33 \end{bmatrix}.$$

Thus, our desired straight line (see Figure 14.2) is given by

$$x_2 = 0.33x_1 - 0.33.$$

Numerical problems can creep into the normal equations of the linear system (14.7). This is particularly so when the $n \times m$ matrix A has many more equations than unknowns, $n \gg m$. In Section 16.1, we will examine the Householder method for finding the least squares

---

[2]For reasons beyond the scope of this text, we have formulated the line in the explicit form rather than the more favored parametric or implicit forms.

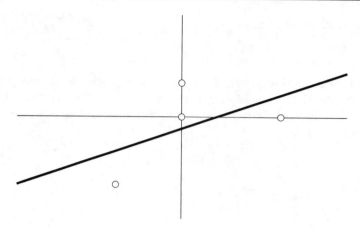

Figure 14.2.
Least squares: fitting a straight line to a set of points.

solution to the linear system $A\mathbf{u} = \mathbf{b}$ directly, without forming the normal equations. Example 14.7 will be revisited in Example 16.2.

14.5 Inverse Matrices

The *inverse* of a square matrix A is the matrix that "undoes" A's action, i.e., the combined action of A and A^{-1} is the identity:

$$AA^{-1} = I. \tag{14.8}$$

Example 14.8

The following scheme shows a matrix A multiplied by its inverse A^{-1}. The matrix A is on the left, A^{-1} is on top, and the result of the multiplication, the identity, is on the lower right:

$$
\begin{array}{ccc|ccc}
 & & & 1 & 0 & -1 \\
 & & & 3 & 1 & -3 \\
 & & & 1 & 2 & -2 \\
\hline
-4 & 2 & -1 & 1 & 0 & 0 \\
-3 & 1 & 0 & 0 & 1 & 0 \\
-5 & 2 & -1 & 0 & 0 & 1
\end{array}.
$$

How do we find the inverse of a matrix? In much the same way as we did in the 2×2 case in Section 5.5, write

$$A \begin{bmatrix} \bar{\mathbf{a}}_1 & \cdots & \bar{\mathbf{a}}_n \end{bmatrix} = \begin{bmatrix} \mathbf{e}_1 & \cdots & \mathbf{e}_n \end{bmatrix}. \qquad (14.9)$$

Here, the matrices are $n \times n$, and the vectors $\bar{\mathbf{a}}_i$ as well as the $\mathbf{e}_i$ are vectors with n components. The vector $\mathbf{e}_i$ has all nonzero entries except for its ith component; it equals 1.

We may now interpret (14.9) as n linear systems:

$$A\bar{\mathbf{a}}_1 = \mathbf{e}_1, \quad \ldots, \quad A\bar{\mathbf{a}}_n = \mathbf{e}_n. \qquad (14.10)$$

Each of these may be solved as per Section 14.2. Most economically, an LU decomposition should be used here (see Section 14.6).

The inverse of a matrix A only exists if the action of A does not reduce dimensionality, as in a projection. This means that all columns of A must be linearly independent. There is a simple way to see if a matrix A is invertible; just perform Gauss elimination for the first of the linear systems in (14.10). If you are able to transform A to upper triangular with all nonzero diagonal elements, then A is invertible. Otherwise, it is said to be *singular*.

A singular matrix reduces dimensionality. An invertible (hence square!) matrix does not do this; it is said to have *rank n*, or *full rank*. If a matrix reduces dimensionality by k, then it has rank $n - k$.

Example 14.9

The 4×4 matrix

$$\begin{bmatrix} 1 & 3 & -3 & 0 \\ 0 & 3 & 3 & 1 \\ 0 & 0 & 0 & 0 \\ 0 & 0 & 0 & 0 \end{bmatrix}$$

has rank 2, while

$$\begin{bmatrix} 1 & 3 & -3 & 0 \\ 0 & 3 & 3 & 1 \\ 0 & 0 & -1 & 0 \\ 0 & 0 & 0 & 0 \end{bmatrix}$$

has rank 3.

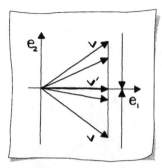

The $n \times n$ identity matrix has rank n; the zero matrix has rank 0. An example of a matrix which does not have full rank is a projection. All of its image vectors $\mathbf{v}'$ may be the result of many different operations $A\mathbf{v}$, as Sketch 14.3 shows.

Example 14.10

Let us compute the inverse of the $n \times n$ matrix G_j as defined in (14.2):

$$
G_j^{-1} =
\begin{bmatrix}
1 & & & & & & \\
 & \ddots & & & & & \\
 & & 1 & & & & \\
 & & & 1 & & & \\
 & & & g_{j+1,j} & 1 & & \\
 & & & \vdots & & \ddots & \\
 & & & g_{n,j} & & & 1
\end{bmatrix}.
$$

That's simple! To make some geometric sense of this, you should realize that G_j is a shear, and so is G_j^{-1}, except it "undoes" G_j.

Here is another interesting property of the inverse of a matrix. Suppose $k \neq 0$ and kA is an invertible matrix, then

$$
(kA)^{-1} = \frac{1}{k} A^{-1}.
$$

And yet another: If two matrices, A and B, are invertible, then the product AB is invertible, too.

Inverse matrices are primarily a theoretical concept. They suggest to solve a linear system $A\mathbf{v} = \mathbf{b}$ by computing A^{-1} and then to set $\mathbf{v} = A^{-1}\mathbf{b}$. Don't do that! It is a very expensive way to solve a linear system; simple Gauss elimination is much cheaper.

14.6   LU Decomposition

Gauss elimination has two major parts: transforming the system to upper triangular form, and then back substitution. The creation of the upper triangular matrix may be written in terms of matrix multiplies using Gauss matrices G_j. If we denote the final upper triangular matrix[3] by U, then we have

$$G_{n-1} \cdot \ldots \cdot G_1 \cdot A = U. \qquad (14.11)$$

It follows that

$$A = G_1^{-1} \cdot \ldots \cdot G_{n-1}^{-1} U.$$

The neat thing about the product $G_1^{-1} \cdot \ldots \cdot G_{n-1}^{-1}$ is that it is a *lower triangular* matrix with elements $g_{i,j}$:

$$G_1^{-1} \cdot \ldots \cdot G_{n-1}^{-1} = \begin{bmatrix} 1 & & & \\ g_{2,1} & 1 & & \\ \vdots & \ddots & \ddots & \\ g_{n,1} & \cdots & g_{n,n-1} & 1 \end{bmatrix}.$$

We denote this product by L (for lower triangular). Thus,

$$A = LU, \qquad (14.12)$$

which is known as the *LU decomposition* of A. It is also called the *factorization* of A. Every invertible matrix A has such a decomposition, although it may be necessary to employ pivoting during its computation.

If we denote the elements of L by $l_{i,j}$ (keeping in mind that $l_{i,i} = 1$) and those of U by $u_{i,j}$, the elements of A may be rewritten as

$$a_{i,j} = l_{i,1}u_{1,i} + \ldots + l_{i,j}u_{j,j}; \quad j < i,$$

i.e., for those $a_{i,j}$ below A's diagonal. For those on or above the diagonal, we get

$$a_{i,j} = l_{i,1}u_{1,j} + \ldots + l_{i,i-1}u_{i-1,j} + l_{i,i}u_{i,j}; \quad j \geq i.$$

This leads to

$$l_{i,j} = \frac{1}{u_{j,j}}(a_{i,j} - l_{i,1}u_{1,j} - \ldots - l_{i,j-1}u_{j-1,j}); \quad j < i \qquad (14.13)$$

and

$$u_{i,j} = a_{i,j} - l_{i,1}u_{1,j} - \ldots - l_{i,i-1}u_{i-1,j}; \quad j \geq i. \qquad (14.14)$$

If A has a decomposition $A = LU$, then the system can be written as

$$LU\mathbf{u} = \mathbf{b}. \qquad (14.15)$$

---

[3]This matrix U was called $A^{\triangle}$ in Section 14.2.

The matrix vector product $U\mathbf{u}$ results in a vector; call this $\mathbf{y}$. Reexamining (14.15), it becomes a two-step problem. First solve

$$Ly = \mathbf{b}, \tag{14.16}$$

then solve

$$U\mathbf{u} = \mathbf{y}. \tag{14.17}$$

Hence, we have if $U\mathbf{u} = \mathbf{y}$, then $LU\mathbf{u} = L\mathbf{y} = \mathbf{b}$. The two systems in (14.16) and (14.17) are triangular and easy to solve. *Forward substitution* is used with the matrix L. (See Exercise 9 and its solution for an algorithm.) Back substitution is used with the matrix U. An algorithm is provided in Section 14.2.

Here is a more direct method for forming L and U, rather than through Gauss elimination. This then is the method of LU decomposition.

LU Decomposition

Given: A coefficient matrix A and a right-hand side $\mathbf{b}$ describing a linear system

$$A\mathbf{u} = \mathbf{b}.$$

Find: The unknowns $u_1, \ldots, u_n$.

Algorithm:

Calculate the nonzero elements of L and U:
For $k = 1, \ldots, n$
$\qquad u_{k,k} = a_{k,k} - l_{k,1}u_{1,k} - \ldots - l_{k,k-1}u_{k-1,k}$
$\qquad$ For $i = k+1, \ldots, n$
$\qquad\qquad l_{i,k} = \frac{1}{u_{k,k}}[a_{i,k} - l_{i,1}u_{1,k} - \ldots - l_{i,k-1}u_{k-1,k}]$
$\qquad$ For $j = k+1, \ldots, n$
$\qquad\qquad u_{k,j} = a_{k,j} - l_{k,1}u_{1,j} - \ldots - l_{k,k-1}u_{k-1,j}$

Using forward substitution solve $L\mathbf{y} = \mathbf{b}$.
Using back substitution solve $U\mathbf{u} = \mathbf{y}$.

The $u_{k,k}$ term must not be zero; we had a similiar situation with Gauss elimination. Again, this situation requires pivoting. Pivoting makes the bookkeeping trickier; see [14] for a detailed description.

Example 14.11

Decompose A into LU, where

$$A = \begin{bmatrix} 2 & 2 & 4 \\ -1 & 2 & -3 \\ 1 & 2 & 2 \end{bmatrix}.$$

Following the steps in the algorithm above, we calculate the following matrix entries:

$$u_{1,1} = 2,$$
$$l_{2,1} = a_{2,1}/u_{1,1} = -1/2,$$
$$l_{3,1} = a_{3,1}/u_{1,1} = 1/2,$$
$$u_{1,2} = a_{1,2} = 2,$$

$$u_{2,2} = a_{2,2} - l_{2,1}u_{1,2} = 2 + 1 = 3,$$
$$l_{3,2} = \frac{1}{u_{2,2}}[a_{3,2} - l_{3,1}u_{1,2}] = \frac{1}{3}[2-1] = 1/3,$$
$$u_{1,3} = 4,$$
$$u_{2,3} = a_{2,3} - l_{2,1}u_{1,3} = -3 + 2 = -1,$$

$$u_{3,3} = a_{3,3} - l_{3,1}u_{1,3} - l_{3,2}u_{2,3} = 2 - 2 + 1/3 = 1/3.$$

Check that this produced valid entries for L and U:

$$\begin{array}{ccc|ccc}
 & & & 2 & 2 & 4 \\
 & & & 0 & 3 & -1 \\
 & & & 0 & 0 & 1/3 \\
\hline
1 & 0 & 0 & 2 & 2 & 4 \\
-1/2 & 1 & 0 & -1 & 2 & -3 \\
1/2 & 1/3 & 1 & 1 & 2 & 2
\end{array}.$$

Finally, the major benefit of the LU-decomposition: speed. In cases where one has to solve multiple linear systems with the same coefficient matrix,[4] the LU-decomposition is a big timesaver. We perform it once, and then perform the forward and backward substitutions (14.16) and (14.17) for each right-hand side. This is significantly less work than performing a complete Gauss elimination every time over!

---

[4]Finding the inverse of a matrix, as described in (14.10), is an example.

- coefficient matrix
- consistent system
- subspace
- back substitution
- solvable system
- unsolvable system
- Gauss elimination
- Gauss matrix
- upper triangular matrix
- row echelon form
- pivoting
- augmented matrix
- inverse matrix
- elementary row operation
- multiplier
- augmented matrix
- determinant
- cofactor expansion
- overdetermined system
- least squares solution
- normal equations
- singular matrix
- matrix rank
- full rank
- LU decomposition
- forward substitution
- lower triangular matrix

14.7   Exercises

1. Solve the linear system $A\mathbf{v} = \mathbf{b}$ where

$$A = \begin{bmatrix} 1 & 0 & -1 & 2 \\ 0 & 0 & 1 & -2 \\ 2 & 0 & 0 & 1 \\ 1 & 1 & 1 & 0 \end{bmatrix}, \quad \text{and} \quad \mathbf{b} = \begin{bmatrix} -1 \\ 2 \\ 1 \\ -3 \end{bmatrix}.$$

2. Solve the linear system $A\mathbf{v} = \mathbf{b}$ where

$$A = \begin{bmatrix} 0 & 0 & 1 \\ 1 & 0 & 0 \\ 1 & 1 & 1 \end{bmatrix}, \quad \text{and} \quad \mathbf{b} = \begin{bmatrix} -1 \\ 0 \\ -1 \end{bmatrix}.$$

3. Find the inverse of the matrix from the previous problem.

4. Let five points be given by

$$\mathbf{p}_2 = \begin{bmatrix} 1 \\ -1 \end{bmatrix}, \mathbf{p}_1 = \begin{bmatrix} 0 \\ 0 \end{bmatrix}, \mathbf{p}_3 = \begin{bmatrix} -3 \\ 0 \end{bmatrix}, \mathbf{p}_4 = \begin{bmatrix} 3 \\ -1 \end{bmatrix}, \mathbf{p}_5 = \begin{bmatrix} 0 \\ 1 \end{bmatrix}.$$

Find the least squares approximation using a straight line.

5. Restate the Gauss elimination algorithm in pseudocode with pivoting that is done without actually exchanging rows, but rather using an ordering vector.

6. Calculate the determinant of

$$A = \begin{bmatrix} 3 & 0 & 1 \\ 1 & 2 & 0 \\ 1 & 1 & 1 \end{bmatrix}.$$

7. Calculate the LU decomposition of the matrix in the previous problem.

8. What is the rank of the matrix in Exercise 6?

9. Write a forward substitution algorithm for solving the lower triangular system (14.16).

10. Apply Cramer's rule to solve the following linear system:

$$\begin{bmatrix} 3 & 0 & 1 \\ 1 & 2 & 0 \\ 1 & 1 & 1 \end{bmatrix} \mathbf{u} = \begin{bmatrix} 8 \\ 6 \\ 6 \end{bmatrix}.$$

   Hint: Reuse your work from Exercise 6.

15

General Linear Spaces

Figure 15.1.
General linear spaces: all cubic polynomials over the interval [0,1] form a linear space.
Some elements of this space are shown.

In this chapter, we will review the theory about vectors and linear spaces. We will provide a framework to characterize linear spaces which are not only of dimension two or three, but of possibly much larger dimension. These spaces tend to be somewhat abstract, but they are a powerful concept in dealing with many real-life problems, such as car crash simulations, weather forecasts, or computer games.

15.1 Basic Properties

We first define a general *linear space* $\mathcal{L}$. We denote its elements—the vectors—by boldface letters such as $\mathbf{u}$, $\mathbf{v}$, etc. One basic operation must be defined. It is the *linear combination* of two vectors $s\mathbf{u} + t\mathbf{v}$ with scalars s and t. With this operation in place, the defining property for a linear space is that any linear combination of vectors results in a vector in the same space. More precisely, if $\mathbf{v}_1, \mathbf{v}_2, \ldots, \mathbf{v}_n$ are in $\mathcal{L}$ then any linear combination $\mathbf{v}$ of the form

$$\mathbf{v} = s_1\mathbf{v}_1 + s_2\mathbf{v}_2 + \ldots + s_n\mathbf{v}_n \tag{15.1}$$

is also in $\mathcal{L}$. Note that all s_i may be zero, asserting that every linear space has a zero vector in it.

A set of vectors $\mathbf{v}_1, \ldots, \mathbf{v}_n$ is called *linearly independent* if it is impossible to express one of them as a linear combination of the others. For example, the equation

$$\mathbf{v}_1 = s_2\mathbf{v}_2 + s_3\mathbf{v}_3 + \ldots + s_n\mathbf{v}_n$$

will not have a solution set $s_2, \ldots, s_n$ in case the vectors $\mathbf{v}_1, \ldots, \mathbf{v}_n$ are linearly independent. As a simple consequence, the zero vector can only be expressed in a trivial manner in terms of linearly independent vectors, namely if

$$\mathbf{0} = s_1\mathbf{v}_1 + \ldots + s_n\mathbf{v}_n$$

then $s_1 = \ldots = s_n = 0$. If the zero vector *can* be expressed as a nontrivial combination of n vectors, then we say these vectors are *linearly dependent*.

If the vectors $\mathbf{v}_1, \ldots, \mathbf{v}_n$ are linearly independent, then the set of all vectors which may be expressed as a linear combination of them is said to form a *subspace* of $\mathcal{L}$ of *dimension* n. We also say this subspace is *spanned* by $\mathbf{v}_1, \ldots, \mathbf{v}_n$. If this subspace equals the whole space $\mathcal{L}$, then we call $\mathbf{v}_1, \ldots, \mathbf{v}_n$ a *basis* for $\mathcal{L}$, and the dimension of $\mathcal{L}$ is n. If we wish to emphasize that a space is of dimension n, we call it $\mathcal{L}_n$. We add that if $\mathcal{L}$ is a linear space of dimension n, then any $n+1$ vectors are linearly dependent.

Example 15.1

Let's start with one very familiar linear space, $\mathbb{R}^3$, which has dimension 3. A commonly used basis for this space is the familiar, linearly independent, $\mathbf{e}_1, \mathbf{e}_2, \mathbf{e}_3$ vectors. A linear combination of this basis, for example,

$$\mathbf{v} = \begin{bmatrix} 3 \\ 4 \\ 7 \end{bmatrix} = 3 \begin{bmatrix} 1 \\ 0 \\ 0 \end{bmatrix} + 4 \begin{bmatrix} 0 \\ 1 \\ 0 \end{bmatrix} + 7 \begin{bmatrix} 0 \\ 0 \\ 1 \end{bmatrix}$$

is also in $\mathbb{R}^3$. This vector forms a one-dimensional subspace of $\mathbb{R}^3$. If we choose another vector, say $\mathbf{w} = 2\mathbf{e}_1 + 0\mathbf{e}_2 + 0\mathbf{e}_3$, then $\mathbf{v}$ and $\mathbf{w}$ together form a two-dimensional subspace of $\mathbb{R}^3$.

For more examples of linear spaces, consult Section 15.6.

Example 15.2

In a 4D space $\mathcal{L}_4$, let three vectors be given by

$$\mathbf{v}_1 = \begin{bmatrix} -1 \\ 0 \\ 0 \\ 1 \end{bmatrix}, \mathbf{v}_2 = \begin{bmatrix} 5 \\ 0 \\ -3 \\ 1 \end{bmatrix}, \mathbf{v}_3 = \begin{bmatrix} 3 \\ 0 \\ -3 \\ 0 \end{bmatrix}.$$

These three vectors are linearly dependent since

$$\mathbf{v}_2 = \mathbf{v}_1 + 2\mathbf{v}_3 \quad \text{or} \quad \mathbf{0} = \mathbf{v}_1 - \mathbf{v}_2 + 2\mathbf{v}_3.$$

Our set $\{\mathbf{v}_1, \mathbf{v}_2, \mathbf{v}_3\}$ contains only two linearly independent vectors, hence they span a subspace of $\mathcal{L}_4$ of dimension two.

Example 15.3

In a 3D space $\mathcal{L}_3$, let four vectors be given by

$$\mathbf{v}_1 = \begin{bmatrix} -1 \\ 0 \\ 0 \end{bmatrix}, \mathbf{v}_2 = \begin{bmatrix} 1 \\ 2 \\ 0 \end{bmatrix}, \mathbf{v}_3 = \begin{bmatrix} 1 \\ 2 \\ -3 \end{bmatrix} \mathbf{v}_4 = \begin{bmatrix} 0 \\ 0 \\ -3 \end{bmatrix}.$$

These four vectors are linearly dependent since

$$\mathbf{v}_3 = -\mathbf{v}_1 + 2\mathbf{v}_2 + \mathbf{v}_4.$$

Any set of three of these vectors is a basis for $\mathcal{L}_3$.

15.2   Linear Maps

A map $\Phi : \mathcal{L} \to \mathcal{M}$ between two linear spaces $\mathcal{L}$ and $\mathcal{M}$ is called a *linear map* if it preserves linear relationships. Let three *preimage* vectors $\mathbf{v}_1, \mathbf{v}_2, \mathbf{v}_3$ be mapped to three *image* vectors $\Phi\mathbf{v}_1, \Phi\mathbf{v}_2, \Phi\mathbf{v}_3$. If there is a linear relationship among the preimages, then the same relationship will hold for the images:

$$\mathbf{v}_1 = \alpha\mathbf{v}_2 + \beta\mathbf{v}_3 \quad \text{implies} \quad \Phi\mathbf{v}_1 = \alpha\Phi\mathbf{v}_2 + \beta\Phi\mathbf{v}_3. \qquad (15.2)$$

Maps that do not have this property are called *nonlinear maps* and are much harder to deal with.

Let us now consider a set of vectors forming a basis of $\mathcal{L}$. Their images span a subspace of $\mathcal{M}$ which is called Φ's *range*, $\Phi\mathcal{L}$. If the dimension of $\mathcal{L}$ is n and if the dimension of $\Phi\mathcal{L}$ is k, then Φ has rank k. Thus, the rank indicates how much Φ reduces dimension: it is reduced from n to k. Of course $n = k$ is a possibility, in which case there is no loss of dimensionality. A linear map can never *increase* dimension. It is possible to map $\mathcal{L}$ (of dimension n) to a higher-dimensional space $\mathcal{M}$. However, the images of $\mathcal{L}$'s n basis vectors will span a subspace of $\mathcal{M}$ of dimension at most n.

Linear maps are conveniently written in terms of *matrices*. If $\mathbf{v}_1, \ldots, \mathbf{v}_n$ are vectors[1] in $\mathcal{L}$, then we may form the linear combination

$$\mathbf{v} = s_1\mathbf{v}_1 + \ldots + s_n\mathbf{v}_n.$$

Note that we do not assume that the vectors $\mathbf{v}_1, \ldots, \mathbf{v}_n$ are linearly independent. The scalars $s_1, \ldots, s_n$ may themselves be interpreted as a vector, leading to the expression

$$\mathbf{v} = \begin{bmatrix} \mathbf{v}_1 & \cdots & \mathbf{v}_n \end{bmatrix} \begin{bmatrix} s_1 \\ \vdots \\ s_n \end{bmatrix}$$

where the matrix $A = \begin{bmatrix} \mathbf{v}_1 & \cdots & \mathbf{v}_n \end{bmatrix}$ consists of the (column) vectors $\mathbf{v}_1, \ldots, \mathbf{v}_n$. We may condense this to

$$\mathbf{v} = A\mathbf{s}.$$

---

[1]More precisely, they are the coordinate columns of vectors expressed with respect to a basis of $\mathcal{L}$.

The matrix A describes a linear map Φ, taking the vector **s** to the vector **v**. It has m rows and n columns. If $m = n$, then A is said to be square. To reiterate, the matrix *describes* a linear map, and may be viewed as that map's coordinates. Sometimes it is convenient to drop the distinction between a map and its matrix form.

The matrix A has a certain rank k—how can we infer this rank from the matrix? First of all, a matrix of size $m \times n$, can be at most rank $k = min\{m,n\}$. This is called *full rank*. A matrix with rank less than this $min\{m,n\}$ is called *rank deficient*. We perform Gauss elimination (possibly with row exchanges) until the matrix is in upper triangular form. If after Gauss elimination there are k nonzero rows,[2] then the rank of A is k. Figure 15.2 gives an illustration of some possible scenarios.

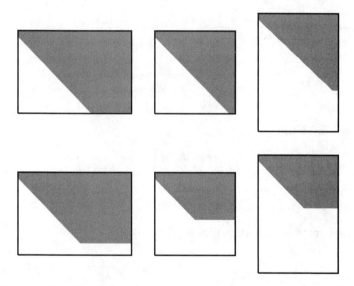

Figure 15.2.
The three types of matrices: from left to right, m < n, m = n, m > n. Examples of full rank matrices are on the top row, and examples of rank deficient matrices are on the bottom row. In each, gray indicates nonzero entries and white indicates zero entries after Gauss elimination was performed.

---

[2]This is equivalent to our earlier definition that the rank is equal to the number of linearly independent column vectors.

Example 15.4

Let us determine the rank of the matrix

$$\begin{bmatrix} 1 & 3 & 4 \\ 0 & 1 & 2 \\ 1 & 2 & 2 \\ -1 & 1 & 1 \end{bmatrix}.$$

We perform Gauss elimination to obtain

$$\begin{bmatrix} 1 & 3 & 4 \\ 0 & 1 & 2 \\ 0 & 0 & -3 \\ 0 & 0 & 0 \end{bmatrix}.$$

There is one row of zeroes, and we conclude that the matrix has rank 3.

Next, let us take the matrix

$$\begin{bmatrix} 1 & 3 & 4 \\ 0 & 1 & 2 \\ 1 & 2 & 2 \\ 0 & 1 & 2 \end{bmatrix}.$$

Gauss elimination yields

$$\begin{bmatrix} 1 & 3 & 4 \\ 0 & 1 & 2 \\ 0 & 0 & 0 \\ 0 & 0 & 0 \end{bmatrix},$$

and we conclude that this matrix has rank 2.

A square $n \times n$ matrix A of rank n is *invertible*, i.e., there is a matrix which undoes A's action. This is the inverse matrix, denoted by A^{-1}. See Section 14.5 on how to compute the inverse.

If a matrix is invertible, then it does not reduce dimension and its *determinant* is nonzero. The determinant of a square matrix measures the volume of the n-dimensional parallelepiped which is defined by its columns vectors. The determinant of a matrix is computed by subjecting its column vectors to a sequence of shears until it is of upper triangular form (Gauss elimination). The value is then the product of the diagonal elements.

15.3   Inner Products

A map from $\mathcal{L}$ to the reals $\mathbb{R}$ is called an *inner product* if it assigns a real number $\langle \mathbf{v}, \mathbf{w} \rangle$ to any pair of vectors $\mathbf{v}, \mathbf{w}$ in $\mathcal{L}$ such that:

$$\langle \mathbf{v}, \mathbf{w} \rangle = \langle \mathbf{w}, \mathbf{v} \rangle, \qquad (15.3)$$

$$\langle \alpha \mathbf{v}_1 + \beta \mathbf{v}_2, \mathbf{w} \rangle = \alpha \langle \mathbf{v}_1, \mathbf{w} \rangle + \beta \langle \mathbf{v}_2, \mathbf{w} \rangle, \qquad (15.4)$$

$$\langle \mathbf{v}, \mathbf{v} \rangle \ \geq \ 0, \qquad (15.5)$$

$$\langle \mathbf{v}, \mathbf{v} \rangle = 0 \ \text{ if } \ \text{ and } \ \text{ only } \ \text{ if } \ \mathbf{v} = \mathbf{0}. \qquad (15.6)$$

A standard example is the dot product where we write $\langle \mathbf{v}, \mathbf{w} \rangle = \mathbf{v} \cdot \mathbf{w}$.

Once an inner product is defined, it may be used to define the *length* $\|\mathbf{v}\|$ of a vector $\mathbf{v}$. It is given by

$$\|\mathbf{v}\| = \sqrt{\langle \mathbf{v}, \mathbf{v} \rangle}.$$

If two vectors $\mathbf{v}$ and $\mathbf{w}$ in a linear space $\mathcal{L}$ satisfy

$$\langle \mathbf{v}, \mathbf{w} \rangle = 0$$

then they are called *orthogonal*. If $\mathbf{v}_1, \dots, \mathbf{v}_n$ form a basis for $\mathcal{L}$ and all $\mathbf{v}_i$ are mutually orthogonal: $\langle \mathbf{v}_i, \mathbf{v}_j \rangle = 0$ for $i \neq j$, then the $\mathbf{v}_i$ are said to form an orthogonal basis. If in addition they are also of unit length: $\|\mathbf{v}_i\| = 1$, they form an *orthonormal basis*. Any basis of a linear space may be transformed to an orthonormal basis by the Gram-Schmidt process, described next.

15.4   Gram-Schmidt Orthonormalization

Let $\mathbf{b}_1, \dots, \mathbf{b}_r$ be a set of orthonormal vectors, forming the basis of an r-dimensional subspace $\mathcal{S}_r$ of $\mathcal{L}$. Let $\mathbf{u}$ be an arbitrary vector in $\mathcal{L}$, but not in $\mathcal{S}_r$. Define a vector $\hat{\mathbf{u}}$ by

$$\hat{\mathbf{u}} = \langle \mathbf{u}, \mathbf{b}_1 \rangle \mathbf{b}_1 + \dots + \langle \mathbf{u}, \mathbf{b}_r \rangle \mathbf{b}_r.$$

This vector is $\mathbf{u}$'s *projection* into $\mathcal{S}_r$. See Section 11.8 for a 3D illustration. This is seen by checking that the difference vector $\mathbf{u} - \hat{\mathbf{u}}$ is orthogonal to each of the $\mathbf{b}_i$. We first check that it is orthogonal to $\mathbf{b}_1$ and observe

$$\langle \mathbf{u} - \hat{\mathbf{u}}, \mathbf{b}_1 \rangle = \langle \mathbf{u}, \mathbf{b}_1 \rangle - \langle \mathbf{u}, \mathbf{b}_1 \rangle \langle \mathbf{b}_1, \mathbf{b}_1 \rangle + \dots + \langle \mathbf{u}, \mathbf{b}_r \rangle \langle \mathbf{b}_1, \mathbf{b}_r \rangle.$$

All terms $\langle \mathbf{b}_1, \mathbf{b}_2 \rangle$, $\langle \mathbf{b}_1, \mathbf{b}_3 \rangle$ etc. vanish since the $\mathbf{b}_i$ are orthogonal. Thus, $\langle \mathbf{u} - \hat{\mathbf{u}}, \mathbf{b}_1 \rangle = 0$. In the same manner, we show that $\mathbf{u} - \hat{\mathbf{u}}$ is orthogonal to the remaining $\mathbf{b}_i$. We now normalize $\mathbf{u} - \hat{\mathbf{u}}$, rename it $\mathbf{b}_{r+1}$ and add it to the existing $\mathbf{b}_1, \ldots, \mathbf{b}_r$. Then the set $\mathbf{b}_1, \ldots, \mathbf{b}_{r+1}$ forms an orthonormal basis for the subspace $\mathcal{S}_{r+1}$ of $\mathcal{L}$. We may repeat this process until we have found an orthonormal basis for all of $\mathcal{L}$.

This process is known as *Gram-Schmidt* orthonormalization. Given any basis $\mathbf{v}_1, \ldots, \mathbf{v}_n$ of $\mathcal{L}$, we can find an orthonormal basis by setting $\mathbf{b}_1 = \mathbf{v}_1 / \|\mathbf{v}_1\|$ and continuing to construct, one by one, vectors $\mathbf{b}_2, \ldots, \mathbf{b}_n$ using the above procedure.

15.5 Higher Dimensional Eigen Things

For any $n \times n$ matrix A we may ask if it has fixed directions, i.e., are there vectors $\mathbf{r}$ which are mapped to multiples λ of themselves by A? Such vectors are characterized by

$$A\mathbf{r} = \lambda\mathbf{r}$$

or

$$[A - \lambda I]\mathbf{r} = \mathbf{0}. \tag{15.7}$$

Since $\mathbf{r} = \mathbf{0}$ trivially performs this way, we will not consider it (the zero vector) from now on. In (15.7), we see that the matrix $[A - \lambda I]$ maps a nonzero vector $\mathbf{r}$ to the zero vector $\mathbf{0}$. Thus, its determinant must vanish (see Section 4.10):

$$\det[A - \lambda I] = 0. \tag{15.8}$$

The term $\det[A - \lambda I]$ is called the *characteristic polynomial* of A. It is a polynomial of degree n in λ, and its zeroes are A's eigenvalues.

Example 15.5

Let

$$A = \begin{bmatrix} 1 & 1 & 0 & 0 \\ 0 & 3 & 1 & 0 \\ 0 & 0 & 4 & 1 \\ 0 & 0 & 0 & 2 \end{bmatrix}.$$

We find the degree four characteristic polynomial $p(\lambda)$:

$$p(\lambda) = \det[A - \lambda I] = \begin{vmatrix} 1-\lambda & 1 & 0 & 0 \\ 0 & 3-\lambda & 1 & 0 \\ 0 & 0 & 4-\lambda & 1 \\ 0 & 0 & 0 & 2-\lambda \end{vmatrix},$$

resulting in

$$p(\lambda) = (1-\lambda)(3-\lambda)(4-\lambda)(2-\lambda).$$

The zeroes of this polynomial are found by solving $p(\lambda) = 0$. In our slightly contrived example, we find $\lambda_1 = 1, \lambda_2 = 3, \lambda_3 = 4, \lambda_4 = 2$.

The bad news is that one does not always have trivial matrices like the above to deal with. A general $n \times n$ matrix has a characteristic polynomial $p(\lambda) = \det[A - \lambda I]$ of degree n, and the eigenvalues are the zeroes of this polynomial. Finding the zeroes of an nth degree polynomial is a nontrivial numerical task.

Needless to say, not all eigenvalues of a matrix are real in general. But the important class of *symmetric* matrices always does have real eigenvalues.

Two more properties of eigenvalues:

- The matrices A and A^{T} have the same eigenvalues.

- If A has eigenvalues λ_i, then A^{-1} has eigenvalues $1/\lambda_i$.

Having found the λ_i, we can now solve linear systems

$$[A - \lambda_i I]\mathbf{r}_i = \mathbf{0}$$

in order to find the eigenvectors $\mathbf{r}_i$. These are homogeneous systems, and thus have no unique solutions. In practice, one would normalize all eigenvectors in order to eliminate this ambiguity. The situation gets trickier if some of the eigenvalues are multiple zeroes of the characteristic polynomial. We will not consider this case.

Symmetric matrices are special again. Not only do they have real eigenvalues, but their eigenvectors are orthonormal. This can be shown in exactly the same way as we did for the 2D case in Section 7.5.

Example 15.6

Let a rotation matrix be given by

$$A = \begin{bmatrix} c & -s & 0 \\ s & c & 0 \\ 0 & 0 & 1 \end{bmatrix}$$

with $c = \cos\alpha$ and $s = \sin\alpha$. It rotates around the $\mathbf{e}_3$-axis by α degrees. We should thus expect that $\mathbf{e}_3$ is an eigenvector—and indeed, one easily verifies that $A\mathbf{e}_3 = \mathbf{e}_3$. Thus, the eigenvalue, corresponding to $\mathbf{e}_3$ is 1.

15.6   A Gallery of Spaces

In this section, we highlight some special linear spaces—but there are many more!

For a first example, consider all polynomials of a fixed degree n. These are functions of the form

$$p(t) = a_0 + a_1 t + a_2 t^2 + \ldots + a_n t^n$$

where t is the independent variable of $p(t)$. It is easy to check that these polynomials have the linearity property (15.1). For example, if $p(t) = 3 - 2t + 3t^2$ and $q(t) = -1 + t + 2t^2$, then $2p(t) + 3q(t) = 3 - t + 12t^2$ is yet another polynomial of the same degree.

Thus we can construct a linear space whose elements (we will not use the term "vectors" for them) are all polynomials of a fixed degree. Addition in this space is addition of polynomials, i.e., coefficient by coefficient; multiplication is multiplication of a polynomial by a real number.

This example also serves to introduce a not-so-obvious linear map. The operation of forming derivatives turns out to be linear! The derivative p' of a degree n polynomial p is a polynomial of degree $n - 1$, given by

$$p'(t) = a_1 + 2a_2 t + \ldots + n a_n t^{n-1}.$$

We set $\Phi p = p'$ and simply check that the linearity condition (15.2) is met! The linear map of forming derivatives thus maps the space of all degree n polynomials into that of all degree $n - 1$ polynomials. The rank of this map is thus $n - 1$.

Example 15.7

Let us consider the two cubic polynomials

$$p(t) = 3 - t + 2t^2 + 3t^3 \quad \text{and} \quad q(t) = 1 + t - t^3.$$

Let

$$r(t) = 2p(t) - q(t) = 5 - 3t + 4t^2 + 7t^3.$$

Now

$$r'(t) = -3 + 8t + 21t^2, \qquad (15.9)$$
$$p'(t) = -1 + 4t + 9t^2, \qquad (15.10)$$
$$q'(t) = 1 - 3t^2. \qquad (15.11)$$

It is now trivial to check that $r'(t) = 2p'(t) - q'(t)$, thus asserting the linearity of the derivative map.

Another linear space is given by the set of all real-valued continuous functions over the interval $[0, 1]$. This space is typically named $C[0, 1]$. Clearly the linearity condition is met: if f and g are elements of $C[0, 1]$, then $\alpha f + \beta g$ is also in $C[0, 1]$. Here we have an example of a linear space which is infinite-dimensional, meaning that no finite set of functions forms a basis for $C[0, 1]$.

For a third example, consider the set of all 3×3 matrices. They form a linear space; this space consists of "vectors" which are matrices. In this space, linear combinations are formed using standard matrix addition and multiplication with a scalar.

And, finally, a more abstract example. The set of all linear maps from a linear space $\mathcal{L}$ into the reals (known as *linear functionals*) forms a linear space itself. It is called the *dual space* $\mathcal{L}^*$ of $\mathcal{L}$. Its dimension equals that of $\mathcal{L}$. Here is an example of linear functionals. Fix some vector $\mathbf{v}$ in $\mathcal{L}$. Then the functionals defined by $\Phi_\mathbf{v}(\mathbf{u}) = \langle \mathbf{u}, \mathbf{v} \rangle$ are in $\mathcal{L}^*$. If $\mathbf{b}_1, \ldots, \mathbf{b}_n$ is an orthonormal basis of $\mathcal{L}$, we can define linear functionals $\Phi_1, \ldots, \Phi_n$ by the property

$$\Phi_i(\mathbf{b}_j) = \langle \mathbf{b}_i \mathbf{b}_j \rangle.$$

These functionals form a basis for $\mathcal{L}^*$.

- dimension
- linear combination
- image
- preimage
- rank
- full rank
- rank deficient
- inverse
- determinant
- subspace
- span
- linear space
- dual space

- linear functional
- functional
- inner product
- range
- Gram-Schmidt orthonormalization
- projection
- linearity
- basis
- linear independence
- orthonormal
- characteristic polynomial

15.7 Exercises

1. Let $p = p_0 + p_1 t + p_2 t^2$ and $q = q_0 + q_1 t + q_2 t^2$ be two quadratic polynomials. Define

$$\langle p, q \rangle = p_0 q_0 + p_1 q_1 + p_2 q_2.$$

   Is this an inner product for the space of all quadratic polynomials?

2. Define an inner product on the space of all 3×3 matrices.

3. Does the set of all polynomials with leading coefficient $a_n = 1$ form a linear space?

4. Find a basis for the linear space formed by all 2×2 matrices.

5. What is the dimension of the linear space formed by all $n \times n$ matrices?

6. Does the set of all 3D vectors with nonnegative components form a subspace of $\mathbb{R}^3$?

7. What is the rank of the matrix

$$\begin{bmatrix} 1 & 2 & 0 \\ -1 & -2 & 1 \\ 0 & 0 & 1 \\ 2 & 4 & -1 \end{bmatrix}?$$

8. What is the rank of the matrix

$$\begin{bmatrix} 1 & 2 & 0 & 0 & 0 \\ -1 & 0 & 0 & -2 & 1 \\ 0 & 0 & 1 & 0 & 1 \end{bmatrix}?$$

9. Let $\mathcal{L}$ be a linear space. Is the map $\Phi(\mathbf{u}) = \|\mathbf{u}\|$ an element of the dual space $\mathcal{L}^*$?

10. Does the set of all monotonically increasing functions over $[0, 1]$ form a linear space?

11. Find the eigenvalues of the matrix

$$\begin{bmatrix} 0 & 0 & 0 & 2 \\ 0 & 0 & 1 & 0 \\ 0 & 1 & 0 & 0 \\ 2 & 0 & 0 & 0 \end{bmatrix}.$$

16

Numerical Methods

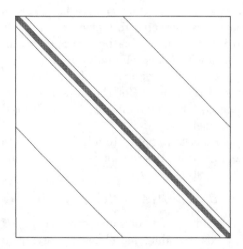

Figure 16.1.
A sparse matrix: all nonzero entries are marked.

We have seen methods for solving linear systems or for finding eigenvalues. In practical applications, these methods may not be very efficient. For example, it is not numerically feasible to compute the zeroes of a 1000th-degree polynomial, as one would have to do for finding the eigenvalues of a 1000×1000 matrix. For such cases, sophisticated numerical methods have been developed; we highlight

a few of them in this chapter. A *sparse matrix* is one that has many
zero entries. An example of such a matrix is illustrated in Figure 16.1.
Efficient handling of matrices such as this one is another motivation
for developing special numerical methods.

16.1 Another Linear System Solver:
The Householder Method

Let's revisit the problem of solving the linear system $A\mathbf{u} = \mathbf{b}$, where
A is an $n \times n$ matrix. We may think of A as n column vectors, each
with n elements,

$$[\mathbf{a}_1 \ldots \mathbf{a}_n]\mathbf{u} = \mathbf{b}.$$

In Section 14.2 we examined the classical method of Gauss elimina-
tion, or the process of applying shears G_i to the vectors $\mathbf{a}_1 \ldots \mathbf{a}_n$ and
$\mathbf{b}$ in order to convert A to upper triangular form, or

$$G_{n-1} \ldots G_1 A\mathbf{u} = G_{n-1} \ldots G_1 \mathbf{b},$$

and then we were able to solve for $\mathbf{u}$ with back substitution. Each
G_i is a shear matrix, constructed to transform the ith column vector
$G_{i-1} \ldots G_1 \mathbf{a}_i$ to a vector with zeroes below the diagonal element, $a_{i,i}$.

A serious problem with Gauss elimination, even with modifications
such as pivoting, is that the application of the shears can change the
condition number of the matrix (for the worse). A more numerically
stable method may be found by replacing shears with *reflections*. This
is the Householder method.

The Householder method applied to a linear system takes the same
form as Gauss elimination. A series of reflections H_i are constructed
and applied to the system,

$$H_{n-1} \ldots H_1 A\mathbf{u} = H_{n-1} \ldots H_1 \mathbf{b},$$

where each H_i transforms the column vector $H_{i-1} \ldots H_1 \mathbf{a}_i$ to a vector with zeroes below the diagonal element. Let's examine how we
construct a *Householder transformation H_i*.

A simple 2×2 matrix

$$\begin{bmatrix} 1 & -2 \\ 1 & 0 \end{bmatrix}$$

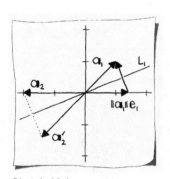

Sketch 16.1.
Householder reflection of
vector $\mathbf{a}_1$.

and Sketch 16.1 will help to illustrate the construction of a Householder transformation. The first transformation, $H_1 A$ reflects $\mathbf{a}_1$ onto
the $\mathbf{e}_1$ axis, to the vector $\mathbf{a}_1' = \|\mathbf{a}_1\|\mathbf{e}_1$, or

$$\begin{bmatrix} 1 \\ 1 \end{bmatrix} \rightarrow \begin{bmatrix} \sqrt{2} \\ 0 \end{bmatrix}.$$

We will reflect about the line L_1 illustrated in Sketch 16.1, so we must construct the normal $\mathbf{n}_1$ to this line:

$$\mathbf{n}_1 = \frac{\mathbf{a}_1 - \|\mathbf{a}_1\|\mathbf{e}_1}{\|\mathbf{a}_1 - \|\mathbf{a}_1\|\mathbf{e}_1\|}, \qquad (16.1)$$

which is simply the normalized difference between the original vector and the target vector (after the reflection). The implicit equation of the line L_1 is

$$\mathbf{n}_1^T\mathbf{x} = 0,$$

and $\mathbf{n}_1^T\mathbf{a}_1$ is the distance of the point $(\mathbf{o} + \mathbf{a}_1)$ to L_1. Therefore, the reflection constitutes moving twice the distance to the plane, in the direction of the normal, so

$$\mathbf{a}_1' = \mathbf{a}_1 - \left[2\mathbf{n}_1^T\mathbf{a}_1\right]\mathbf{n}_1. \qquad (16.2)$$

To write this reflection in matrix form, we rearrange the terms from (16.2),

$$\mathbf{a}_1' = \mathbf{a}_1 - \mathbf{n}_1\left[2\mathbf{n}_1^T\mathbf{a}_1\right]$$
$$= \left(I - \mathbf{n}_1 2\mathbf{n}_1^T\right)\mathbf{a}_1$$

Notice that $\mathbf{n}_1 2\mathbf{n}_1^T$ is a dyadic matrix, as introduced in Section 11.5. We now have the matrix H_1 defining one Householder transformation:

$$H_1 = I - \mathbf{n}_1 2\mathbf{n}_1^T. \qquad (16.3)$$

This is precisely the reflection we constructed in (11.11)!

The 2×2 example only serves to illustrate the underlying geometry of a reflection matrix. The construction of a general Householder transformation H_i is a little more complicated. Suppose now that we have the following matrix

$$A = \begin{bmatrix} a_{1,1} & a_{1,2} & a_{1,3} & a_{1,4} \\ 0 & a_{2,2} & a_{2,3} & a_{2,4} \\ 0 & 0 & a_{3,3} & a_{3,4} \\ 0 & 0 & a_{4,3} & a_{4,4} \end{bmatrix}.$$

We need to construct H_3 to zero the element $a_{4,3}$ while preserving A's upper triangular nature. In other words, we want to preserve the elimination done by previous transformations H_1 and H_2. To achieve this construct

$$\bar{\mathbf{a}}_3 = \begin{bmatrix} 0 \\ 0 \\ a_{3,3} \\ a_{4,3} \end{bmatrix},$$

or in general,

$$\bar{\mathbf{a}}_i = \begin{bmatrix} 0 \\ \vdots \\ 0 \\ a_{i,i} \\ \vdots \\ a_{n,i} \end{bmatrix}.$$

Then

$$\mathbf{n}_i = \frac{\bar{\mathbf{a}}_i - ||\bar{\mathbf{a}}_i||\mathbf{e}_i}{||\bar{\mathbf{a}}_i - ||\bar{\mathbf{a}}_i||\mathbf{e}_i||}, \tag{16.4}$$

results in a Householder transformation

$$H_i = I - 2\mathbf{n}_i\mathbf{n}_i^{\mathrm{T}}. \tag{16.5}$$

For any n-vector $\mathbf{v}$, $H_i\mathbf{v}$ and $\mathbf{v}$ coincide in the first $i-1$ components, and if the components of $\mathbf{v}$ from i to n are zero, they will remain so. Properties of H_i include being:

- symmetric: $H_i = H_i^{\mathrm{T}}$,

- involutary: $H_iH_i = I$ and thus $H_i = H_i^{-1}$,

- unitary (orthogonal): $H_i^{\mathrm{T}}H_i = I$, and thus $H_i\mathbf{v}$ has the same length as $\mathbf{v}$.

Implementation of Householder transformations doesn't involve explicit construction and multiplication by the Householder matrix in (16.5). A numerically and computationally more efficient algorithm is easy to achieve since we know quite a bit about how each H_i acts on the column vectors. Additionally, if $\mathbf{a}_i$ is nearly parallel to $\mathbf{e}_i$ then numerical problems can creep into our construction due to loss of *significant digits* in subtraction of nearly equal numbers. Therefore, it is better to reflect onto the direction of the $\mathbf{e}_i$-axis which represents the largest reflection. That is the purpose of the factor γ in the algorithm below. For a more detailed discussion, see [2] or [14].

In the Householder algorithm that follows, as we work on the jth column vector, we use

$$\bar{\mathbf{a}}_k = \begin{bmatrix} a_{j,k} \\ \vdots \\ a_{n,k} \end{bmatrix}$$

to indicate that only elements $j, \ldots, n$ of the kth column vector $\mathbf{a}_k$ (with $k \geq j$) are involved in a calculation.

Householder's Method

Algorithm:

Input:
    $n \times m$ matrix A, where $n \geq m$ and rank of A is m;
    n vector $\mathbf{b}$, augmented to A as the $(m+1)^{st}$ column.
Output:
    Upper triangular matrix HA written over A;
    $H\mathbf{b}$ written over $\mathbf{b}$ in the augmented $(m+1)^{st}$
        column of A.

If $n = m$ then $p = n - 1$; Else $p = m$
For $j = 1, 2, \ldots, p$
    $a = \bar{\mathbf{a}}_j \cdot \bar{\mathbf{a}}_j$
    $\gamma = -\text{sign}(a_{j,j})\sqrt{a}$
    $\alpha = a - a_{j,j}\gamma$
    Temporarily set $a_{j,j} = a_{j,j} - \gamma$
    For $k = j + 1, \ldots, m + 1$
        $s = \frac{1}{\alpha}(\bar{\mathbf{a}}_j \cdot \bar{\mathbf{a}}_k)$
        $\bar{\mathbf{a}}_k = \bar{\mathbf{a}}_k - s\bar{\mathbf{a}}_j$
    Set $\bar{\mathbf{a}}_j = \begin{bmatrix} \gamma & 0 & \cdots & 0 \end{bmatrix}$

Example 16.1

Let's apply the Householder algorithm to the linear system

$$\begin{bmatrix} 1 & 1 & 0 \\ 1 & -1 & 0 \\ 0 & 0 & 1 \end{bmatrix} \mathbf{u} = \begin{bmatrix} -1 \\ 0 \\ 1 \end{bmatrix}.$$

For $j = 1$ in the algorithm, we calculate the following values: $\gamma = -\sqrt{2}$, $\alpha = 2 + \sqrt{2}$, and then we temporarily set

$$\bar{\mathbf{a}}_1 = \begin{bmatrix} 1 + \sqrt{2} \\ 1 \\ 0 \end{bmatrix}.$$

It's a little tricky, but notice for $k = 2$, $s = \sqrt{2} - 1$. This results in

$$\bar{\mathbf{a}}_2 = \begin{bmatrix} 0 \\ -\sqrt{2} \\ 0 \end{bmatrix}.$$

For $k = 3$, $s = 0$, and $\bar{\mathbf{a}}_3$ remains unchanged. For $k = 4$, $s = -\sqrt{2}/2$, and then the right-hand side vector becomes

$$\bar{\mathbf{a}}_4 = \begin{bmatrix} \sqrt{2}/2 \\ \sqrt{2}/2 \\ 0 \end{bmatrix}.$$

Now we set $\bar{\mathbf{a}}_1$, and the reflection H_1 results in the linear system

$$\begin{bmatrix} -\sqrt{2} & 0 & 0 \\ 0 & -\sqrt{2} & 0 \\ 0 & 0 & 1 \end{bmatrix} \mathbf{u} = \begin{bmatrix} \sqrt{2}/2 \\ \sqrt{2}/2 \\ 1 \end{bmatrix}.$$

Notice that $\mathbf{a}_3$ was not effected because we were reflecting in the $\mathbf{e}_1, \mathbf{e}_2$-plane, and the length of each column vector was not changed. Since the matrix is upper triangular, we may now use back substitution to find the solution vector

$$\mathbf{u} = \begin{bmatrix} -1/2 \\ -1/2 \\ 1 \end{bmatrix}.$$

Householder's algorithm is computationally more involved than Gauss elimination. It is the method of choice, however, if one is dealing with ill-conditioned systems as they arise, for example, in least squares problems. The algorithm above is set-up for such overdetermined problems, as discussed in Section 14.4: The input matrix A is of dimension $n \times m$ where $n \geq m$. The following example illustrates that the Householder method will result in the least squares solution to an overdetermined system.

Example 16.2

Let's revisit a least squares line fitting problem from Example 14.7. See that example for a problem description, and see Figure 14.2 for an illustration. The overdetermined linear system for this problem is

$$\begin{bmatrix} -2 & 1 \\ 0 & 1 \\ 3 & 1 \\ 0 & 1 \end{bmatrix} \begin{bmatrix} a \\ b \end{bmatrix} = \begin{bmatrix} -2 \\ 0 \\ 0 \\ 1 \end{bmatrix}.$$

After the first Householder reflection ($j = 1$), the linear system becomes

$$\begin{bmatrix} 3.605 & 0.278 \\ 0 & 1 \\ 0 & 1.387 \\ 0 & 1 \end{bmatrix} \begin{bmatrix} a \\ b \end{bmatrix} = \begin{bmatrix} 1.103 \\ 0 \\ -1.662 \\ 1 \end{bmatrix}.$$

For the second Householder reflection ($j = 2$), we see the application of the modified column vectors, $\bar{a}_i$. For example, the calculation of a takes the form

$$a = \begin{bmatrix} 1 \\ 1.387 \\ 1 \end{bmatrix} \cdot \begin{bmatrix} 1 \\ 1.387 \\ 1 \end{bmatrix},$$

thus the dot product is applied to elements $a_{2,2}, a_{3,2}, a_{4,2}$ only. This reflection results in the linear system

$$\begin{bmatrix} 3.605 & 0.278 \\ 0 & -1.981 \\ 0 & 0 \\ 0 & 0 \end{bmatrix} \begin{bmatrix} a \\ b \end{bmatrix} = \begin{bmatrix} 1.103 \\ 0.659 \\ -1.362 \\ 1.221 \end{bmatrix}.$$

Notice that the second reflection did not modify the elements $a_{1,2}$ or $a_{1,3}$ (on the right-hand side).

We can now solve the system with back substitution, and the solution is

$$\begin{bmatrix} a \\ b \end{bmatrix} = \begin{bmatrix} 0.33 \\ -0.33 \end{bmatrix}.$$

Excluding numerical round-off, this is the same solution found using the normal equations in Example 14.7.

16.2   Vector Norms and Sequences

You are all familiar with sequences of real numbers such as

$$1, \frac{1}{2}, \frac{1}{4}, \frac{1}{8}, \dots$$

or

$$1, 2, 4, 8, \dots$$

The first of these has the *limit* 0 while the second one does not have a limit. If you are not familiar with the concept of a limit, you should

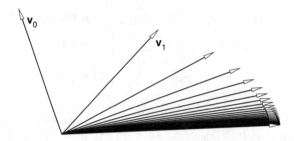

Figure 16.2.

Vector sequences: a sequence which converges.

now consult your favorite calculus text. One way of saying a sequence of real numbers a_i has a limit a is that beyond some index i, all a_i differ from the limit a by an arbitrarily small amount ϵ.

Vector sequences

$$\mathbf{v}^{(0)}, \mathbf{v}^{(1)}, \mathbf{v}^{(2)}, \ldots$$

are not all that different. Here we say the sequence has a limit $\mathbf{v}$ if from some index i on, the distance of any $\mathbf{v}^{(i)}$ from $\mathbf{v}$ is smaller than an arbitrarily small amount ϵ. By "distance" of two vectors, we are referring to the usual *Euclidean norm*: if $\mathbf{w} = \mathbf{a} - \mathbf{b}$, then the length or magnitude of $\mathbf{w}$ is given by

$$\|\mathbf{w}\| = \sqrt{\mathbf{w} \cdot \mathbf{w}}.$$

See Figure 16.2 for an example. The Euclidean norm is also known as the L^2 *norm*, and sometimes it is written $\|\mathbf{w}\|_2$.

Different measures for vectors exist, for example the *Manhattan norm* or L^1 *norm*

$$\|\mathbf{w}\|_1 = |w_1| + \ldots + |w_n|.$$

We could define other norms as well; (nonnegative) vector norms must satisfy the following criteria:

- $\|\mathbf{w}\| > 0$ when $\mathbf{w} \neq \mathbf{0}$,

- $\|\mathbf{w}\| = 0$ when $\mathbf{w} = \mathbf{0}$,

- $\|c\mathbf{w}\| = |c|\|\mathbf{w}\|$ for a scalar c,

- $\|\mathbf{v} + \mathbf{w}\| \leq \|\mathbf{v}\| + \|\mathbf{w}\|$ (triangle inequality).

For our purposes, the Euclidean norm will do, so unless otherwise noted, this is the norm of choice, and the subscript 2 will be omitted.

Example 16.3

Let a vector sequence be given by

$$\mathbf{v}^{(i)} = \begin{bmatrix} 1/i \\ 1/i^2 \\ 1/i^3 \end{bmatrix}.$$

This sequence has the limit

$$\mathbf{v} = \begin{bmatrix} 0 \\ 0 \\ 0 \end{bmatrix}.$$

Now take the sequence

$$\mathbf{v}^{(i)} = \begin{bmatrix} i \\ 1/i^2 \\ 1/i^3 \end{bmatrix}.$$

It does not have a limit: even though the last two components each have a limit, the first component diverges.

16.3   Iterative System Solvers: Gauss-Jacobi and Gauss-Seidel

In applications such as *Finite Element Methods* (*FEM*) in the context of the solution of fluid flow problems, scientists are faced with linear systems with many thousands of equations. Gauss elimination would work, but would be far too slow. Typically, huge linear systems have one advantage: the coefficient matrix is *sparse* meaning it only has very few (such as ten) nonzero entries per row. Thus, a 100,000 × 100,000 system would only have 1,000,000 nonzero entries, compared to 10,000,000,000 matrix elements! In these cases, one does not store the whole matrix, but only its nonzero entries, together with their i, j location. An example of a sparse matrix is shown in Figure 16.1. The solution to such systems is typically obtained by *iterative methods* which we will discuss next.

Example 16.4

Let the system be given by

$$\begin{bmatrix} 4 & 1 & 0 \\ 2 & 5 & 1 \\ -1 & 2 & 4 \end{bmatrix} \begin{bmatrix} u_1 \\ u_2 \\ u_3 \end{bmatrix} = \begin{bmatrix} 1 \\ 0 \\ 3 \end{bmatrix}.$$

(This example was taken from Johnson and Riess [14].) An iterative method starts from a guess for the solution and then refines it until it *is* the solution. Let's take

$$\mathbf{u}^{(1)} = \begin{bmatrix} u_1^{(1)} \\ u_2^{(1)} \\ u_3^{(1)} \end{bmatrix} = \begin{bmatrix} 1 \\ 1 \\ 1 \end{bmatrix}$$

for our first guess, and note that it clearly is not the solution to our system: $A\mathbf{u}^{(1)} \neq \mathbf{b}$.

A better guess ought to be obtained by using the current guess and solving the first equation for a new $u_1^{(2)}$, the second for a new $u_2^{(2)}$, and so on. This gives us

$$4u_1^{(2)} + 1 = 1$$
$$2 + 5u_2^{(2)} + 1 = 0$$
$$-1 + 2 + 4u_3^{(2)} = 3$$

and thus

$$\mathbf{u}^{(2)} = \begin{bmatrix} 0 \\ -0.6 \\ 0.5 \end{bmatrix}.$$

The next iteration becomes

$$4u_1^{(3)} - 0.6 = 1$$
$$5u_2^{(3)} + 0.5 = 0$$
$$-1.2 + 4u_3^{(3)} = 3$$

and thus

$$\mathbf{u}^{(3)} = \begin{bmatrix} 0.4 \\ -0.1 \\ 1.05 \end{bmatrix}.$$

After a few more iterations, we will be close enough to the true solution

$$\mathbf{u} = \begin{bmatrix} 0.333 \\ -0.333 \\ 1.0 \end{bmatrix}.$$

Try one more iteration for yourself.

This iterative method is known as *Gauss-Jacobi iteration*. Let us now formulate this process for the general case. We are given a linear system with n equations and n unknowns u_i, written in matrix form as

$$A\mathbf{u} = \mathbf{b}.$$

Let us also assume that we have an initial guess $\mathbf{u}^{(1)}$ for the solution vector $\mathbf{u}$.

We now define two matrices D and R as follows: D is the diagonal matrix whose diagonal elements are those of A and R is the matrix obtained from A by setting all its diagonal elements to zero. Clearly then

$$A = D + R$$

and our linear system becomes

$$D\mathbf{u} + R\mathbf{u} = \mathbf{b}$$

or

$$\mathbf{u} = D^{-1}[\mathbf{b} - R\mathbf{u}].$$

In the spirit of our previous development, we now write this as

$$\mathbf{u}^{(k+1)} = D^{-1}\left[\mathbf{b} - R\mathbf{u}^{(k)}\right],$$

meaning that we attempt to compute a new estimate $\mathbf{u}^{(k+1)}$ from an existing one $\mathbf{u}^{(k)}$. Note that D must not contain zeroes on the diagonal; this can be achieved by row or column interchanges.

Example 16.5

With this new framework, let us reconsider our last example. We have

$$A = \begin{bmatrix} 4 & 1 & 0 \\ 2 & 5 & 1 \\ -1 & 2 & 4 \end{bmatrix}, \quad R = \begin{bmatrix} 0 & 1 & 0 \\ 2 & 0 & 1 \\ -1 & 2 & 0 \end{bmatrix}, \quad D^{-1} = \begin{bmatrix} 0.25 & 0 & 0 \\ 0 & 0.2 & 0 \\ 0 & 0 & 0.25 \end{bmatrix}.$$

Then

$$\mathbf{u}^{(2)} = \begin{bmatrix} 0.25 & 0 & 0 \\ 0 & 0.2 & 0 \\ 0 & 0 & 0.25 \end{bmatrix} \left(\begin{bmatrix} 1 \\ 0 \\ 3 \end{bmatrix} - \begin{bmatrix} 0 & 1 & 0 \\ 2 & 0 & 1 \\ -1 & 2 & 0 \end{bmatrix} \begin{bmatrix} 1 \\ 1 \\ 1 \end{bmatrix} \right) = \begin{bmatrix} 0 \\ -0.6 \\ 0.5 \end{bmatrix}$$

Will the Gauss-Jacobi method succeed, i.e., will the sequence of vectors $\mathbf{u}^{(k)}$ converge? The answer is: sometimes yes, and sometimes no. It will *always* succeed if A is diagonally dominant,[1] and then it will succeed no matter what our initial guess $\mathbf{u}^{(1)}$ was. Many practical problems result in diagonally dominant systems.

In a practical setting, how do we determine if convergence is taking place? Ideally, we would like $\mathbf{u}^{(k)} = \mathbf{u}$, the true solution, after a number of iterations. Equality will most likely not happen, but the length of the *residual vector*

$$\|A\mathbf{u}^{(k)} - \mathbf{b}\|$$

should become small (i.e., less than some preset tolerance). Thus, we check the size of the residual vector after each iteration, and stop once it is smaller than some preset tolerance.

A modification of the Gauss-Jacobi method is known as *Gauss-Seidel* iteration. When we compute $\mathbf{u}^{(k+1)}$ in the Gauss-Jacobi method, we can observe the following: the second element, $u_2^{(k+1)}$, is computed using $u_1^{(k)}, u_3^{(k)}, \ldots, u_n^{(k)}$. We had just computed $u_1^{(k+1)}$. It stands to reason that using it instead of $u_1^{(k)}$ would be advantageous. This idea gives rise to the Gauss-Seidel method: as soon as a new element $u_i^{(k+1)}$ is computed, the estimate vector $\mathbf{u}^{(k+1)}$ is updated.

In summary, Gauss-Jacobi updates the new estimate vector once all of its elements are computed, Gauss-Seidel updates as soon as a new element is computed. Typically, Gauss-Seidel iteration converges faster than Gauss-Jacobi iteration.

16.4 Finding Eigenvalues: the Power Method

Let A be a symmetric $n \times n$ matrix.[2] Further, let the eigenvalues of A be ordered such that $|\lambda_1| \geq |\lambda_2| \geq \ldots \geq |\lambda_n|$. Then λ_1 is called

[1] Recall that a matrix is diagonally dominant if for every row, the absolute value of its diagonal element is larger that the sum of the absolute values of its remaining elements.

[2] The method discussed in this section may be extended to nonsymmetric matrices, but since those eigenvalues may be complex, we will avoid them here.

the *dominant eigenvalue* of A. To simplify the notation to come, refer to this dominant eigenvalue as λ, and let $\mathbf{r}$ be its corresponding eigenvector. In Section 7.6, we considered repeated applications of a matrix; we restricted ourselves to the 2D case. We encountered an equation of the form

$$A^i \mathbf{r} = \lambda^i \mathbf{r} \qquad (16.6)$$

which clearly holds for matrices of arbitrary size. This equation may be interpreted as follows: if A has an eigenvalue λ, then A^i has an eigenvalue λ^i.

This property may be used to *find* the dominant eigenvalue. Consider the vector sequence

$$\mathbf{r}^{(i+1)} = A\mathbf{r}^{(i)}; \qquad i = 1, 2, \ldots$$

where $\mathbf{r}^{(1)}$ is an arbitrary (nonzero) vector.

After a sufficiently large i, the $\mathbf{r}^{(i)}$ will begin to line up with $\mathbf{r}$, as illustrated in the two left-most examples in Figure 16.3. Here is how to utilize that fact for finding λ: for sufficiently large i, we will approximately have

$$\mathbf{r}^{(i+1)} = \lambda \mathbf{r}^{(i)}.$$

This means that all components of $\mathbf{r}^{(i+1)}$ and $\mathbf{r}^{(i)}$ are (approximately) related by

$$\frac{r_j^{(i+1)}}{r_j^{(i)}} = \lambda \quad \text{for } j = 1, \ldots, n. \qquad (16.7)$$

Hence, all we need to do is to increase i until all ratios in (16.7) approach a common value of λ.

Some remarks on using this method:

- If $|\lambda|$ is either "large" or "close" to zero, the $\mathbf{r}^{(i)}$ will either become unbounded or approach zero in length, respectively. This has the potential to cause numerical problems. It is prudent, therefore, to *normalize* the $\mathbf{r}^{(i)}$ every few steps. A reasonable number here might be five, so that after every fifth iteration we start over with a unit vector.

- Convergence seems impossible if $\mathbf{r}^{(1)}$ is perpendicular to $\mathbf{r}$, the eigenvector of λ. Theoretically, no convergence will kick in, but for once numerical roundoff is on our side: after a few iterations, $\mathbf{r}^{(i)}$ will not be perpendicular to $\mathbf{r}$ and we will converge—if slowly!

- Very slow convergence will also be observed if $|\lambda_1| \approx |\lambda_2|$.

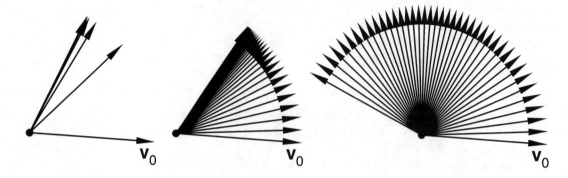

Figure 16.3.
The power method: three examples whose matrices are given in Example 16.6.

- The power method as described here is limited to symmetric matrices with one dominant eigenvalue. It may be generalized to more cases, but for the purpose of this exposition, we decided to outline the principle rather than to cover all details. For more of those, see [14].

Example 16.6

Figure 16.3 illustrates three cases, A_1, A_2, A_3, from left to right. The three matrices and their eignevalues are as follows:

$$A_1 = \begin{bmatrix} 2 & 1 \\ 1 & 2 \end{bmatrix}, \qquad \lambda_1 = 3, \qquad \lambda_2 = 1,$$

$$A_2 = \begin{bmatrix} 2 & 0.1 \\ 0.1 & 2 \end{bmatrix}, \quad \lambda_1 = 2.1, \qquad \lambda_2 = 1.9,$$

$$A_3 = \begin{bmatrix} 2 & -.1 \\ 0.1 & 2 \end{bmatrix}, \quad \lambda_1 = 2 + 0.1i, \quad \lambda_2 = 2 - 0.1i,$$

and for all of them, we used

$$\mathbf{r}^{(1)} = \begin{bmatrix} 1.5 \\ -0.1 \end{bmatrix}.$$

In all three examples, the vectors $\mathbf{r}^{(i)}$ were normalized for display. The first matrix, A_1 is symmetric and the dominant eigenvalue is reasonably separated from λ_2, hence the rapid convergence. The second matrix, A_2, is also symmetric, however λ_1 is close in value to λ_2, hence the slow convergence. The third matrix has complex eigenvalues, hence no convergence.

A more realistic numerical example is next.

Example 16.7

Let a matrix A be given by

$$A = \begin{bmatrix} 0 & 1 & 1 \\ 1 & 1 & 0 \\ 1 & 0 & 1 \end{bmatrix}.$$

We start with an initial vector

$$\mathbf{r}^{(1)} = \begin{bmatrix} 1 & 0 & 0 \end{bmatrix}.$$

Iterations 10 and 11 yield

$$\mathbf{r}^{10} = \begin{bmatrix} 342 \\ 341 \\ 341 \end{bmatrix}, \qquad \mathbf{r}^{11} = \begin{bmatrix} 682 \\ 683 \\ 683 \end{bmatrix},$$

from which we obtain the three ratios 2.0, 2.003, 2.003. The true dominant eigenvalue is $\lambda = 2$.

- reflection matrix
- Householder method
- overdetermined system
- symmetric matrix
- involutary matrix
- unitary matrix
- orthogonal matrix
- significant digits
- vector sequence
- vector norm
- convergence
- Euclidean norm
- L^2 norm
- Manhattan norm
- iterative method
- sparse matrix
- Gauss-Jacobi method
- Gauss-Seidel method
- residual vector
- dominant eigenvalue
- power method

16.5   Exercises

1. Let a linear system be given by

$$A = \begin{bmatrix} 4 & 0 & -1 \\ 2 & 8 & 2 \\ 1 & 0 & 2 \end{bmatrix} \quad \text{and} \quad \mathbf{b} = \begin{bmatrix} 2 \\ -2 \\ 0 \end{bmatrix}.$$

Carry out three iterations of the Gauss-Jacobi iteration starting with an initial guess

$$\mathbf{u}^{(1)} = \begin{bmatrix} 0 & 0 & 0 \end{bmatrix}.$$

2. Let a linear system be given by

$$A = \begin{bmatrix} 4 & 0 & 1 \\ 2 & -8 & 2 \\ 1 & 0 & 2 \end{bmatrix} \quad \text{and} \quad \mathbf{b} = \begin{bmatrix} 0 \\ 2 \\ 0 \end{bmatrix}.$$

Carry out three iterations of the Gauss-Jacobi iteration starting with an initial guess

$$\mathbf{u}^{(1)} = \begin{bmatrix} 0 & 0 & 0 \end{bmatrix}.$$

3. Let A be the matrix

$$A = \begin{bmatrix} 4 & 0 & 1 \\ 0 & 1 & 2 \\ 1 & 2 & 2 \end{bmatrix}.$$

Starting with the vector

$$\mathbf{r}^{(1)} = \begin{bmatrix} 1 & 1 & 1 \end{bmatrix}$$

carry out three steps of the power method to find A's dominant eigenvalue.

4. Let A be the matrix

$$A = \begin{bmatrix} -8 & 0 & 8 \\ 0 & 1 & -2 \\ 8 & -2 & 0 \end{bmatrix}.$$

Starting with the vector

$$\mathbf{r}^{(1)} = \begin{bmatrix} 1 & 1 & 1 \end{bmatrix}$$

carry out three steps of the power method to find A's dominant eigenvalue.

5. Carry out three Gauss-Jacobi iterations for the linear system

$$A = \begin{bmatrix} 4 & 1 & 0 & 1 \\ 1 & 4 & 1 & 0 \\ 0 & 1 & 4 & 1 \\ 1 & 0 & 1 & 4 \end{bmatrix} \quad \text{and} \quad \mathbf{b} = \begin{bmatrix} 0 \\ 1 \\ 1 \\ 0 \end{bmatrix},$$

starting with the initial guess

$$\mathbf{u}^{(1)} = \begin{bmatrix} 1 \\ 1 \\ 1 \\ 1 \end{bmatrix}.$$

6. Define a vector sequence by

$$\mathbf{u}^i = \begin{bmatrix} 0 & 0 & 1/i \\ 0 & 1 & 0 \\ -1/i & 0 & 0 \end{bmatrix} \begin{bmatrix} 1 \\ 1 \\ 1 \end{bmatrix}; \qquad i = 1, 2, \ldots.$$

Does it have a limit? If so, what is it?

7. Let a vector sequence be given by

$$\mathbf{u}^{(1)} = \begin{bmatrix} 1 \\ -1 \end{bmatrix} \quad \text{and} \quad \mathbf{u}^{(i+1)} = \begin{bmatrix} 1/2 & 1/4 \\ 0 & 3/4 \end{bmatrix} \mathbf{u}^{(i)}.$$

Will this sequence converge? If so, to what vector?

8. Carry out three iterations of Gauss-Seidel for Example 16.4. Which method, Gauss-Jacobi or Gauss-Seidel, is converging to the solution faster? Why?

9. If you graph all 2D vectors of unit length with respect to the Euclidean norm, the outline of the set forms a circle. What form is the outline for the Manhattan norm?

10. The infinity norm is defined as

$$\|\mathbf{w}\|_\infty = \max_i |w_i|.$$

What is the outline of the graph of all 2D vectors of unit length in this norm?

11. Examining the L^1 and L^2 norms defined in Section 16.2, how would you define an L^3 norm?

12. Use the Householder method to solve the following linear system

$$\begin{bmatrix} 1 & 1.1 & 1.1 \\ 1 & 0.9 & 0.9 \\ 0 & -0.1 & 0.2 \end{bmatrix} \mathbf{u} = \begin{bmatrix} 1 \\ 1 \\ 0.3 \end{bmatrix}.$$

Notice that the columns are almost linearly dependent.

17

Putting Lines Together:
Polylines and Polygons

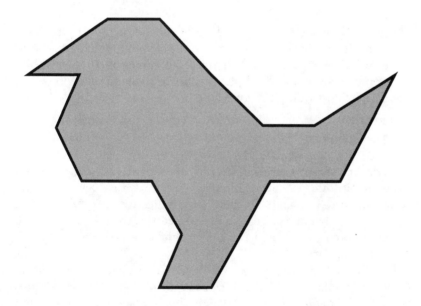

Figure 17.1.

Polygon: straight line segments forming a bird shape.

Figure 17.1 shows a *polygon*. It is the outline of a shape, drawn with straight line segments. Since such shapes are all a printer or plotter can draw, just about every computer-generated drawing consists of

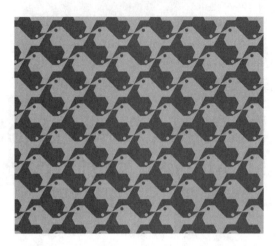

Figure 17.2.
Mixing maps: a pattern is created by composing rotations and translations.

polygons. If we add an "eye" to the bird-shaped polygon from Figure 17.1, and if we apply a sequence of rotations and translations to it, then we arrive at Figure 17.2. It turns out copies of our special bird polygon can cover the whole plane! This technique is also present in the Escher illustration in Figure 6.8.

17.1 Polylines

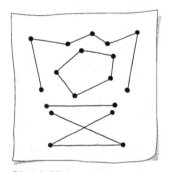

Sketch 17.1.
2D polyline examples.

Straight line segments, called *edges*, connecting an ordered set of *vertices* constitute a *polyline*. The first and last vertices are not necessarily connected. Some 2D examples are illustrated in Sketch 17.1, however, a polyline can be 3D, too. Since the vertices are ordered, the edges are oriented and can be thought of as vectors. Let's call them *edge vectors*.

Polylines are a primary output primitive, and thus they are included in graphics standards. One example of a graphics standard is the *GKS (Graphical Kernel System)*; this is a specification of what belongs in a graphics package. The development of Postscript was based on GKS. Polylines have many uses in computer graphics and modeling. Whether in 2D or 3D, they are typically used to outline a shape. The power of polylines to reveal a shape is illustrated in Figure 17.3 in the display of a 3D surface. The surface is *evaluated* in an organized fashion so that the points can be logically connected as

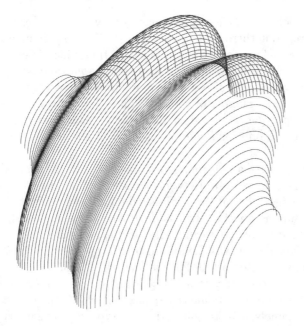

Figure 17.3.
Polylines: the display of a 3D surface.

polylines, giving the observer a feeling of the "flow" of the surface. In modeling, a polyline is often used to approximate a complex curve or data, which in turn makes analysis easier and less costly. An example of this is illustrated in Figure 14.2.

17.2 Polygons

When the first and last vertices of a polyline are connected, it is called a *polygon*. Normally, a polygon is thought to enclose an area. For this reason, unless a remark is made, we will consider planar polygons only. Just as with polylines, polygons constitute an ordered set of vertices and we will continue to use the term edge vectors. Thus, a polygon with n edges is given by an ordered set of 2D points

$$\mathbf{p}_1, \mathbf{p}_2, \ldots, \mathbf{p}_n$$

and has edge vectors $\mathbf{v}_i = \mathbf{p}_{i+1} - \mathbf{p}_i; i = 1, \ldots, n$. Note that the edge vectors sum to zero![1]

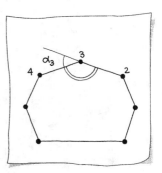

Sketch 17.2.
Interior and exterior angles.

---

[1]To be precise, they sum to the *zero vector*.

Sketch 17.3.

A minmax box is a polygon.

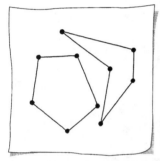

Sketch 17.4.

Convex (left) and nonconvex (right) polygons.

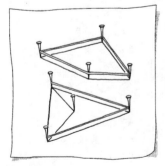

Sketch 17.5.

Rubberband test for convexity of a polygon.

If you look at the edge vectors carefully, you'll discover that $\mathbf{v}_n = \mathbf{p}_{n+1} - \mathbf{p}_n$, but there is no vertex $\mathbf{p}_{n+1}$! This apparent problem is resolved by defining $\mathbf{p}_{n+1} = \mathbf{p}_1$, a convention which is called *cyclic numbering*. We'll use this convention throughout, and will not mention it every time. We also add one more, topological, characterization of polygons: the number of vertices equals the number of edges.

Since a polygon is closed, it divides the plane into two parts: a finite part, the polygon's *interior*, and an infinite part, the polygon's *exterior*.

As you traverse a polygon, you follow the path determined by the vertices and edge vectors. Between vertices, you'll move along straight lines (the edges), but at the vertices, you'll have to perform a rotation before resuming another straight line path. The angle α_i by which you rotate at vertex $\mathbf{p}_i$ is called the *turning angle* or *exterior angle* at $\mathbf{p}_i$. The *interior angle* is then given by $\pi - \alpha_i$ (see Sketch 17.2).

Polygons are used a lot! For instance, in Chapter 1 we discussed the *extents* of geometry in a 2D coordinate system. Another name for these extents is a *minmax box* (see Sketch 17.3). It is a special polygon, namely a rectangle. Another type of polygon studied in Chapter 8 is the triangle. This type of polygon is often used to define a *polygonal mesh* of a 3D model. The triangles may then be filled with color to produce a shaded image. A polygonal mesh from triangles is illustrated in Figures 8.3 and 8.4, and one from rectangles is illustrated in Figure 10.3.

17.3 Convexity

Polygons are commonly classified by their shape. There are many ways of doing this. One important classification is as *convex* or *nonconvex*. The latter is also referred to as *concave*. Sketch 17.4 gives an example of each. How do you describe the shape of a convex polygon? As in Sketch 17.5, stick nails into the paper at the vertices. Now take a rubberband and stretch it around the nails, then let go. If the rubberband shape follows the outline of the polygon, it is convex. Another definition: take any two points in the polygon (including on the edges) and connect them with a straight line. If the line never leaves the polygon, then it is convex. This must work for *all* possible pairs of points!

The issue of convexity is important, because algorithms which involve polygons can be simplified if the polygons are known to be convex. This is true for algorithms for the problem of *polygon clipping*. This problem starts with two polygons, and the goal is to find the intersection of the polygon areas. Some examples are illustrated in Sketch 17.6. The intersection area is defined in terms of one or more polygons. If both polygons are convex, then the result will be just one convex polygon. However, if even one polygon is not convex then the result might be two or more, possibly disjoint or nonconvex, polygons. Thus, nonconvex polygons need more record keeping in order to properly define the intersection area(s). Not all algorithms are designed for nonconvex polygons. See [11] for a detailed description of clipping algorithms.

An n-sided convex polygon has a sum of interior angles I equal to

$$I = (n-2)\pi. \tag{17.1}$$

To see this, take one polygon vertex and form triangles with the other vertices, as illustrated in Sketch 17.7. This forms $n-2$ triangles. The sum of interior angles of a triangle is known to be π. Thus, we get the above result.

The sum of the exterior angles of a convex polygon is easily found with this result. Each interior and exterior angle sums to π. Suppose the ith interior angle is α_i radians, then the exterior angle is $\pi - \alpha_i$ radians. Sum over all angles, and the exterior angle sum E is

$$E = n\pi - (n-2)\pi = 2\pi. \tag{17.2}$$

To test if an n-sided polygon is convex, we'll use the *barycenter* of the vertices $\mathbf{p}_1, \ldots, \mathbf{p}_n$. The barycenter $\mathbf{b}$ is a special barycentric combination (see Sections 8.1 and 8.3):

$$\mathbf{b} = \frac{1}{n}(\mathbf{p}_1 + \ldots + \mathbf{p}_n).$$

It is the center of gravity of the vertices. We need to construct the implicit line equation for each edge vector in a consistent manner. If the polygon is convex, then the point $\mathbf{b}$ will be on the "same" side of every line. The implicit equation will result in all positive or all negative values. (This will not work for some unusual, nonsimple, polygons. See Section 17.5 and the Exercises.)

Another test for convexity is to check if there is a *re-entrant angle*. This is an interior angle which is greater than π.

17.4 Types of Polygons

There are a variety of special polygons. First, we introduce two terms to help describe these polygons:

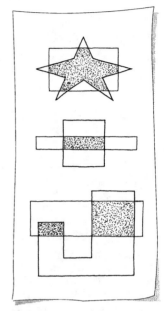

Sketch 17.6.
Polygon clipping.

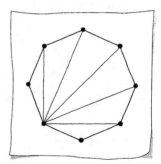

Sketch 17.7.
Sum of interior angles using triangles.

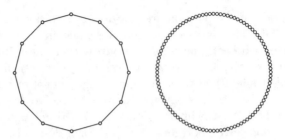

Figure 17.4.
Circle approximation: using an n-gon to represent a circle.

Sketch 17.8.
Regular polygons.

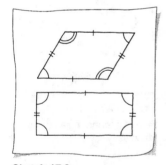

Sketch 17.9.
Rhombus and rectangle.

- *equilateral* means that all sides are of equal length, and

- *equiangular* means that all *interior angles* at the vertices are equal.

In the following illustrations, edges with the same number of tick marks are of equal length and angles with the same number of arc markings are equal.

A very special polygon is the *regular polygon*: it is equilateral and equiangular. Examples are illustrated in Sketch 17.8. This polygon is also referred to as an *n-gon*, indicating it has n edges. We list the names of the "classical" n-gons:

- a 3-gon is an equilateral triangle,

- a 4-gon is a square,

- a 5-gon is a pentagon,

- a 6-gon is a hexagon, and

- an 8-gon is an octagon

An n-gon is commonly used to approximate a circle in computer graphics, as illustrated in Figure 17.4.

A *rhombus* is equilateral but not equiangular, whereas a *rectangle* is equiangular but not equilateral. These are illustrated in Sketch 17.9.

17.5   Unusual Polygons

Most often, applications deal with *simple* polygons, as opposed to *nonsimple* polygons. A nonsimple polygon, as illustrated in Sketch

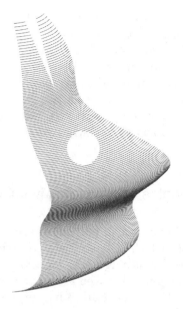

Figure 17.5.
Trimmed surface: an application of polygons with holes.

17.10, is characterized by edges intersecting other than at the vertices. Topology is the reason nonsimple polygons can cause havoc in some algorithms. For convex and nonconvex simple polygons, as you traverse along the boundary of the polygon, the interior remains on one side. This is not the case for nonsimple polygons. At the mid-edge intersections, the interior switches sides. In more concrete terms, recall how the implicit line equation could be used to determine if a point is on the line, and more generally, which side it is on. Suppose you have developed an algorithm that associates the + side of the line with being inside the polygon. This rule will work fine if the polygon is simple, but not otherwise.

Sometimes nonsimple polygons can arise due to an error. The polygon clipping algorithms, as discussed in Section 17.3, involve sorting vertices to form the final polygon. If this sorting goes haywire, then you could end up with a nonsimple polygon rather than a simple one as desired.

In applications, it is not uncommon to encounter polygons with holes. Such a polygon is illustrated in Sketch 17.11. As you see, this is actually more than one polygon. An example of this, illustrated in Figure 17.5, is a special CAD/CAM surface called a *trimmed surface*.

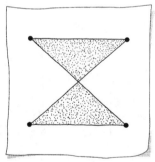

Sketch 17.10.
Nonsimple polygon.

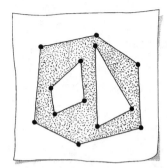

Sketch 17.11.
Polygon with holes.

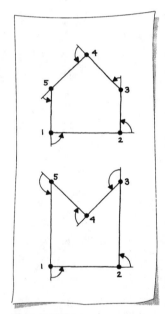

Sketch 17.12.
Turning angles.

The polygons define parts of the material to be cut or punched out. This allows other parts to fit to this one.

For trimmed surfaces and other CAD/CAM applications, a certain convention is accepted in order to make more sense out of this multi-polygon geometry. The polygons must be oriented a special way. The *visible region*, or the region that is not cut out, is to the "left." As a result, the outer boundary is oriented counterclockwise and the inner boundaries are oriented clockwise. More on the visible region in Section 17.9.

17.6 Turning Angles and Winding Numbers

The *turning angle* of a polygon or polyline is essentially another name for the *exterior angle*, which is illustrated in Sketch 17.2. Notice that this sketch illustrates the turning angles for a convex and a nonconvex polygon. Here the difference between a turning angle and an exterior angle is illustrated. The turning angle has an orientation as well as an angle measure. All turning angles for a convex polygon have the same orientation, which is not the case for a nonconvex polygon. This fact will allow us to easily differentiate the two types of polygons.

Here is an application of the turning angle. Suppose for now that a given polygon is 2D and lives in the $[\mathbf{e}_1, \mathbf{e}_2]$-plane. Its n vertices are labeled

$$\mathbf{p}_1, \mathbf{p}_2, \cdots \mathbf{p}_n.$$

We want to know if the polygon is convex. We only need to look at the orientation of the turning angles, not the actual angles. First, let's embed the 2D vectors in 3D by adding a zero third coordinate, for example:

$$\mathbf{p}_1 = \begin{bmatrix} p_{1,1} \\ p_{2,1} \\ 0 \end{bmatrix}.$$

Recall that the cross product of two vectors in the $\mathbf{e}_1, \mathbf{e}_2$-plane will produce a vector which points "in" or "out," that is in the $+\mathbf{e}_3$ or $-\mathbf{e}_3$ direction. Therefore, by taking the cross product of successive edge vectors

$$\mathbf{u}_i = (\mathbf{p}_{i+1} - \mathbf{p}_i) \wedge (\mathbf{p}_{i+2} - \mathbf{p}_{i+1}), \qquad (17.3)$$

we'll encounter $\mathbf{u}_i$ of the form

$$\begin{bmatrix} 0 \\ 0 \\ u_{3,i} \end{bmatrix}.$$

If the sign of the $u_{3,i}$ value is the *same* for all angles, then the polygon is convex. A mathematical way to describe this is by using the scalar triple product (see Section 10.6). The turning angle orientation is determined by the scalar

$$u_{3,i} = \mathbf{e}_3 \cdot ((\mathbf{p}_{i+1} - \mathbf{p}_i) \wedge (\mathbf{p}_{i+2} - \mathbf{p}_{i+1})).$$

Notice that the sign is dependent upon the traversal direction of the polygon, but only a change of sign is important. The determinant of the 2D vectors would have worked just as well, but the 3D approach is more useful for what follows.

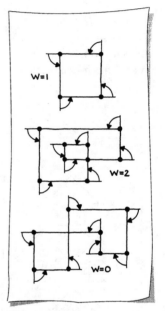

    If the polygon lies in an arbitrary plane, having a normal $\mathbf{n}$, then the above convex/concave test is changed only a bit. The cross product in (17.3) produces a vector $\mathbf{u}_i$ that has direction $\pm\mathbf{n}$. Now we need the dot product, $\mathbf{n} \cdot \mathbf{u}_i$ to extract a signed scalar value.

    If we actually computed the turning angle at each vertex, we could form an accumulated value called the *total turning angle*. Recall from (17.2) that the total turning angle for a convex polygon is 2π. For a polygon that is not known to be convex, assign a sign using the scalar triple product as above, to each angle measurement. The sum E will then be used to compute the *winding number* of the polygon. The winding number W is

$$W = \frac{E}{2\pi}.$$

Thus, for a convex polygon, the winding number is one. Sketch 17.13 illustrates a few examples. A non-self-intersecting polygon is essentially one loop. A polygon can have more than one loop, with different orientation: clockwise versus counterclockwise. The winding number gets decremented for each clockwise loop and incremented for each counterclockwise loop, or vice versa depending on how you assign signs to your angles.

Sketch 17.13.
Winding numbers.

17.7 Area

A simple method for calculating the area of a 2D polygon is to use the *signed area* of a triangle as in Section 4.10. First, *triangulate* the polygon. For example, choose one vertex of the polygon and form all triangles from it and successive pairs of vertices, as is illustrated in Sketch 17.14. The sum of the signed areas of the triangles results in the area of the polygon. For this method to work, we must form the

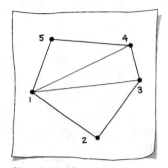

triangles with a consistent orientation. For example, in Sketch 17.14, triangles $(\mathbf{p}_1, \mathbf{p}_2, \mathbf{p}_3)$, $(\mathbf{p}_1, \mathbf{p}_3, \mathbf{p}_4)$, and $(\mathbf{p}_1, \mathbf{p}_4, \mathbf{p}_5)$ are all counterclockwise or right-handed, and therefore have positive area. More precisely, if we form $\mathbf{v}_i = \mathbf{p}_i - \mathbf{p}_1$, then the area of the polygon in Sketch 17.14 is

$$A = \frac{1}{2}(\det[\mathbf{v}_2, \mathbf{v}_3] + \det[\mathbf{v}_3, \mathbf{v}_4] + \det[\mathbf{v}_4, \mathbf{v}_5]).$$

In general, if a polygon has n vertices, then this area calculation becomes

$$A = \frac{1}{2}(\det[\mathbf{v}_2, \mathbf{v}_3] + \ldots + \det[\mathbf{v}_{n-1}, \mathbf{v}_n]). \tag{17.4}$$

The use of signed area makes this idea work for non-convex polygons as in Sketch 17.15. As illustrated in the sketch, the negative areas cancel duplicate and extraneous areas.

Interestingly, (17.4) takes an interesting form if its terms are expanded. We observe that the determinants that represent edges of triangles within the polygon cancel. So this leaves us with

$$A = \frac{1}{2}(\det[\mathbf{p}_1, \mathbf{p}_2] + \ldots + \det[\mathbf{p}_{n-1}, \mathbf{p}_n] + \det[\mathbf{p}_n, \mathbf{p}_1]). \tag{17.5}$$

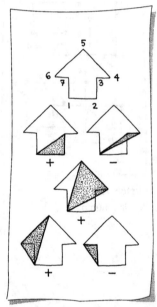

Equation (17.5) seems to have lost all geometric meaning because it involves the determinant of point pairs, but we can recapture geometric meaning if we consider each point to be $\mathbf{p}_i - \mathbf{o}$.

Is (17.4) or (17.5) the preferred form? The amount of computation for each equation is similar; however, there is one drawback of (17.5). If the polygon is far from the origin then numerical problems can occur because the vectors $\mathbf{p}_i$ and $\mathbf{p}_{i+1}$ will be close to parallel. The form in (17.4) essentially builds a local frame in which to compute the area. For debugging and making sense of intermediate computations, (17.4) is easier to work with. This is a nice example of how reducing an equation to its "simplest" form is not always "optimal"!

An interesting observation is that (17.5) may be written as a *generalized determinant*. The coordinates of the vertices are

$$\mathbf{p}_i = \begin{bmatrix} p_{1,i} \\ p_{2,i} \end{bmatrix}.$$

The area is computed as follows,

$$A = \frac{1}{2}\begin{vmatrix} p_{1,1} & p_{1,2} & \cdots & p_{1,n} & p_{1,1} \\ p_{2,1} & p_{2,2} & \cdots & p_{2,n} & p_{2,1} \end{vmatrix},$$

which is computed by adding the products of all "downward" diagonals, and subtracting the products of all "upward" diagonals.

Example 17.1

Let

$$\mathbf{p}_1 = \begin{bmatrix} 0 \\ 0 \end{bmatrix}, \quad \mathbf{p}_2 = \begin{bmatrix} 1 \\ 0 \end{bmatrix}, \quad \mathbf{p}_3 = \begin{bmatrix} 1 \\ 1 \end{bmatrix}, \quad \mathbf{p}_4 = \begin{bmatrix} 0 \\ 1 \end{bmatrix}.$$

We have

$$A = \frac{1}{2} \begin{vmatrix} 0 & 1 & 1 & 0 & 0 \\ 0 & 0 & 1 & 1 & 0 \end{vmatrix} = \frac{1}{2}[0 + 1 + 1 + 0 - 0 - 0 - 0 - 0] = 1.$$

Since our polygon was a square, this is as expected.

But now take

$$\mathbf{p}_1 = \begin{bmatrix} 0 \\ 0 \end{bmatrix}, \quad \mathbf{p}_2 = \begin{bmatrix} 1 \\ 1 \end{bmatrix}, \quad \mathbf{p}_3 = \begin{bmatrix} 0 \\ 1 \end{bmatrix}, \quad \mathbf{p}_4 = \begin{bmatrix} 1 \\ 0 \end{bmatrix}.$$

This is a nonsimple polygon! Its area computes to

$$A = \frac{1}{2} \begin{vmatrix} 0 & 1 & 0 & 1 & 0 \\ 0 & 1 & 1 & 1 & 0 \end{vmatrix} = \frac{1}{2}[0 + 1 + 0 + 0 - 0 - 0 - 1 - 0] = 0.$$

Draw a sketch and convince yourself this is correct!

Planar polygons are sometimes specified by 3D points. We can adjust the area formula (17.4) accordingly. Recall that the cross product of two vectors results in a vector whose length equals the area of the parallelogram spanned by the two vectors. So we will replace each determinant by a cross product

$$\mathbf{u}_i = \mathbf{v}_i \wedge \mathbf{v}_{i+1} \qquad \text{for} \quad i = 2, n - 1 \qquad (17.6)$$

and as before, each $\mathbf{v}_i = \mathbf{p}_i - \mathbf{p}_1$. Suppose the (unit) normal to the polygon is $\mathbf{n}$. Notice that $\mathbf{n}$ and all $\mathbf{u}_i$ share the same direction, therefore $\|\mathbf{u}\| = \mathbf{u} \cdot \mathbf{n}$. Now we can rewrite (17.5) for 3D points,

$$A = \frac{1}{2}\mathbf{n} \cdot (\mathbf{u}_2 + \ldots + \mathbf{u}_{n-1}). \qquad (17.7)$$

with the $\mathbf{u}_i$ defined in (17.6). Notice that (17.7) is a sum of scalar triple products, which were introduced in Section 10.6.

Example 17.2

Take the four coplanar 3D points

$$\mathbf{p}_1 = \begin{bmatrix} 0 \\ 2 \\ 0 \end{bmatrix}, \quad \mathbf{p}_2 = \begin{bmatrix} 2 \\ 0 \\ 0 \end{bmatrix}, \quad \mathbf{p}_3 = \begin{bmatrix} 2 \\ 0 \\ 3 \end{bmatrix}, \quad \mathbf{p}_4 = \begin{bmatrix} 0 \\ 2 \\ 3 \end{bmatrix}.$$

Compute the area with (17.7), and note that the normal

$$\mathbf{n} = \begin{bmatrix} -1/\sqrt{2} \\ -1/\sqrt{2} \\ 0 \end{bmatrix}.$$

First compute

$$\mathbf{v}_2 = \begin{bmatrix} 2 \\ -2 \\ 0 \end{bmatrix} \mathbf{v}_3 = \begin{bmatrix} 2 \\ -2 \\ 3 \end{bmatrix} \mathbf{v}_4 = \begin{bmatrix} 0 \\ 0 \\ 3 \end{bmatrix},$$

and then compute the cross products:

$$\mathbf{u}_2 = \mathbf{v}_2 \wedge \mathbf{v}_3 = \begin{bmatrix} -6 \\ -6 \\ 0 \end{bmatrix} \quad \text{and} \quad \mathbf{u}_3 = \mathbf{v}_3 \wedge \mathbf{v}_4 = \begin{bmatrix} -6 \\ -6 \\ 0 \end{bmatrix}.$$

Then the area is

$$A = \frac{1}{2}\mathbf{n} \cdot (\mathbf{u}_2 + \mathbf{u}_3) = 6\sqrt{2}.$$

(The equality $\sqrt{2}/2 = 1/\sqrt{2}$ was used to eliminate the denominator.) You may also realize that our simple example polygon is just a rectangle, and so you have a another way to check the area!

In Section 10.5, we looked at calculating normals to a polygonal mesh specifically for computer graphics lighting models. The results of this section are useful for this task as well. By removing the dot product with the normal, (17.7) provides us with a method of computing a good average normal to a non-planar polygon:

$$\mathbf{n} = \frac{(\mathbf{u}_2 + \mathbf{u}_2 + \ldots + \mathbf{u}_{n-2})}{\|\mathbf{u}_2 + \mathbf{u}_2 + \ldots + \mathbf{u}_{n-2}\|}.$$

This normal estimation method is a weighted average based on the areas of the triangles. To eliminate this weighting, normalize each $\mathbf{u}_i$ before summing them.

Example 17.3

What is an estimate normal to the non-planar polygon

$$\mathbf{p}_1 = \begin{bmatrix} 0 \\ 0 \\ 1 \end{bmatrix}, \quad \mathbf{p}_2 = \begin{bmatrix} 1 \\ 0 \\ 0 \end{bmatrix}, \quad \mathbf{p}_3 = \begin{bmatrix} 1 \\ 1 \\ 1 \end{bmatrix}, \quad \mathbf{p}_4 = \begin{bmatrix} 0 \\ 1 \\ 0 \end{bmatrix}?$$

Calculate

$$\mathbf{u}_2 = \begin{bmatrix} 1 \\ 0 \\ 1 \end{bmatrix}, \quad \mathbf{u}_3 = \begin{bmatrix} 0 \\ 1 \\ 1 \end{bmatrix},$$

and the normal is

$$\mathbf{n} = \frac{1}{\sqrt{6}} \begin{bmatrix} 1 \\ 1 \\ 2 \end{bmatrix}.$$

17.8 Planarity Test

Suppose someone sends you a CAD file that contains a polygon. For your application, the polygon must be 2D; however, it is oriented arbitrarily in 3D. How do you verify that the data points are *coplanar*?

There are many ways to solve this problem, although some solutions have clear advantages over the others. Some considerations when comparing algorithms include:

- numerical stability,

- speed,

- ability to define a meaningful tolerance,

- size of data set, and

- maintainability of the algorithm.

The order of importance is arguable.

Let's look at three possible methods to solve this planarity test and then compare them.

- **Volume test:** Choose the first polygon vertex as a base point. Form vectors to the next three vertices. Use the scalar triple product to calculate the volume spanned by these three vectors. If it is less than a given tolerance, then the four points are coplanar. Continue for all other sets.

- **Plane test:** Construct the plane through the first three vertices. Check if all of the other vertices lie in this plane, within a given tolerance.

- **Average normal test:** Find the centroid $\mathbf{c}$ of all points. Compute all normals $\mathbf{n}_i = [\mathbf{p}_i - \mathbf{c}] \wedge [\mathbf{p}_{i+1} - \mathbf{c}]$. Check if all angles formed by two subsequent normals are below a given angle tolerance.

If we compare these three methods, we see that they employ different kinds of tolerances: for volumes, distances, and angles. Which of these is preferable must depend on the application at hand. Clearly the plane test is the fastest of the three; yet it has a problem if the first three vertices are close to being collinear.

17.9 Inside or Outside?

Another important concept for 2D polygons is the *inside/outside test* or *visibility test*. The problem is this: Given a polygon in the $[\mathbf{e}_1, \mathbf{e}_2]$-plane and a point $\mathbf{p}$, determine if the point lies inside the polygon.

One obvious application for this is polygon fill for raster device software, e.g., as with PostScript. Each pixel must be checked to see if it is in the polygon and should be colored. The inside/outside test is also encountered in CAD with *trimmed surfaces*, which were introduced in Section 17.5. With both applications it is not uncommon to have a polygon with one or more holes, as illustrated in Sketch 17.11. In the PostScript fill application, nonsimple polygons are not unusual either.[2]

We will present two similar algorithms, producing different results in some special cases. However, we choose to solve this problem, we will want to incorporate a *trivial reject* test. This simply means that if a point is "obviously" not in the polygon, then we output that result immediately, i.e., with a minimal amount of calculation. In this problem, trivial reject refers to constructing a *minmax box* around the polygon. If a point lies outside of this minmax box then it may be

---

[2]As an example, take Figure 3.5. The lines inside the bounding rectangle are the edges of a nonsimple polygon.

trivially rejected. As you see, this involves simple comparison of $\mathbf{e}_1$- and $\mathbf{e}_2$-coordinates.

17.9.1   Even-Odd Rule

From a point $\mathbf{p}$, construct a line in parametric form with vector $\mathbf{r}$ in any direction. The parametric line is

$$\mathbf{l}(t) = \mathbf{p} + t\mathbf{r}.$$

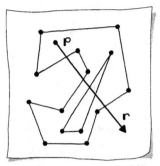

This is illustrated in Sketch 17.16. Count the number of intersections this line has with the polygon edges for $t \geq 0$ only. This is why the vector is sometimes referred to as a *ray*. The number of intersection will be odd if $\mathbf{p}$ is inside and even if $\mathbf{p}$ is outside. Figure 17.6 illustrates the results of this rule with the polygon fill application.

It can happen that $\mathbf{l}(t)$ coincides with an edge of the polygon or passes through a vertex. Either a more elaborate counting scheme must be developed, or you can choose a different $\mathbf{r}$. As a rule, it is better to not choose $\mathbf{r}$ parallel to the $\mathbf{e}_1$- or $\mathbf{e}_2$-axis, because the polygons often have edges parallel to these axes.

Sketch 17.16.
Even-Odd rule.

Figure 17.6.
Even-Odd Rule: applied to polygon fill.

17.9.2   Nonzero Winding Number

In Section 17.6 the winding number was introduced. Here is another use for it. This method proceeds similarly to the even-odd rule. Construct a parametric line at a point **p** and intersect the polygon edges. Again, only consider those intersections for $t \geq 0$. The counting method depends on the orientation of the polygon edges. Start with a winding number of zero. Following Sketch 17.17, if a polygon edge is oriented "right to left" then add one to the winding number. If a polygon edge is oriented "left to right" then subtract one from the winding number. If the final result is zero then the point is outside the polygon. Figure 17.7 illustrates the results of this rule with the same polygons used in the even-odd rule. As with the previous rule, if you encounter edges head-on, then choose a different ray.

The differences in the algorithms are interesting. The PostScript language uses the nonzero winding number rule as the default. The authors (of the PostScript language) feel that this produces better results for the polygon fill application, but the even-odd rule is available with a special command. PostScript must deal with the most general

Sketch 17.17.
Nonzero winding number rule.

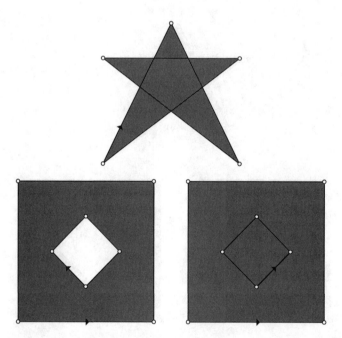

Figure 17.7.
Nonzero Winding Rule: applied to polygon fill.

(and crazy) polygons. In the trimmed surface application, the polygons must be simple and polygons cannot intersect; therefore, either algorithm is suitable.

If you happen to know that you are dealing only with *convex polygons*, another inside/outside test is available. Check which side of the edges the point **p** is on. If it is on the same side for all edges, then **p** is inside the polygon. All you have to do is to compute all determinants of the form

$$\left| (\mathbf{p}_i - \mathbf{p}) \quad (\mathbf{p}_{i+1} - \mathbf{p}) \right|.$$

If they are all of the same sign, **p** is inside the polygon.

- polygon
- polyline
- cyclic numbering
- turning angle
- exterior angle
- interior angle
- polygonal mesh
- convex
- concave
- polygon clipping
- sum of interior angles
- sum of exterior angles
- re-entrant angle
- equilateral polygon
- equiangular polygon
- regular polygon
- n-gon
- rhombus
- simple polygon
- trimmed surface
- visible region
- total turning angle
- winding number
- polygon area
- planarity test
- trivial reject
- inside/outside test
- scalar triple product

17.10 Exercises

1. What is the sum of the interior angles of a six-sided polygon? What is the sum of the exterior angles?

2. What type of polygon is equiangular and equilateral?

3. Which polygon is equilateral but not equiangular?

4. Develop an algorithm which determines whether or not a polygon is simple.

5. Calculate the winding number of the polygon with the following vertices.

$$\mathbf{p}_1 = \begin{bmatrix} 0 \\ 0 \end{bmatrix} \quad \mathbf{p}_2 = \begin{bmatrix} -2 \\ 0 \end{bmatrix} \quad \mathbf{p}_3 = \begin{bmatrix} -2 \\ 2 \end{bmatrix}$$

$$\mathbf{p}_4 = \begin{bmatrix} 0 \\ 2 \end{bmatrix} \quad \mathbf{p}_5 = \begin{bmatrix} 2 \\ 2 \end{bmatrix} \quad \mathbf{p}_6 = \begin{bmatrix} 2 \\ -2 \end{bmatrix}$$

$$\mathbf{p}_7 = \begin{bmatrix} 3 \\ -2 \end{bmatrix} \quad \mathbf{p}_8 = \begin{bmatrix} 3 \\ -1 \end{bmatrix} \quad \mathbf{p}_9 = \begin{bmatrix} 0 \\ -1 \end{bmatrix}$$

6. Compute the area of the polygon with the following vertices.

$$\mathbf{p}_1 = \begin{bmatrix} -1 \\ 0 \end{bmatrix} \quad \mathbf{p}_2 = \begin{bmatrix} 0 \\ 1 \end{bmatrix} \quad \mathbf{p}_3 = \begin{bmatrix} 1 \\ 0 \end{bmatrix}$$

$$\mathbf{p}_4 = \begin{bmatrix} 1 \\ 2 \end{bmatrix} \quad \mathbf{p}_5 = \begin{bmatrix} -1 \\ 2 \end{bmatrix}$$

Use both methods from Section 17.7.

7. Give an example of a nonsimple polygon that will pass the test for convexity, which uses the barycenter from Section 17.3.

8. Find an estimate normal to the non-planar polygon

$$\mathbf{p}_1 = \begin{bmatrix} 0 \\ 0 \\ 1 \end{bmatrix}, \quad \mathbf{p}_2 = \begin{bmatrix} 1 \\ 0 \\ 0 \end{bmatrix}, \quad \mathbf{p}_3 = \begin{bmatrix} 1 \\ 1 \\ 1 \end{bmatrix}, \quad \mathbf{p}_4 = \begin{bmatrix} 0 \\ 1 \\ 0 \end{bmatrix}.$$

9. All points but one of the following making the vertices of a polygon lie in a plane:

$$\mathbf{p}_1 = \begin{bmatrix} 2 \\ 0 \\ -2 \end{bmatrix}, \quad \mathbf{p}_2 = \begin{bmatrix} 3 \\ 2 \\ -4 \end{bmatrix}, \quad \mathbf{p}_3 = \begin{bmatrix} 2 \\ 4 \\ -4 \end{bmatrix},$$

$$\mathbf{p}_4 = \begin{bmatrix} 0 \\ 3 \\ -1 \end{bmatrix}, \quad \mathbf{p}_5 = \begin{bmatrix} 0 \\ 2 \\ -1 \end{bmatrix}, \quad \mathbf{p}_6 = \begin{bmatrix} 0 \\ 0 \\ 0 \end{bmatrix}.$$

Which point is the *outlier*?[3] Which planarity test is the most suited to this problem?

---

[3]This term is frequently used to refer to noisy, inaccurate data from a laser scanner.

18

Curves

Figure 18.1.
Car design: curves are used to design cars such as the the Ford Synergy 2010 concept car. (Source http://www.ford.com.)

Earlier in this book, we mentioned that all letters that you see here were designed by a font designer, and then put into a font library. The font designer's main tool is a cubic curve, also called a cubic Bézier curve. Such curves are handy for font design, but they were initially invented for car design. This happened in France in the early 1960s at Rénault and Citroën in Paris. These techniques are still in use today, as illustrated in Figure 18.1. We will briefly outline this kind of curve, and also apply previous geometric concepts to the study of curves in general. This type of work is called *Geometric Modeling* or *Computer Aided Geometric Design*, see and introductory text such as [7]. Please keep in mind: this chapter just scratches the surface of the modeling field!

18.1   Application: Parametric Curves

You will recall that one way to write a straight line is the *parametric* form:

$$\mathbf{x}(t) = (1-t)\mathbf{a} + t\mathbf{b}.$$

If we interpret t as time, then this says at time $t = 0$ a moving point is at $\mathbf{a}$. It moves towards $\mathbf{b}$, and reaches it at time $t = 1$. You might have observed that the coefficients $(1-t)$ and t are linear, or degree 1, polynomials, which explains another name for this: *linear interpolation*. So we have a simple example of a *parametric curve*: a curve that can be written as

$$\mathbf{x}(t) = \begin{bmatrix} f(t) \\ g(t) \end{bmatrix},$$

where $f(t)$ and $g(t)$ are functions of the parameter t. For the linear interpolant above, $f(t) = (1-t)a_1 + tb_1$ and $g(t) = (1-t)a_2 + tb_2$ In general, f and g can be any functions, e.g., polynomial, trigonometric, or exponential. However, in this chapter we will be looking at polynomial f and g.

Let us be a bit more ambitious now and study motion along *curves*, i.e., paths that do not have to be straight. The simplest example is that of driving a car along a road. At time $t = 0$, you start, you follow the road, and at time $t = 1$, you have arrived somewhere. It does not really matter what kind of units we use to measure time; the $t = 0$ and $t = 1$ may just be viewed as a normalization of an arbitrary time interval.

We will now attack the problem of modeling curves, and we will choose a particularly simple way of doing this, namely *cubic Bézier curves*. We start with four points in 2D or 3D, $\mathbf{b}_0, \mathbf{b}_1, \mathbf{b}_2$, and $\mathbf{b}_3$, called *Bézier (control) points*. Connect them with straight lines as shown in Sketch 18.1. The resulting polygon is called a *Bézier (control) polygon*.[1]

The four control points, $\mathbf{b}_0, \mathbf{b}_1, \mathbf{b}_2, \mathbf{b}_3$, define a cubic curve, and some examples are illustrated in Figure 18.2. To create these plots, we *evaluate* the cubic curve at many t-parameters that range between zero and one, $t \in [0, 1]$. If we evaluate at 50 points, then we would find points on the curve associated with

$$t = 0, 1/50, 2/50, \ldots, 49/50, 1,$$

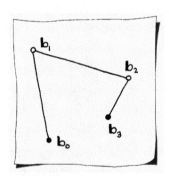

Sketch 18.1.
A Bézier polygon.

---

[1]Bézier polygons are not assumed to be closed as were the polygons of Section 17.2.

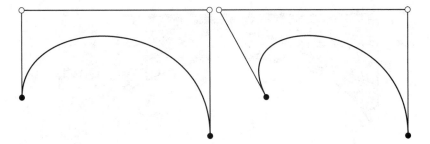

Figure 18.2.
Bézier curves: two examples that differ in the location of one control point, $\mathbf{b}_0$ only.

and then these points are connected by straight line segments to make the curve look smooth. The points are so close together that you cannot detect the line segments. In other words, we plot a *discrete approximation* of the curve.

Here is how you generate one point on a cubic Bézier curve. Pick a parameter value t between 0 and 1. Find the corresponding point on each polygon leg by linear interpolation:

$$\mathbf{b}_0^1(t) = (1 - t)\mathbf{b}_0 + t\mathbf{b}_1,$$
$$\mathbf{b}_1^1(t) = (1 - t)\mathbf{b}_1 + t\mathbf{b}_2,$$
$$\mathbf{b}_2^1(t) = (1 - t)\mathbf{b}_2 + t\mathbf{b}_3.$$

These three points form a polygon themselves. Now repeat the linear interpolation process, and you get two points:

$$\mathbf{b}_0^2 = (1 - t)\mathbf{b}_0^1(t) + t\mathbf{b}_1^1(t),$$
$$\mathbf{b}_1^2 = (1 - t)\mathbf{b}_1^1(t) + t\mathbf{b}_2^1(t).$$

Repeat one more time,

$$\mathbf{b}_0^3(t) = (1 - t)\mathbf{b}_0^2(t) + t\mathbf{b}_1^2(t), \tag{18.1}$$

and you have a point on the Bézier curve defined by $\mathbf{b}_0, \mathbf{b}_1, \mathbf{b}_2, \mathbf{b}_3$ at the parameter value t. The recursive process of applying linear interpolation is called the *de Casteljau algorithm*, and the steps above are shown in Sketch 18.2. Figure 18.3 illustrates all de Casteljau steps for 33 evaluations. The points $\mathbf{b}_i^j$ are often called *intermediate Bézier*

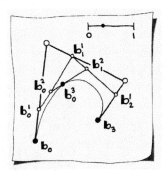

Sketch 18.2.
The de Casteljau algorithm.

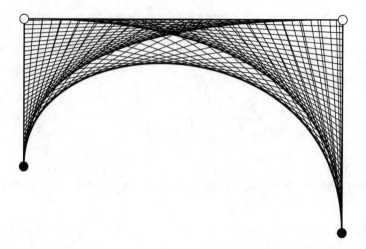

Figure 18.3.
Bézier curves: all intermediate Bézier points generated by the de Casteljau algorithm for 33 evaluations.

points, and the following schematic is helpful in keeping track of how each point is generated.

$$
\begin{array}{llll}
\mathbf{b}_0 & & & \\
\mathbf{b}_1 & \mathbf{b}_0^1 & & \\
\mathbf{b}_2 & \mathbf{b}_1^1 & \mathbf{b}_0^2 & \\
\mathbf{b}_3 & \mathbf{b}_2^1 & \mathbf{b}_1^2 & \mathbf{b}_0^3 \\
\text{stage}: & 1 & 2 & 3
\end{array}
$$

Except for the (input) Bézier polygon, each point in the schematic is a function of t.

Example 18.1

A numerical counterpart to Sketch 18.3 follows. Let the polygon be given by

$$
\mathbf{b}_0 = \begin{bmatrix} 4 \\ 4 \end{bmatrix}, \quad \mathbf{b}_1 = \begin{bmatrix} 0 \\ 8 \end{bmatrix}, \quad \mathbf{b}_2 = \begin{bmatrix} 8 \\ 8 \end{bmatrix}, \quad \mathbf{b}_3 = \begin{bmatrix} 8 \\ 0 \end{bmatrix}.
$$

For simplicity, let $t = 1/2$. Linear interpolation is then nothing but finding midpoints, and we have the following intermediate Bézier

points,

$$\mathbf{b}_0^1 = \frac{1}{2}\mathbf{b}_0 + \frac{1}{2}\mathbf{b}_1 = \begin{bmatrix} 2 \\ 6 \end{bmatrix},$$

$$\mathbf{b}_1^1 = \frac{1}{2}\mathbf{b}_1 + \frac{1}{2}\mathbf{b}_2 = \begin{bmatrix} 4 \\ 8 \end{bmatrix},$$

$$\mathbf{b}_2^1 = \frac{1}{2}\mathbf{b}_2 + \frac{1}{2}\mathbf{b}_3 = \begin{bmatrix} 8 \\ 4 \end{bmatrix}.$$

Next,

$$\mathbf{b}_0^2 = \frac{1}{2}\mathbf{b}_0^1 + \frac{1}{2}\mathbf{b}_1^1 = \begin{bmatrix} 3 \\ 7 \end{bmatrix},$$

$$\mathbf{b}_1^2 = \frac{1}{2}\mathbf{b}_1^1 + \frac{1}{2}\mathbf{b}_2^1 = \begin{bmatrix} 6 \\ 6 \end{bmatrix},$$

and finally

$$\mathbf{b}_0^3 = \frac{1}{2}\mathbf{b}_0^2 + \frac{1}{2}\mathbf{b}_1^2 = \begin{bmatrix} \frac{9}{2} \\ \frac{13}{2} \end{bmatrix}.$$

This is the point on the curve corresponding to $t = 1/2$.

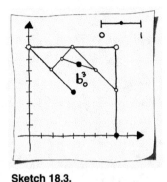

Sketch 18.3.
Evaluation via the de Casteljau algorithm at t = 1/2.

The equation for a cubic Bézier curve is found by expanding (18.1):

$$\begin{aligned} \mathbf{b}_0^3 &= (1-t)\mathbf{b}_0^2 + t\mathbf{b}_1^2 \\ &= (1-t)\left[(1-t)\mathbf{b}_0^1 + t\mathbf{b}_1^1\right] + t\left[(1-t)\mathbf{b}_1^1 + t\mathbf{b}_2^1\right] \\ &= (1-t)\left[(1-t)\left[(1-t)\mathbf{b}_0 + t\mathbf{b}_1\right] + t\left[(1-t)\mathbf{b}_1 + t\mathbf{b}_2\right]\right] \\ &\quad + t\left[(1-t)\left[(1-t)\mathbf{b}_1 + t\mathbf{b}_2\right] + t\left[(1-t)\mathbf{b}_2 + t\mathbf{b}_3\right]\right]. \end{aligned}$$

After collecting terms with the same $\mathbf{b}_i$, this becomes

$$\mathbf{b}_0^3(t) = (1-t)^3\mathbf{b}_0 + 3(1-t)^2 t\mathbf{b}_1 + 3(1-t)t^2\mathbf{b}_2 + t^3\mathbf{b}_3. \qquad (18.2)$$

This is the general form of a cubic Bézier curve. As t traces out values between 0 and 1, the point $\mathbf{b}_0^3(t)$ traces out a curve.

The polynomials in (18.2) are called the *Bernstein basis functions*:

$$\begin{aligned} B_0^3(t) &= (1-t)^3 \\ B_1^3(t) &= 3(1-t)^2 t \\ B_2^3(t) &= 3(1-t)t^2 \\ B_3^3(t) &= t^3, \end{aligned}$$

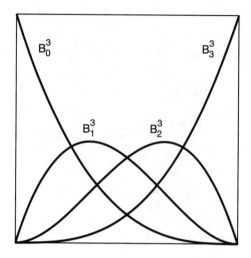

Figure 18.4.

Bernstein polynomials: a plot of the four cubic polynomials for $t \in [0,1]$.

and they are illustrated in Figure 18.4. Now $\mathbf{b}_0^3(t)$ can be written as

$$\mathbf{b}_0^3(t) = B_0^3(t)\mathbf{b}_0 + B_1^3(t)\mathbf{b}_1 + B_2^3(t)\mathbf{b}_2 + B_3^3(t)\mathbf{b}_3. \qquad (18.3)$$

As we see, the Bernstein basis functions are cubic, or degree 3, polynomials. This set polynomials is a bit different than the cubic *monomials*, $1, t, t^2, t^3$ that we are used to from calculus. However, either set allow us to write all cubic polynomials. We say that the Bernstein (or monomial) polynomials form a basis for all cubic polynomials, hence the name basis function. See Section 15.1 for more on bases. We'll look at the relationship between these sets of polynomials in Section 18.3.

The original curve was given by the control polygon

$$\mathbf{b}_0, \mathbf{b}_1, \mathbf{b}_2, \mathbf{b}_3.$$

Inspection of Sketch 18.3 suggests that a subset of the intermediate Bézier points form two cubic polygons which also mimic the curve's shape. The curve segment from $\mathbf{b}_0$ to $\mathbf{b}_0^3(t)$ has the polygon

$$\mathbf{b}_0, \mathbf{b}_0^1, \mathbf{b}_0^2, \mathbf{b}_0^3,$$

and the other from $\mathbf{b}_0^3(t)$ to $\mathbf{b}_3$ has the polygon

$$\mathbf{b}_0^3, \mathbf{b}_1^2, \mathbf{b}_2^1, \mathbf{b}_3.$$

This process of generating two Bézier curves from one, is called *sub-division*.

From now on, we will also use the shorter $\mathbf{b}(t)$ instead of $\mathbf{b}_0^3(t)$.

18.2   Properties of Bézier Curves

From inspection of the examples, but also from (18.2), we see that the curve passes through the first and last control points:

$$\mathbf{b}(0) = \mathbf{b}_0 \quad \text{and} \quad \mathbf{b}(1) = \mathbf{b}_3. \qquad (18.4)$$

Another way to say this: the curve *interpolates* to $\mathbf{b}_0$ and $\mathbf{b}_3$.

If we map the control polygon using an affine map, then the curve undergoes the same transformation, as shown in Figure 18.5. This is called *affine invariance*, and it is due to the fact that the cubic coefficients of the control points, the Bernstein polynomials, in (18.3) sum to one. This can be seen as follows:

$$(1-t)^3 + 3(1-t)^2 t + 3(1-t)t^2 + t^3 = [(1-t)+t]^3 = 1.$$

Thus, every point on the curve is a *barycentric combination* of the control points. Such relationships are not changed under affine maps, as per Section 6.2.

Figure 18.5.
Affine invariance: as the control polygon rotates, so does the curve.

Figure 18.6.
Convex hull property: the curve lies in the convex hull of the control polygon.

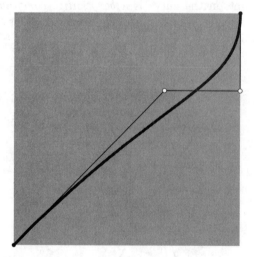

Figure 18.7.
Convex hull property: as a result of the convex hull property, the curve lies inside the minmax box of the control polygon.

The curve, for $t \in [0, 1]$, lies in the convex hull of the control polygon—a fact called the *convex hull property*. This can be seen by observing in Figure 18.4 that the Bernstein polynomials in (18.2) are nonnegative for $t \in [0, 1]$. It follows that every point on the curve is a *convex combination* of the control points, and hence is inside their *convex hull*. For a definition, see Section 8.4; for an illustration, see Figure 18.6. If we evaluate the curve for t-values outside of $[0, 1]$, this is called *extrapolation*. We can no longer predict the shape of the curve, so this procedure is normally not recommended.

Clearly, the control polygon is inside its minmax box.[2] Because of the convex hull property, we also know that the curve is inside this box—a property that has numerous applications. See Figure 18.7 for an illustration.

18.3 The Matrix Form

As a preparation for what is to follow, let us rewrite (18.2) using the formalism of dot products. It then looks like this:

$$b(t) = \begin{bmatrix} b_0 & b_1 & b_2 & b_3 \end{bmatrix} \begin{bmatrix} (1-t)^3 \\ 3(1-t)^2 t \\ 3(1-t)t^2 \\ t^3 \end{bmatrix}. \tag{18.5}$$

Instead of the Bernstein polynomials as above, most people think of polynomials as combinations of the *monomials*; they are $1, t, t^2, t^3$ for the cubic case. We can related the Bernstein and monomial forms by rewriting our expression (18.2):

$$b(t) = b_0 + 3t(b_1 - b_0) + 3t^2(b_2 - 2b_1 + b_0) + t^3(b_3 - 3b_2 + 3b_1 - b_0). \tag{18.6}$$

This allows a concise formulation using matrices:

$$b(t) = \begin{bmatrix} b_0 & b_1 & b_2 & b_3 \end{bmatrix} \begin{bmatrix} 1 & -3 & 3 & -1 \\ 0 & 3 & -6 & 3 \\ 0 & 0 & 3 & -3 \\ 0 & 0 & 0 & 1 \end{bmatrix} \begin{bmatrix} 1 \\ t \\ t^2 \\ t^3 \end{bmatrix}. \tag{18.7}$$

This is the matrix form of a Bézier curve.

Equation (18.6) or (18.7) shows how to write a Bézier curve in monomial form. A curve in monomial form looks like this:

$$b(t) = a_0 + a_1 t + a_2 t^2 + a_3 t^3.$$

Geometrically, the four control "points" for the curve in monomial form are now a mix of points and vectors: $a_0 = b_0$ is a point, but

---

[2]Recall that the minmax box of a polygon is the smallest rectangle with edges parallel to the coordinate axes that contains the polygon.

$\mathbf{a}_1, \mathbf{a}_2, \mathbf{a}_3$ are vectors. Using the dot product form, this becomes

$$\mathbf{b}(t) = \begin{bmatrix} \mathbf{a}_0 & \mathbf{a}_1 & \mathbf{a}_2 & \mathbf{a}_3 \end{bmatrix} \begin{bmatrix} 1 \\ t \\ t^2 \\ t^3 \end{bmatrix}.$$

Thus, the monomial $\mathbf{a}_i$ are defined as

$$\begin{bmatrix} \mathbf{a}_0 & \mathbf{a}_1 & \mathbf{a}_2 & \mathbf{a}_3 \end{bmatrix} = \begin{bmatrix} \mathbf{b}_0 & \mathbf{b}_1 & \mathbf{b}_2 & \mathbf{b}_3 \end{bmatrix} \begin{bmatrix} 1 & -3 & 3 & -1 \\ 0 & 3 & -6 & 3 \\ 0 & 0 & 3 & -3 \\ 0 & 0 & 0 & 1 \end{bmatrix}.$$

$$(18.8)$$

How about the inverse process: If we are given a curve in monomial form, how can we write it as a Bézier curve? Simply rearrange (18.8) to solve for the $\mathbf{b}_i$:

$$\begin{bmatrix} \mathbf{b}_0 & \mathbf{b}_1 & \mathbf{b}_2 & \mathbf{b}_3 \end{bmatrix} = \begin{bmatrix} \mathbf{a}_0 & \mathbf{a}_1 & \mathbf{a}_2 & \mathbf{a}_3 \end{bmatrix} \begin{bmatrix} 1 & -3 & 3 & -1 \\ 0 & 3 & -6 & 3 \\ 0 & 0 & 3 & -3 \\ 0 & 0 & 0 & 1 \end{bmatrix}^{-1}.$$

A matrix inversion is all that is needed here! Notice that the square matrix in (18.8) is nonsingular, therefore we can conclude that any cubic curve can be written in either the Bézier or the monomial form.

18.4 Derivatives

Equation (18.2) consists of two (in 2D) or three (in 3D) cubic equations in t. We can take the derivative in each of the components:

$$\frac{d\mathbf{b}(t)}{dt} = -3(1-t)^2\mathbf{b}_0 - 6(1-t)t\mathbf{b}_1 + 3(1-t)^2\mathbf{b}_1$$
$$- 3t^2\mathbf{b}_2 + 6(1-t)t\mathbf{b}_2 + 3t^2\mathbf{b}_3.$$

Rearranging, and using the abbreviation $\frac{d\mathbf{b}(t)}{dt} = \dot{\mathbf{b}}(t)$, we have

$$\dot{\mathbf{b}}(t) = 3(1-t)^2[\mathbf{b}_1 - \mathbf{b}_0] + 6(1-t)t[\mathbf{b}_2 - \mathbf{b}_1] + 3t^2[\mathbf{b}_3 - \mathbf{b}_2]. \quad (18.9)$$

As expected, the derivative of a degree three curve is one of degree two.[3]

---

[3]Note that the derivative curve does not have control *points* anymore, but rather *control vectors*!

One very nice feature of the de Casteljau algorithm is that the intermediate Bézier points generated by it allow us to express the derivative very simply. A simpler expression than (18.9) for the derivative is

$$\dot{\mathbf{b}}(t) = 3\left[\mathbf{b}_1^2(t) - \mathbf{b}_0^2(t)\right]. \tag{18.10}$$

Getting the the derivative for free makes the de Casteljau algorithm more efficient than evaluating (18.2) and (18.9) directly to get a point and a derivative.

At the endpoints, the derivative formula is very simple. For $t = 0$, we obtain

$$\dot{\mathbf{b}}(0) = 3[\mathbf{b}_1 - \mathbf{b}_0],$$

and, similarly, for $t = 1$,

$$\dot{\mathbf{b}}(1) = 3[\mathbf{b}_3 - \mathbf{b}_2].$$

In words, the control polygon is tangent to the curve at the curve's endpoints. This is not a surprising statement if you check Figure 18.2!

Example 18.2

Let us compute the derivative of the curve from Example 18.1 for $t = 1/2$. First, let's evaluate the direct equation (18.9). We obtain

$$\dot{\mathbf{b}}\left(\frac{1}{2}\right) = 3 \cdot \frac{1}{4}\left[\begin{bmatrix}0\\8\end{bmatrix} - \begin{bmatrix}4\\4\end{bmatrix}\right] + 6 \cdot \frac{1}{4}\left[\begin{bmatrix}8\\8\end{bmatrix} - \begin{bmatrix}0\\8\end{bmatrix}\right] + 3 \cdot \frac{1}{4}\left[\begin{bmatrix}8\\0\end{bmatrix} - \begin{bmatrix}8\\8\end{bmatrix}\right],$$

which yields

$$\dot{\mathbf{b}}\left(\frac{1}{2}\right) = \begin{bmatrix}9\\-3\end{bmatrix}.$$

If instead, we used (18.10), and thus the intermediate control points calculated in Example 18.1, we get

$$\dot{\mathbf{b}}\left(\frac{1}{2}\right) = 3\left[\mathbf{b}_1^2\left(\frac{1}{2}\right) - \mathbf{b}_0^2\left(\frac{1}{2}\right)\right]$$

$$= 3\left[\begin{bmatrix}6\\6\end{bmatrix} - \begin{bmatrix}3\\7\end{bmatrix}\right]$$

$$= \begin{bmatrix}9\\-3\end{bmatrix},$$

which is the same answer but with less work! See Sketch 18.4 for an illustration.

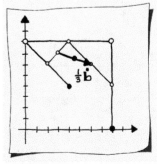

Sketch 18.4.
A derivative vector.

Note that the derivative of a curve is a *vector*. It is tangent to the curve—apparent from our example, but nothing we want to prove here. A convenient way to think about the derivative is by interpreting it as a *velocity vector*. If you interpret the parameter t as time, and you think of traversing the curve such that at time t you have reached $\mathbf{b}(t)$, then the derivative measures your velocity. The larger the magnitude of the tangent vector, the faster you move.

If we rotate the control polygon, the curve will follow, and so will all of its derivative vectors. In calculus, a "horizontal tangent" has a special meaning; it indicates an extreme value of a function. Not here: the very notion of an extreme value is meaningless for parametric curves since the term "horizontal tangent" depends on the curve's orientation and is not a property of the curve itself.

We may take the derivative of (18.9) with respect to t. We then have the *second derivative*. It is given by

$$\ddot{\mathbf{b}}(t) = -6(1-t)[\mathbf{b}_1 - \mathbf{b}_0] - 6t[\mathbf{b}_2 - \mathbf{b}_1] + 6(1-t)[\mathbf{b}_2 - \mathbf{b}_1] + 6t[\mathbf{b}_3 - \mathbf{b}_2]$$

and may be rearranged to

$$\ddot{\mathbf{b}}(t) = 6(1-t)[\mathbf{b}_2 - 2\mathbf{b}_1 + \mathbf{b}_0] + 6t[\mathbf{b}_3 - 2\mathbf{b}_2 + \mathbf{b}_1]. \qquad (18.11)$$

The de Casteljau algorithm supplies a simple way to write the second derivative, too:

$$\ddot{\mathbf{b}}(t) = 6\left[\mathbf{b}_2^1(t) - 2\mathbf{b}_1^1(t) + \mathbf{b}_0^1(t)\right]. \qquad (18.12)$$

Loosely speaking, we may interpret the second derivative $\ddot{\mathbf{b}}(t)$ as acceleration when traversing the curve.

The second derivative at $\mathbf{b}_0$ (see Sketch 18.5) is particularly simple— it is given by

$$\ddot{\mathbf{b}}(0) = 6[\mathbf{b}_2 - 2\mathbf{b}_1 + \mathbf{b}_0].$$

Notice that this is a scaling of the difference of the vectors $\mathbf{b}_1 - \mathbf{b}_0$ and $\mathbf{b}_2 - \mathbf{b}_1$. Recalling the *parallelogram rule*, the second derivative at $\mathbf{b}_0$ is easy to sketch. A similar equation holds for the second derivative at $t = 1$; now the points involved are $\mathbf{b}_1, \mathbf{b}_2, \mathbf{b}_3$.

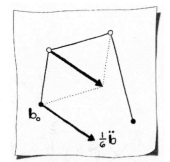

Sketch 18.5.

A second derivative vector.

18.5 Composite Curves

A Bézier curve is a handsome tool, but one such curve would rarely suffice for describing much of any shape! For "real" shapes, we have to be able to line up many cubic Bézier curves. In order to define a smooth overall curve, these pieces must join smoothly.

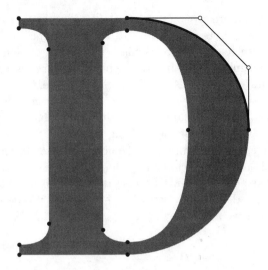

Figure 18.8.
Composite Bézier curves: the letter "D" as a collection of cubic Bézier curves. Only one Bézier polygon of many is shown.

This is easily achieved. Let $\mathbf{b}_0, \mathbf{b}_1, \mathbf{b}_2, \mathbf{b}_3$ and $\mathbf{c}_0, \mathbf{c}_1, \mathbf{c}_2, \mathbf{c}_3$ be the control polygons of two Bézier curves with a common point $\mathbf{b}_3 = \mathbf{c}_0$ (see Sketch 18.6). If the two curves are to have the same tangent vector direction at $\mathbf{b}_3 = \mathbf{c}_0$, then all that is required is

$$\mathbf{c}_1 - \mathbf{c}_0 = c[\mathbf{b}_3 - \mathbf{b}_2] \qquad (18.13)$$

for some positive real number c, meaning that the three points $\mathbf{b}_2, \mathbf{b}_3 = \mathbf{c}_0, \mathbf{c}_1$ are collinear.

If we use this rule to piece curve segments together, we can design many 2D and 3D shapes. Figure 18.8 gives an example.

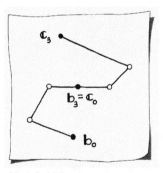

Sketch 18.6.
Smoothly joining Bézier curves.

18.6   The Geometry of Planar Curves

The geometry of planar curves is centered around one concept: their *curvature.* It is easily understood if you imagine driving a car along a road. For simplicity, let's assume you are driving with constant speed. If the road does not curve, i.e., it is straight, you will not have to turn your steering wheel. When the road does curve, you will have to turn the steering wheel, and more so if the road curves rapidly. The curviness of the road (our model of a curve) is thus proportional to the turning of the steering wheel.

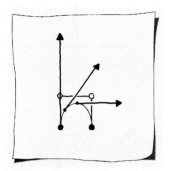

Returning to the more abstract concept of a curve, let us sample its tangents at various points (see Sketch 18.7). Where the curve bends sharply, i.e., where its curvature is high, successive tangents differ from each other significantly. In areas where the curve is relatively flat, or where its curvature is low, successive tangents are almost identical. Curvature may thus be defined as *rate of change of tangents*. (In terms of our car example, the rate of change of tangents is proportional to the turning of the steering wheel.)

Since the tangent is determined by the curve's first derivative, its rate of change should be determined by the second derivative. This is indeed so, but the actual formula for curvature is a bit more complex than can be derived in the context of this book. We denote the curvature of the curve at $\mathbf{b}(t)$ by κ; it is given by

$$\kappa(t) = \frac{\|\dot{\mathbf{b}} \wedge \ddot{\mathbf{b}}\|}{\|\dot{\mathbf{b}}\|^3}. \tag{18.14}$$

This formula holds for both 2D and 3D curves. In the 2D case, it may be rewritten as

$$\kappa(t) = \frac{|\dot{\mathbf{b}} \quad \ddot{\mathbf{b}}|}{\|\dot{\mathbf{b}}\|^3} \tag{18.15}$$

with the use of a 2×2 determinant. Since determinants may be positive or negative, curvature in 2D is *signed*. A point where $\kappa = 0$ is called an *inflection point*: the 2D curvature changes sign here. In Figure 18.9, the inflection point is marked. In calculus, you learned that a curve has an inflection point if the second derivative vanishes. For parametric curves, the situation is different. An inflection point occurs when the first and second derivative vectors are parallel, or linearly dependent. This can lead to the curious effect of a cubic with *two* inflection points. It is illustrated in Figure 18.10.

Figure 18.9.

Inflection point: an inflection point, a point where the curvature changes sign, is marked on the curve.

Figure 18.10.
Inflection point: a cubic with two inflection points.

Figure 18.11.
Curve motions: a letter is moved along a curve.

18.7 Moving along a Curve

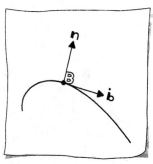

Sketch 18.8.
Sliding along a curve.

Take a look at Figure 18.11. You will see the letter "B" sliding along a curve. If the curve is given in Bézier form, how can that effect be achieved? The answer can be seen in Sketch 18.8. If you want to position an object, such as the letter "B," at a point on a curve, all you need to know is the point and the curve's tangent there. If $\dot{\mathbf{b}}$ is the tangent, then simply define $\mathbf{n}$ to be a vector perpendicular to it.[4] Using the local coordinate system with origin $\mathbf{b}(t)$ and $[\dot{\mathbf{b}}, \mathbf{n}]$-axes, you can position any object as in Section 4.1.

The same story is far trickier in 3D! If you had a point on the curve and its tangent, the exact location of your object would not be fixed; it could still rotate around the tangent. Yet there is a unique way to position objects along a 3D curve. At every point on the curve, we may define a *local coordinate system* as follows.

Let the point on the curve be $\mathbf{b}(t)$; we now want to set up a local coordinate system defined by three vectors $\mathbf{f}_1, \mathbf{f}_2, \mathbf{f}_3$. Following the 2D example, we set $\mathbf{f}_1$ to be in the tangent direction; $\mathbf{f}_1 = \dot{\mathbf{b}}(t)$. If the curve does not have an inflection point at t, then $\dot{\mathbf{b}}(t)$ and $\ddot{\mathbf{b}}(t)$ will not be collinear. This means that they span a plane, and that

---

[4]If $\dot{\mathbf{b}} = \begin{bmatrix} \dot{b}_1 \\ \dot{b}_2 \end{bmatrix}$ then $\mathbf{n} = \begin{bmatrix} -\dot{b}_2 \\ \dot{b}_1 \end{bmatrix}$.

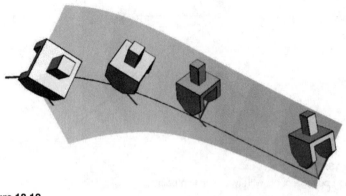

Figure 18.12.

Curve motions: a robot arm is moved along a curve. (Courtesy of M. Wagner, Arizona State University.)

plane's normal is given by $\dot{\mathbf{b}}(t) \wedge \ddot{\mathbf{b}}(t)$. See Sketch 18.9 for some visual information. We make the plane's normal one of our local coordinate axes, namely $\mathbf{f_3}$. The plane, by the way, has a name: it is called the *osculating plane* at $\mathbf{x}(t)$. Since we have two coordinate axes, namely $\mathbf{f_1}$ and $\mathbf{f_3}$, it is not hard to come up with the remaining axis, we just set $\mathbf{f_2} = \mathbf{f_1} \wedge \mathbf{f_3}$. Thus, for every point on the curve (as long as it is not an inflection point), there exists an orthogonal coordinate system. It is customary to use coordinate axes of unit length, and then we have

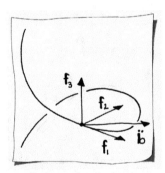

Sketch 18.9.
A Frenet frame.

$$\mathbf{f_1} = \frac{\dot{\mathbf{b}}(t)}{\|\dot{\mathbf{b}}(t)\|}, \qquad (18.16)$$

$$\mathbf{f_3} = \frac{\dot{\mathbf{b}}(t) \wedge \ddot{\mathbf{b}}(t)}{\|\dot{\mathbf{b}}(t) \wedge \ddot{\mathbf{b}}(t)\|}, \qquad (18.17)$$

$$\mathbf{f_2} = \mathbf{f_1} \wedge \mathbf{f_3}. \qquad (18.18)$$

This system with local origin $\mathbf{b}(t)$ and normalized axes $\mathbf{f_1}, \mathbf{f_2}, \mathbf{f_3}$ is called the *Frenet frame* of the curve at $\mathbf{b}(t)$. Equipped with the tool of Frenet frames, we may now position objects along a 3D curve! See Figure 18.12. Since we are working in 3D, we use the cross product to form the orthogonal frame; however, we could have equally as well used the Gram-Schmidt process from Section 11.8.

Let us now work out exactly how to carry out our object-positioning plan. The object is given in some local coordinate system with axes $\mathbf{u_1}, \mathbf{u_2}, \mathbf{u_3}$. Any point of the object has coordinates

$$\mathbf{u} = \begin{bmatrix} u_1 \\ u_2 \\ u_3 \end{bmatrix}.$$

It is mapped to

$$\mathbf{x}(t, \mathbf{u}) = \mathbf{x}(t) + u_1\mathbf{f}_1 + u_2\mathbf{f}_2 + u_3\mathbf{f}_3.$$

A typical application is *robot motion*. Robots are used extensively in automotive assembly lines; one job is to grab a part and move it to its destination inside the car body. This movement happens along well-defined curves. While the car part is being moved, it has to be oriented into its correct position—exactly the process described in this section!

- linear interpolation
- parametric curve
- de Casteljau algorithm
- cubic Bézier curve
- subdivision
- affine invariance
- convex hull property
- Bernstein polynomials
- basis function
- barycentric combination
- matrix form
- cubic monomial curve

- Bernstein and monomial conversion
- nonsingular
- first derivative
- second derivative
- parallelogram rule
- composite Bézier curves
- curvature
- inflection point
- Frenet frame
- osculating plane

18.8   Exercises

Let a cubic Bézier curve $\mathbf{b}(t)$ be given by the control polygon

$$\mathbf{b}_0 = \begin{bmatrix} 0 \\ 0 \\ 0 \end{bmatrix}, \quad \mathbf{b}_1 = \begin{bmatrix} 4 \\ 0 \\ 0 \end{bmatrix}, \quad \mathbf{b}_2 = \begin{bmatrix} 4 \\ 4 \\ 0 \end{bmatrix}, \quad \mathbf{b}_3 = \begin{bmatrix} 4 \\ 4 \\ 4 \end{bmatrix}.$$

1. Sketch this 3D curve manually.

2. Using the de Casteljau algorithm, evaluate it for $t = 1/4$.

3. Evaluate the first and second derivative for $t = 1/4$. Add these vectors to the sketch from Exercise 1.

4. What is the control polygon for the curve defined from $t = 0$ to $t = 1/4$ and the curve defined over $t = 1/4$ to $t = 1$?

5. Rewrite it in monomial form.

6. Find its minmax box.

7. Find its curvature at $t = 0$ and $t = 1/2$.

8. Find its Frenet frame for $t = 1$.

9. Using the de Casteljau algorithm, evaluate it for $t = 2$. This is extrapolation.

10. Attach another curve $\mathbf{c}(t)$ at $\mathbf{b}_0$, creating a composite curve with tangent continuity. What are the constraints on $\mathbf{c}(t)$'s polygon?

PS1 Postscript has a utility for drawing (2D) Bézier curves. You must first move to $\mathbf{b}_0$ and then call the utility with the next three control points as follows.

$b_{0,0}$ $b_{1,0}$ moveto
$b_{0,1}$ $b_{1,1}$ $b_{0,2}$ $b_{1,2}$ $b_{0,3}$ $b_{1,3}$ curveto
Create a PostScript program that draws a Bézier curve and its polygon.

A

PostScript Tutorial

The figures in this book are created using the PostScript language. This is a mini-guide to PostScript with the purpose of giving you enough information so that you can alter these images or create similar ones yourself.

PostScript is a page description language.[1] A PostScript program tells the output device (printer or previewer) how to format a printed page. Most laser printers today understand the PostScript language. A previewer, such as Ghostview, allows you to view your document without printing—a great way to save paper. Ghostview can be found at

> http://www.cs.wisc.edu/~ghost/index.html.

In case you would like more in-depth information about Post-Script, see [13, 12].

PostScript deals with 2D objects only. A 3D package exists at this site:

> http://www.math.ubc.ca/people/faculty/cass/graphics/text/www/.

Before you attempt to go there, be sure that you understand this tutorial!

A.1   A Warm-Up Example

Let's go through the PostScript file that generates Figure A.1.

---

[1]Its origins are in the late seventies, when it was developed by Evans & Sutherland, by Xerox, and finally by Adobe Systems, who now own it.

Figure A.1.
PostScript example: drawing the outline of a box.

```
%!
newpath
        200  200  moveto
        300  200  lineto
        300  300  lineto
        200  300  lineto
        200  200  lineto
stroke

showpage
```

Figure A.1 shows the result of this program: we have drawn a box on a standard size page. (In this figure, the outline of the page is shown for clarity; everything is reduced to fit here.)

The first line of the file is %!. For now, let's say that nothing else belongs on this line, and there are no extra spaces on the line. This command tells the printer that what follows is in the PostScript language.

The actual drawing begins with the `newpath` command. Move to the starting position with `moveto`. Record a path with `lineto`. Finally, indicate that the path should be drawn with `stroke`. These commands simulate the movement of a "virtual pen" in a fairly obvious way.

PostScript uses *prefix notation* for its commands. So if you want to move your "pen" to position $(100, 200)$, the command is `100 200 moveto`.

Finally, you need to invoke the `showpage` command to cause PostScript to actually print your figure. This is important!

Now for some variation:

```
%!
% These are comment lines because they
% begin with a percent sign

% draw the box
newpath
     200 200 moveto
     300 200 lineto
     300 300 lineto
     200 300 lineto
     200 200 lineto
stroke

80 80 translate
0.5 setgray

newpath
     200 200 moveto
     300 200 lineto
     300 300 lineto
     200 300 lineto
     200 200 lineto
fill

showpage
```

This is illustrated in Figure A.2. The `80 80 translate` command moves the origin of your current coordinate system by 80 units in both the e_1- and e_2-directions.

The `0.5 setgray` command causes the next item to be drawn in a gray shade (0 is black, 1 is white).

What follows is the same square as before; instead of `stroke`, however, there's `fill`. This fills the square with the specified gray scale. Note how the second square is drawn on top of the first one!

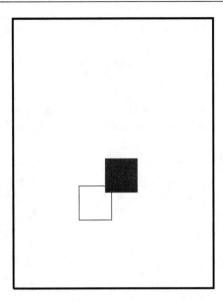

Figure A.2.
PostScript example: drawing two boxes. The outlined box is drawn first and the filled one second.

Another handy basic geometry element is a circle. To create a circle, use the following commands:

```
250 250 50 0 360 arc stroke
```

This generates a circle centered at $(250, 250)$ with radius 50. The 0 360 `arc` indicates we want the whole circle. To create a white filled circle change the command to

```
1.0 setgray
250 250 50 0 360 arc fill
```

Figure A.3 illustrates all these additions.

A.2 Overview

The PostScript language has several basic graphics operators: shapes such as line or arc, painting, text, bit mapped images, and coordinate transformations. Variables and calculations are also in its repertoire; it is a powerful tool. Nearly all the figures for this book are available as PostScript (text) files at the book's website. To illustrate the aspects of the file format, we'll return to some of these figures.

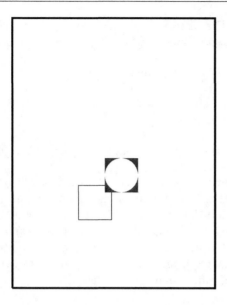

Figure A.3.
PostScript example: building on Figure A.2, a white-filled circle is added.

We use three basic scenarios for generating PostScript files.

1. A program (such as C) generates geometry, then opens a file and writes PostScript commands to plot the geometry. This type of file is typically filled simply with move, draw, and style (gray scale or line width) commands. Example file: Bez_ex.ps which is displayed in Section 18.2.

2. The PostScript language is used to generate geometry. Using a text editor, the PostScript program is typed in. This program might even use control structures such as "for loops." Example file: D_trans.ps which is displayed in Section 6.3.

3. A print screen creating a bit mapped image is made part of a PostScript file. Example file: Flight.ps which is the introductory figure of Chapter 12.

There are also figures which draw from more than one of these scenarios.

Since PostScript tells the printer how to format a page, it is necessary for the move and draw commands to indicate locations on the piece of paper. For historical reasons, printers use a coordinate system based on *points*, abreviated as pt.

$$1 \text{ inch} \equiv 72 \text{ pt}.$$

PostScript follows suit. This means that an $8\frac{1}{2} \times 11$-inch page has the following extents:

$$\text{lower left}: \ (0,0) \qquad \text{upper right}: \ (612, 792).$$

Whichever scenario from above is followed, it is always necessary to be sure that the PostScript commands are drawing in the appropriate area. Keep in mind that you probably want a margin. Chapter 1 covers the basics of setting up the dimensions of the "target box" (on the paper) and those of the "source box" to avoid unwanted distortions. Getting the geometry from the source box to the target box is simply an application of affine maps!

A.3 Affine Maps

In our simple examples above, we created the geometry with respect to the paper coordinates. Sometimes this is inconvenient, so let's discuss the options.

If you create your geometry in, say, a C program, then it is an easy task to apply the appropriate affine map to put it in paper coordinates. However, if this is not the case, the PostScript language has the affine map tools built-in.[2]

There is a matrix (which describes an affine map, i.e., a linear map and a translation) in PostScript which assumes the task of taking "user coordinates" to "device coordinates." In accordance to the terminology of this book, this is our source box to target box transformation. This matrix is called the *current transformation matrix*, or *CTM*.

In other words, the coordinates in your PostScript file are always multiplied by the CTM. If you choose not to change the CTM, then your coordinates must live within the page coordinates, or "device coordinates." Here we give a few details on how to alter the CTM.

There are two "levels" of changing the CTM. The simplest and most basic ones use the `scale`, `rotate`, and `translate` commands.

---

[2]There are many techniques for displaying 3D geometry in 2D; some "realistic" methods are not in the realms of affine geometry. A graphics text should be consulted to determine what is best.

As we know from Chapter 6, these are essentially the most basic affine operations.

Unless you have a complete understanding of the CTM, it is probably a good idea to use each of these commands only once in a PostScript file. It can get confusing! (See Section A.6.)

A scale command such as

```
72 72 scale
```

automatically changes one "unit" from being a *point* to being an inch. A translate command such as

```
2 2 translate
```

will translate the origin to coordinates $(2, 2)$. If this **translate** was preceded by the **scale** command, the effect is different. Try both options for yourself!

A rotate command such as

```
45 rotate
```

will cause a rotation of the coordinate system by 45 degrees in a counterclockwise direction.

These commands are used in the website files **D_scale.ps**, **D_trans.ps**, and **D_rot.ps**.

Instead of the commands above, a more general manipulation of the CTM is available (see Section A.6).

A.4   Variables

The figures in this book use variables quite often. This is a powerful tool allowing a piece of geometry to be defined once, and then affine maps can be applied to it to change its appearance on the page.

Let's use an example to illustrate. Take the file for Figure A.2. We can rewrite this as

```
%!
% define the box
/box {
        200 200 moveto
        300 200 lineto
        300 300 lineto
```

```
                    200 300 lineto
                    200 200 lineto
} def

newpath
box
stroke

80 80 translate
0.5 setgray

newpath
box
fill

showpage
```

The figure does not change. The box that was repeated is defined only once now. Notice the /box {...} **def** structure. This defines the variable box. It is then used without the forward slash.

A.5 Loops

The ability to create "for loops" is another very powerful tool. If you are not familiar with the prefix notation this might look odd. Let's look at the file D_trans.ps, which is displayed in Figure 6.3.

```
%!PS-Adobe-2.0
%%BoundingBox: 90 80 350 270
/Times-Bold findfont
70 scalefont setfont
/printD {
  0 0 moveto
  (D) show
} def

100 100 translate

2.5 2.5  scale
1 -.05 0.35 {setgray printD 3 1 translate } for

showpage
```

Before getting to the loop, let's look at a few other new things in the file. The `BoundingBox` command defines a minmax box that is intended to contain the geometry. This command is not used by PostScript, although it is necessary for correct placement of the figure in a LATEX file. The `PS-Adobe-2.0` string on the first line will allow you to see the "BoundingBox" in Ghostview. The closer the bounding box is to the geometry, the less white space there will be around the figure. The `/Times-Bold findfont` command allows us to access a particular set of fonts—we want to draw the letter D.

Now for the "for loop." The command

```
1 -.05 0.35 { . . . } for
```

tells PostScript to start with the value 1.0 and decrement by 0.05 until it reaches 0.35. At each step it will execute the commands within the parenthesis. This allows the D to be printed 13 times in different gray scales and translated each time. The gray scale is the same as the "for loop" parameter.

A.6   CTM

PostScript always has a CTM which is currently active. If we want to apply a specific matrix to our geometry, then we need to concatenate that matrix with the CTM. The following PostScript code (which generates Figure A.4) gives an example.

```
%!PS-Adobe-2.0
%%BoundingBox: 40 80 400 160
% Demonstrates a shear in the e1-direction
% by modifying the CTM

/shear[1 0  1 1  0 0] def %the shear matrix

/pla {/Helvetica-Bold findfont
    70 scalefont setfont (PLA) show
    } def                  %command to print "PLA"

    70 100 translate
    0 0 moveto
    pla                    %print the left PLA

    150 0 translate        %go to new position to the right
    0 0 moveto
```

PLA *PLA*

Figure A.4.
PostScript example: draw "PLA" and then concatenate the CTM with a shear matrix and redraw "PLA."

```
    shear concat            %apply shear matrix
    pla                     %print PLA again, now sheared

showpage
```

This piece of code shows you how to use a matrix in PostScript. In this book, we write transformations in the form $\mathbf{p}' = A\mathbf{p}$, which is called "right multiply." PostScript, however, uses a transformation matrix for "left multiply." Specifically, this book applies a transformation to a point $\mathbf{p}$ as

$$\mathbf{p}' = \begin{bmatrix} a & c \\ b & d \end{bmatrix} \begin{bmatrix} p_1 \\ p_2 \end{bmatrix},$$

whereas PostScript would apply the transformation as

$$\mathbf{p}' = \begin{bmatrix} p_1 & p_2 \end{bmatrix} \begin{bmatrix} a & b \\ c & d \end{bmatrix}.$$

So the PostScript CTM is the transpose of the matrix as used in this book.

A matrix definition in PostScript has the translation vector included. The matrix has the form

$$[a, b, c, d, t_x, t_y].$$

The matrix *shear* in our example is defined as $[1\ 0\ 1\ 1\ 0\ 0]$ and corresponds to "our" matrix form

$$\mathbf{p}' = \begin{bmatrix} 1 & 1 \\ 0 & 1 \end{bmatrix} + \begin{bmatrix} 0 \\ 0 \end{bmatrix}.$$

B

Selected Problem Solutions

1(a). The triangle vertex with coordinates $(0.1, 0.1)$ in the $[\mathbf{d}_1, \mathbf{d}_2]$-system is mapped to

$$x_1 = 0.9 \times 1 + 0.1 \times 3 = 1.2,$$
$$x_2 = 0.9 \times 2 + 0.1 \times 3 = 2.1.$$

The triangle vertex $(0.9, 0.2)$ in the $[\mathbf{d}_1, \mathbf{d}_2]$-system is mapped to

$$x_1 = 0.1 \times 1 + 0.9 \times 3 = 2.8,$$
$$x_2 = 0.8 \times 2 + 0.2 \times 3 = 2.2.$$

The triangle vertex $(0.4, 0.7)$ in the $[\mathbf{d}_1, \mathbf{d}_2]$-system is mapped to

$$x_1 = 0.6 \times 1 + 0.4 \times 3 = 1.8,$$
$$x_2 = 0.3 \times 2 + 0.7 \times 3 = 2.7.$$

1(b). The coordinates $(2, 2)$ in the $[\mathbf{e}_1, \mathbf{e}_2]$-system are mapped to

$$u_1 = \frac{2-1}{3-1} = \frac{1}{2},$$
$$u_2 = \frac{2-2}{3-2} = 0.$$

2. The local coordinates $(0.5, 0, 0.7)$ are mapped to

$$x_1 = 1 + 0.5 \times 1 = 1.5,$$
$$x_2 = 1 + 0 \times 2 = 1,$$
$$x_3 = 1 + 0.7 \times 4 = 3.8.$$

5. Each coordinate will follow similarly, so let's work out the details for x_1. First, construct the ratio which defines the relationship between a coordinate u_1 in the NDC system and coordinate x_1 in the viewport

$$\frac{u_1 - (-1)}{1 - (-1)} = \frac{x_1 - \min_1}{\max_1 - \min_1}.$$

Solve for x_1, and the equations for x_2 follow similarly, namely

$$x_1 = \frac{(\max_1 - \min_1)}{2}(u_1 + 1) + \min_1,$$

$$x_2 = \frac{(\max_2 - \min_2)}{2}(u_2 + 1) + \min_2.$$

Chapter 2

4. A triangle.

5. The length of the vector $\mathbf{v} = \begin{bmatrix} -4 \\ -3 \end{bmatrix}$ is

$$\|\mathbf{v}\| = \sqrt{4^2 + 3^2} = 5.$$

7. The normalized vector

$$\frac{\mathbf{v}}{\|\mathbf{v}\|} = \begin{bmatrix} -4/5 \\ -3/5 \end{bmatrix}.$$

9. The angle between the vectors

$$\begin{bmatrix} 5 \\ 5 \end{bmatrix} \text{ and } \begin{bmatrix} 3 \\ -3 \end{bmatrix}$$

is 90° by inspection! Sketch it and this will be clear. Additionally, notice $5 \times 3 + 5 \times -3 = 0$.

10. The orthogonal projection $\mathbf{u}$ of $\mathbf{w}$ onto $\mathbf{v}$ is determined by (2.21), or specifically,

$$\mathbf{u} = \frac{\begin{bmatrix} 1 \\ -1 \end{bmatrix} \cdot \begin{bmatrix} 3 \\ 2 \end{bmatrix}}{\sqrt{2}^2} \cdot \begin{bmatrix} 1 \\ -1 \end{bmatrix} = \begin{bmatrix} 1/2 \\ -1/2 \end{bmatrix}.$$

Draw a sketch to verify. Therefore, the $\mathbf{u}^{\perp}$ that completes the orthogonal decomposition of $\mathbf{w}$ is

$$\mathbf{u}^{\perp} = \mathbf{w} - \mathbf{u} = \begin{bmatrix} 3 \\ 2 \end{bmatrix} - \begin{bmatrix} 1 \\ -1 \end{bmatrix} = \begin{bmatrix} 2 \\ 3 \end{bmatrix}.$$

Add $\mathbf{u}^{\perp}$ to your sketch to verify.

2. Form a vector

$$\mathbf{a} = \begin{bmatrix} 1 \\ 2 \end{bmatrix}$$

perpendicular to the line, then $c = 2$, which makes the equation of the line $x_1 + 2x_2 + 2 = 0$.

3. We'll only do one of the four points. The point

$$\begin{bmatrix} 0 \\ 0 \end{bmatrix}$$

is not on the line $x_1 + 2x_2 + 2 = 0$, since $0 + 2 \times 0 + 2 = 2 \neq 0$.

4. First compute $\|\mathbf{a}\| = \sqrt{5}$. The distance d of point

$$\begin{bmatrix} 0 \\ 0 \end{bmatrix}$$

from the line $x_1 + 2x_2 + 2 = 0$ is

$$d = \frac{0 + 2 \times 0 + 2}{\sqrt{5}} \approx 0.9.$$

Sketch the line and the point to convince yourself that this is reasonable.

7. The lines are identical!

10. The line is defined by the equation $\mathbf{l}(t) = \mathbf{p} + t\mathbf{q}$. We can check that $\mathbf{l}(0) = \mathbf{p}$.

11. The point $\mathbf{r}$ is $1 : 1$ with respect to the points $\mathbf{p}$ and $\mathbf{q}$. This means that it has parameter value $t = 1/(1+1) = 1/2$ with respect to the line $\mathbf{l}(t) = \mathbf{p} + t\mathbf{q}$. Therefore,

$$\mathbf{r} = \begin{bmatrix} 2 \\ 2 \end{bmatrix},$$

and the vector

$$\mathbf{v}^{\perp} = \begin{bmatrix} -1 \\ 4 \end{bmatrix}$$

is perpendicular to $\mathbf{q} - \mathbf{p}$. The line $\mathbf{m}(t)$ is then defined as

$$\mathbf{m}(t) = \begin{bmatrix} 2 \\ 2 \end{bmatrix} + t \begin{bmatrix} -1 \\ 4 \end{bmatrix}.$$

2. The product $A\mathbf{v}$:

$$\begin{array}{cc|c}
 & & 2 \\
 & & 3 \\
\hline
0 & -1 & -3 \\
1 & 0 & 2
\end{array}.$$

The product $B\mathbf{v}$:

$$
\begin{array}{cc|c}
 & & 2 \\
 & & 3 \\
\hline
1 & -1 & -1 \\
-1 & 1/2 & -1/2
\end{array}.
$$

3. The sum

$$
A + B = \begin{bmatrix} 1 & -2 \\ 0 & 1/2 \end{bmatrix}.
$$

The product $(A + B)\mathbf{v}$:

$$
\begin{array}{cc|c}
 & & 2 \\
 & & 3 \\
\hline
1 & -2 & -4 \\
0 & 1/2 & 3/2
\end{array}
$$

and

$$
A\mathbf{v} + B\mathbf{v} = \begin{bmatrix} -4 \\ 3/2 \end{bmatrix}.
$$

7. The determinant of A:

$$
|A| = \begin{vmatrix} 0 & -1 \\ 1 & 0 \end{vmatrix} = 0 \times 0 - (-1) \times 1 = 1.
$$

8. The rank, or the number of linearly independent column vectors of B is two. We cannot find a scalar α such that

$$
\begin{bmatrix} 1 \\ -1 \end{bmatrix} = \alpha \begin{bmatrix} -1 \\ 1/2 \end{bmatrix},
$$

therefore, these vectors are linearly independent.

9. Yes, B is a symmetric matrix because

$$
B^T = \begin{bmatrix} 1 & -1 \\ -1 & 1/2 \end{bmatrix} = B.
$$

10. This one is a little tricky! It is a rotation; notice that the determinant is one. See the discussion surrounding (4.15).

12. The matrix $A^2 = A \cdot A$ equals

$$
\begin{bmatrix} -1 & 0 \\ 0 & -1 \end{bmatrix}.
$$

Does this look familiar? Reflect on the discussion surrounding (4.15).

13. We find the kernel of the matrix

$$
A = \begin{bmatrix} 2 & 6 \\ 4 & 12 \end{bmatrix}
$$

by first establishing the fact that $\mathbf{a}_2 = 3\mathbf{a}_1$. Therefore, all vectors of the form $v_1 = -3v_2$ are a part of the kernel. For example,

$$\begin{bmatrix} -3 \\ 1 \end{bmatrix}$$

is in the kernel.

14. No. Simply check that the matrix, multiplied by itself, does not result in the matrix again.

1. The transformation takes the form

Chapter 5

$$\begin{bmatrix} 2 & 6 \\ -3 & 0 \end{bmatrix} \begin{bmatrix} x_1 \\ x_2 \end{bmatrix} = \begin{bmatrix} 6 \\ 3 \end{bmatrix}.$$

2. The linear system

$$\begin{bmatrix} 2 & 6 \\ -3 & 0 \end{bmatrix} \begin{bmatrix} x_1 \\ x_2 \end{bmatrix} = \begin{bmatrix} 6 \\ 3 \end{bmatrix},$$

has the solution

$$x_1 = \frac{\begin{vmatrix} 6 & 6 \\ 3 & 0 \end{vmatrix}}{\begin{vmatrix} 2 & 6 \\ -3 & 0 \end{vmatrix}} = -1 \qquad x_1 = \frac{\begin{vmatrix} 2 & 6 \\ -3 & 3 \end{vmatrix}}{\begin{vmatrix} 2 & 6 \\ -3 & 0 \end{vmatrix}} = \frac{4}{3}.$$

5. The inverse of the matrix A:

$$A^{-1} = \begin{bmatrix} 0 & -3/9 \\ 1/6 & 1/9 \end{bmatrix}.$$

Check that $AA^{-1} = I$!

11. The matrix is

$$A = \begin{bmatrix} 1 & 0 \\ 0 & -1 \end{bmatrix}.$$

12. The kernel is the solution space of the homogeneous system $A\mathbf{v} = \mathbf{0}$. The kernel for the given matrix includes all vectors of the form $v_1 = -3v_2$. For example,

$$\begin{bmatrix} -3 \\ 1 \end{bmatrix}$$

is in the kernel.

Chapter 6

1. The point $\mathbf{q} = \begin{bmatrix} 2/3 \\ 4/3 \end{bmatrix}$. The transformed points:

$$\mathbf{r}' = \begin{bmatrix} 3 \\ 4 \end{bmatrix} \quad \mathbf{s}' = \begin{bmatrix} 11/2 \\ 6 \end{bmatrix} \quad \mathbf{q}' = \begin{bmatrix} 14/3 \\ 16/3 \end{bmatrix}.$$

The point $\mathbf{q}'$ is in fact equal to $1/3\mathbf{r}' + 2/3\mathbf{s}'$.

3. In order to rotate a point $\mathbf{x}$ around another point $\mathbf{r}$, construct the affine map

$$\mathbf{x}' = A(\mathbf{x} - \mathbf{r}) + \mathbf{r}.$$

In this exercise, we rotate $90°$,

$$\mathbf{x} = \begin{bmatrix} -2 \\ -2 \end{bmatrix}, \text{ and } \mathbf{r} = \begin{bmatrix} -2 \\ 2 \end{bmatrix}.$$

The matrix

$$A = \begin{bmatrix} 0 & -1 \\ 1 & 0 \end{bmatrix},$$

thus

$$\mathbf{x}' = \begin{bmatrix} 2 \\ 2 \end{bmatrix}.$$

Be sure to draw a sketch!

5. The point $\mathbf{x}' = \begin{bmatrix} 0 \\ 0 \end{bmatrix}.$

6. Mapping a point $\mathbf{x}$ in NDC coordinates to $\mathbf{x}'$ in the viewport involves the following steps.

1. Translate $\mathbf{x}$ by an amount which translates $\mathbf{l}_n$ to the origin.
2. Scale the resulting $\mathbf{x}$ so that the sides of the NDC box are of unit length.
3. Scale the resulting $\mathbf{x}$ so that the unit box is scaled to match the viewports dimensions.
4. Translate the resulting $\mathbf{x}$ by $\mathbf{l}_v$ to align the scaled box with the viewport.

The sides of the viewport have the lengths $\Delta_1 = 20$ and $\Delta_2 = 10$. We can then express the affine map as

$$\mathbf{x}' = \begin{bmatrix} \Delta_1/2 & 0 \\ 0 & \Delta_2/2 \end{bmatrix} \left(\mathbf{x} - \begin{bmatrix} -1 \\ -1 \end{bmatrix} \right) + \begin{bmatrix} 10 \\ 10 \end{bmatrix}.$$

Applying this affine map to the NDC points in the question yields

$$\mathbf{x}_1' = \begin{bmatrix} 10 \\ 10 \end{bmatrix}, \quad \mathbf{x}_2' = \begin{bmatrix} 30 \\ 20 \end{bmatrix}, \quad \mathbf{x}_3' = \begin{bmatrix} 15 \\ 17\frac{1}{2} \end{bmatrix}.$$

Of course we could easily "eyeball" $\mathbf{x}_1'$ and $\mathbf{x}_2'$, so they provide a good check for our affine map. Be sure to make a sketch.

7. No, affine maps do not transform perpendicular lines to perpendicular lines. A simple shear is a counterexample.

1. Form the matrix $A - \lambda I$. The characteristic equation is

$$\lambda^2 - 2\lambda - 3 = 0.$$

The eigenvalues are $\lambda_1 = -1$ and $\lambda_2 = 3$. The eigenvectors are

$$\mathbf{r}_1 = \begin{bmatrix} 1/\sqrt{2} \\ 1/\sqrt{2} \end{bmatrix} \quad \text{and} \quad \mathbf{r}_2 = \begin{bmatrix} 1/\sqrt{2} \\ -1/\sqrt{2} \end{bmatrix}.$$

3. Form $A^T A$ to be
$$\begin{bmatrix} 1.49 & -0.79 \\ -0.79 & 1.09 \end{bmatrix}.$$

The characteristic equation $|A^T A - \lambda I|$ is

$$\lambda^2 - 2.58\lambda + 1 = 0,$$

and its roots are

$$\lambda_1 = 2.10 \quad \text{and} \quad \lambda_2 = 0.475.$$

Thus, the condition number of the matrix A is $\lambda_1/\lambda_2 \approx 4.42$.

7. With the given projection matrix A, we form

$$A^T A = \begin{bmatrix} 1 & 0 \\ 0 & 0 \end{bmatrix}.$$

This symmetric matrix has eigenvalues $\lambda_1' = 1$ and $\lambda_2' = 0$, therefore, the condition number $c_A = 1/0 = \infty$. This confirms what we already know from Section 5.5: a projection matrix is not invertible.

8. Repeated linear maps do not change the eigenvectors, as demonstrated by (7.15), and the eigenvalues are easily computed. The eigenvalues of A are $\lambda_1 = -1$ and $\lambda_2 = 3$, therefore, for A^2 we have $\lambda_1 = 1$ and $\lambda_2 = 9$, and for A^2 we have $\lambda_1 = -1$ and $\lambda_2 = 27$.

1(a) First of all, draw a sketch. Before calculating the barycentric coordinates (u, v, w) of the point

$$\begin{bmatrix} 0 \\ 1.5 \end{bmatrix},$$

notice that this point is on the edge formed by $\mathbf{p}_1$ and $\mathbf{p}_3$. Thus, the barycentric coordinate $v = 0$.

The problem now is simply to find u and w such that

$$\begin{bmatrix} 0 \\ 1.5 \end{bmatrix} = u \begin{bmatrix} 1 \\ 1 \end{bmatrix} + w \begin{bmatrix} -1 \\ 2 \end{bmatrix}$$

and $u + w = 1$. This is simple enough to see, without computing! The barycentric coordinates are $(1/2, 0, 1/2)$.

1(b) Add the point

$$\mathbf{p} = \begin{bmatrix} 0 \\ 0 \end{bmatrix}$$

to the sketch from the previous exercise. Notice that $\mathbf{p}$, $\mathbf{p}_1$, and $\mathbf{p}_2$ are collinear. Thus, we know that $w = 0$.

The problem now is to find u and v such that

$$\begin{bmatrix} 0 \\ 0 \end{bmatrix} = u \begin{bmatrix} 1 \\ 1 \end{bmatrix} + v \begin{bmatrix} 2 \\ 2 \end{bmatrix}$$

and $u + v = 1$. This is easy to see: $u = 2$ and $v = -1$. If this wasn't obvious, you would calculate

$$u = \frac{\|\mathbf{p}_2 - \mathbf{p}\|}{\|\mathbf{p}_2 - \mathbf{p}_1\|},$$

then $v = 1 - u$.

Thus, the barycentric coordinates of $\mathbf{p}$ are $(2, -1, 0)$.

If you were to write a subroutine to calculate the barycentric coordinates, you would not proceed as we did here. Instead, you would calculate the area of the triangle and two of the three sub-triangle areas. The third barycentric coordinate, say w, can be calculated as $1 - u - v$.

1(c) To calculate the barycentric coordinates (i_1, i_2, i_3) of the incenter, we need the lengths of the sides of the triangles:

$$s_1 = 3 \quad s_2 = \sqrt{5} \quad s_3 = \sqrt{2}.$$

The sum of the lengths, or circumference c, is approximately $c = 6.65$. Thus the barycentric coordinates are

$$(\frac{3}{6.65}, \frac{\sqrt{5}}{6.65}, \frac{\sqrt{2}}{6.65}) = (0.45, 0.34, 0.21).$$

Always double-check that the barycentric coordinates sum to one. Additionally, check that these barycentric coordinates result in a point in the correct location:

$$0.45 \begin{bmatrix} 1 \\ 1 \end{bmatrix} + 0.34 \begin{bmatrix} 2 \\ 2 \end{bmatrix} + 0.21 \begin{bmatrix} -1 \\ 2 \end{bmatrix} = \begin{bmatrix} 0.92 \\ 1.55 \end{bmatrix}.$$

Plot this point on your sketch, and this looks correct! Recall the incenter is the intersection of the three angle bisectors.

1(d) Referring to the circumcenter equations from Section 8.3, first calculate the dot products

$$d_1 = \begin{bmatrix} 1 \\ 1 \end{bmatrix} \cdot \begin{bmatrix} -2 \\ 1 \end{bmatrix} = -1$$

$$d_2 = \begin{bmatrix} -1 \\ -1 \end{bmatrix} \cdot \begin{bmatrix} -3 \\ 0 \end{bmatrix} = 3$$

$$d_3 = \begin{bmatrix} 2 \\ -1 \end{bmatrix} \cdot \begin{bmatrix} 3 \\ 0 \end{bmatrix} = 6,$$

then $D = 18$. The barycentric coordinates (cc_1, cc_2, cc_3) of the circumcenter are

$$cc_1 = -1 \times 9/18 = -1/2$$
$$cc_2 = 3 \times 5/18 = 5/6$$
$$cc_3 = 6 \times 2/18 = 2/3.$$

The circumcenter is

$$\frac{-1}{2}\begin{bmatrix} 1 \\ 1 \end{bmatrix} + \frac{5}{6}\begin{bmatrix} 2 \\ 2 \end{bmatrix} + \frac{2}{3}\begin{bmatrix} -1 \\ 2 \end{bmatrix} = \begin{bmatrix} 0.5 \\ 2.5 \end{bmatrix}.$$

Plot this point on your sketch. Construct the perpendicular bisectors of each edge to verify.

1(e) The centroid of the triangle is simple:

$$\mathbf{c} = \frac{1}{3}\left(\begin{bmatrix} 1 \\ 1 \end{bmatrix} + \begin{bmatrix} 2 \\ 2 \end{bmatrix} + \begin{bmatrix} -1 \\ 2 \end{bmatrix}\right) = \begin{bmatrix} 2/3 \\ 1\frac{2}{3} \end{bmatrix}.$$

Chapter 9

1. The matrix

$$A = \begin{bmatrix} 1 & 1 \\ 1 & 0 \end{bmatrix},$$

thus the characteristic equation is

$$\lambda^2 - \lambda - 1 = 0,$$

which has roots $\lambda_1 = 1.62$ and $\lambda_2 = -0.62$. Since there are two distinct roots of opposite sign, the conic is a hyperbola.

2. Translate by $\mathbf{v} = \begin{bmatrix} 3 \\ -1 \end{bmatrix}$ and scale with the matrix

$$\begin{bmatrix} 1/4 & 0 \\ 0 & 1/8 \end{bmatrix}.$$

4. An ellipse is the result because a conic's type is invariant under affine maps!

5. The given conic is an ellipse. The standard form of this ellipse was rotated by $-45°$ and translated by

$$\mathbf{v} = \begin{bmatrix} -2 \\ 1 \end{bmatrix}.$$

Chapter 10

1. For the given vector

$$\mathbf{r} = \begin{bmatrix} 4 \\ 2 \\ 4 \end{bmatrix},$$

we have $\|\mathbf{r}\| = 6$. Then $\|2\mathbf{r}\| = 2\|\mathbf{r}\| = 12$.

3. The cross product of the vectors $\mathbf{v}$ and $\mathbf{w}$:

$$\mathbf{v} \wedge \mathbf{w} = \begin{bmatrix} 0 \\ -1 \\ 1 \end{bmatrix}.$$

5. The sine of the angle between $\mathbf{v}$ and $\mathbf{w}$:

$$\sin\theta = \frac{\|\mathbf{v} \wedge \mathbf{w}\|}{\|\mathbf{v}\|\|\mathbf{w}\|} = \frac{\sqrt{2}}{1 \times \sqrt{3}} = 0.82$$

This means that $\theta = 55°$. Draw a sketch to double-check this for yourself.

7. The point normal form of the plane through $\mathbf{p}$ with normal direction $\mathbf{r}$ is found by first defining the normal $\mathbf{n}$ by normalizing $\mathbf{r}$:

$$\mathbf{n} = \begin{bmatrix} 2/3 \\ 1/3 \\ 2/3 \end{bmatrix}.$$

The point normal form of the plane is

$$\mathbf{n}(\mathbf{x} - \mathbf{p}) = 0,$$

or

$$\frac{2}{3}x_1 + \frac{1}{3}x_2 + \frac{2}{3}x_3 - \frac{2}{3} = 0.$$

9. A parametric form of the plane P through the points $\mathbf{p}$, $\mathbf{q}$, and $\mathbf{r}$ is

$$P(s,t) = \mathbf{p} + s(\mathbf{q} - \mathbf{p}) + t(\mathbf{r} - \mathbf{p})$$
$$= (1 - s - t)\mathbf{p} + s\mathbf{q} + t\mathbf{r}.$$

11. The volume V formed by the vectors $\mathbf{v}, \mathbf{w}, \mathbf{u}$ can be computed as the scalar triple product

$$V = \mathbf{v} \cdot (\mathbf{w} \wedge \mathbf{u}).$$

This is invariant under cyclic permutations, thus we can also compute V as

$$V = \mathbf{u} \cdot (\mathbf{v} \wedge \mathbf{w}),$$

which allows us to reuse the cross product from Exercise 1. Thus,

$$V = \begin{bmatrix} 0 \\ 0 \\ 1 \end{bmatrix} \cdot \begin{bmatrix} 0 \\ -1 \\ 1 \end{bmatrix} = 1.$$

12. The length d of the projection of $\mathbf{w}$ bound to $\mathbf{q}$ is calculated by using two definitions of $\cos(\theta)$:

$$\cos(\theta) = \frac{d}{\|\mathbf{w}\|} \quad \text{and} \quad \cos(\theta) = \frac{\mathbf{v} \cdot \mathbf{w}}{\|\mathbf{v}\|\|\mathbf{w}\|}.$$

Solve for d,

$$d = \frac{\mathbf{v} \cdot \mathbf{w}}{\|\mathbf{v}\|} = \frac{\begin{bmatrix} 1 \\ 0 \\ 0 \end{bmatrix} \cdot \begin{bmatrix} 1 \\ 1 \\ 1 \end{bmatrix}}{\left\| \begin{bmatrix} 1 \\ 0 \\ 0 \end{bmatrix} \right\|} = 1.$$

The projection length has nothing to do with $\mathbf{q}$!

13. The distance h may be found using two definitions of $\sin(\theta)$:

$$\sin(\theta) = \frac{h}{\|\mathbf{w}\|} \quad \text{and} \quad \sin(\theta) = \frac{\|\mathbf{v} \wedge \mathbf{w}\|}{\|\mathbf{v}\|\|\mathbf{w}\|}.$$

Solve for h,

$$h = \frac{\|\mathbf{v} \wedge \mathbf{w}\|}{\|\mathbf{v}\|} = \frac{\left\| \begin{bmatrix} 0 \\ -1 \\ 1 \end{bmatrix} \right\|}{\left\| \begin{bmatrix} 1 \\ 0 \\ 0 \end{bmatrix} \right\|} = \sqrt{2}.$$

14. The cross product of parallel vectors results in the zero vector.

17. We use barycentric coordinates to determine the color at a point inside the triangle given the colors at the vertices:

$$\mathbf{i_c} = \frac{1}{3}\mathbf{i_p} + \frac{1}{3}\mathbf{i_q} + \frac{1}{3}\mathbf{i_r}$$

$$= \frac{1}{3}\begin{bmatrix}1\\0\\0\end{bmatrix} + \frac{1}{3}\begin{bmatrix}0\\1\\0\end{bmatrix} + \frac{1}{3}\begin{bmatrix}1\\0\\0\end{bmatrix}$$

$$= \begin{bmatrix}2/3\\1/3\\0\end{bmatrix}$$

which is reddish-yellow.

18. The vector $\mathbf{u}$ is *not* an element of the subspace defined by $\mathbf{w}$ and $\mathbf{v}$ because we cannot find scalars s and t such that $\mathbf{u} = s\mathbf{w} + t\mathbf{v}$.

19. This solution is easy enough to determine without formulas, but let's practice using the equations. First, project $\mathbf{w}$ onto $\mathbf{v}$, forming

$$\mathbf{u}_1 = \frac{\mathbf{v}\cdot\mathbf{w}}{\|\mathbf{v}\|^2}\mathbf{v} = \begin{bmatrix}1\\0\\0\end{bmatrix}.$$

Next form $\mathbf{u}_2$ so that $\mathbf{w} = \mathbf{u}_1 + \mathbf{u}_2$:

$$\mathbf{u}_2 = \mathbf{w} - \mathbf{u}_1 = \begin{bmatrix}0\\1\\1\end{bmatrix}.$$

To complete the orthogonal frame, we compute

$$\mathbf{u}_3 = \mathbf{u}_1 \wedge \mathbf{u}_2 = \begin{bmatrix}0\\-1\\1\end{bmatrix}.$$

Draw a sketch to see what you have created. This frame is not orthonormal, but that would be easy to do.

Chapter 11

1. Make a sketch! You will find that the line is parallel to the plane. The actual calculations would involve finding the parameter t on the line for the intersection:

$$t = \frac{(\mathbf{p}-\mathbf{q})\cdot\mathbf{n}}{\mathbf{v}\cdot\mathbf{n}}.$$

In this exercise, $\mathbf{v}\cdot\mathbf{n} = 0$.

5. The vector $\mathbf{a}$ is projected to the vector

$$\mathbf{a}' = \mathbf{a} - \frac{\mathbf{a}\cdot\mathbf{n}}{\mathbf{v}\cdot\mathbf{n}}\mathbf{v}.$$

6. The reflected direction is

$$\mathbf{v}' = \begin{bmatrix} 1/3 \\ 1/3 \\ -2/3 \end{bmatrix} + \frac{4}{3} \begin{bmatrix} 0 \\ 0 \\ 1 \end{bmatrix} = \begin{bmatrix} 1/3 \\ 1/3 \\ 2/3 \end{bmatrix}.$$

7. With a sketch, you can find the intersection point without calculation, but let's practice calculating the point. The linear system is:

$$\begin{bmatrix} 1 & 1 & 0 \\ 1 & 0 & 0 \\ 0 & 0 & 1 \end{bmatrix} \mathbf{x} = \begin{bmatrix} 1 \\ 1 \\ 4 \end{bmatrix},$$

and the solution is $\begin{bmatrix} 1 \\ 0 \\ 4 \end{bmatrix}$.

8. Just like the previous Exercise, with a sketch you can find the intersection line without calculation, but let's practice calculating the line. The planes $x_1 = 1$ and $x_3 = 4$ have normal vectors

$$\mathbf{n}_1 = \begin{bmatrix} 1 \\ 0 \\ 0 \end{bmatrix} \quad \text{and} \quad \mathbf{n}_2 = \begin{bmatrix} 0 \\ 0 \\ 1 \end{bmatrix},$$

respectively. Form the vector

$$\mathbf{v} = \mathbf{n}_1 \wedge \mathbf{n}_2 = \begin{bmatrix} 0 \\ 1 \\ 0 \end{bmatrix},$$

and then a third plane is defined by $\mathbf{vx} = 0$. Set up a linear system to intersect these three planes:

$$\begin{bmatrix} 1 & 0 & 0 \\ 0 & 0 & 1 \\ 0 & 1 & 0 \end{bmatrix} \mathbf{p} = \begin{bmatrix} 1 \\ 4 \\ 0 \end{bmatrix},$$

and the solution is

$$\begin{bmatrix} 1 \\ 0 \\ 4 \end{bmatrix}.$$

The intersection of the given two planes is the line

$$\mathbf{l}(t) = \begin{bmatrix} 1 \\ 0 \\ 4 \end{bmatrix} + t \begin{bmatrix} 0 \\ 1 \\ 0 \end{bmatrix}.$$

9. Set

$$\mathbf{b}_1 = \begin{bmatrix} 1 \\ 0 \\ 0 \end{bmatrix},$$

and then calculate

$$\mathbf{b}_2 = \begin{bmatrix} 1 \\ 1 \\ 0 \end{bmatrix} - \begin{bmatrix} 1 \\ 0 \\ 0 \end{bmatrix} = \begin{bmatrix} 0 \\ 1 \\ 0 \end{bmatrix},$$

$$\mathbf{b}_3 = \begin{bmatrix} -1 \\ -1 \\ 1 \end{bmatrix} - (-1) \begin{bmatrix} 1 \\ 0 \\ 0 \end{bmatrix} - (-1) \begin{bmatrix} 0 \\ 1 \\ 0 \end{bmatrix} = \begin{bmatrix} 0 \\ 0 \\ 1 \end{bmatrix}.$$

Each vector is unit length, so we have an orthonormal frame.

10. We'll see that using cross products in 3D is less work and conceptually easier than Gram-Schmidt. Set

$$\mathbf{b}_1 = \mathbf{v}_1,$$
$$\mathbf{b}_3 = \mathbf{v}_1 \wedge \mathbf{v}_2,$$
$$\mathbf{b}_2 = \mathbf{b}_3 \wedge \mathbf{b}_1.$$

For this simple example, the vectors were unit length, however in general they will need normalized. If $\mathbf{b}_3$ and $\mathbf{b}_1$ are unit length, then $\mathbf{b}_2$ will be unit length automatically!

Chapter 12

2. The scale matrix is
$$\begin{bmatrix} 2 & 0 & 0 \\ 0 & 1/4 & 0 \\ 0 & 0 & -4 \end{bmatrix}.$$
This matrix changes the volume of a unit cube to be $2 \times 1/4 \times -4 = -2$.

4. The shear matrix is
$$\begin{bmatrix} 1 & 0 & -a/c \\ 0 & 1 & -b/c \\ 0 & 0 & 1 \end{bmatrix}.$$
Shears do not change volume, therefore the volume of the mapped unit cube is still 1.

6. To rotate about the vector
$$\begin{bmatrix} -1 \\ 0 \\ -1 \end{bmatrix},$$
first form the unit vector
$$\mathbf{a} = \begin{bmatrix} -1/\sqrt{2} \\ 0 \\ -1/\sqrt{2} \end{bmatrix}.$$

Then, following (12.9), the rotation matrix is

$$\begin{bmatrix} \frac{1}{2}\left(1+\frac{\sqrt{2}}{2}\right) & 1/2 & \frac{1}{2}\left(1-\frac{\sqrt{2}}{2}\right) \\ -1/2 & \sqrt{2}/2 & 1/2 \\ \frac{1}{2}\left(1-\frac{\sqrt{2}}{2}\right) & -1/2 & \frac{1}{2}\left(1+\frac{\sqrt{2}}{2}\right) \end{bmatrix}.$$

The matrices for rotating about an arbitrary vector are difficult to verify by inspection. One test is to check the vector about which we rotated. You'll find that

$$\begin{bmatrix} -1 \\ 0 \\ -1 \end{bmatrix} \rightarrow \begin{bmatrix} -1 \\ 0 \\ -1 \end{bmatrix},$$

which is precisely correct.

7. The resulting matrix is

$$\begin{bmatrix} 2 & 3 & -4 \\ 3 & 9 & -4 \\ -1 & -9 & 4 \end{bmatrix}.$$

9. The transpose of the given matrix is

$$\begin{bmatrix} 1 & -1 & 2 \\ 5 & -2 & 3 \\ -4 & 0 & -4 \end{bmatrix}.$$

10. The matrix $(A^{\mathrm{T}})^{\mathrm{T}}$ is simply A.

11. To find the projection direction, we solve the homogeneous system $A\mathbf{d} = \mathbf{0}$. This system gives us two equations to satisfy:

$$d_1 - d_3 = 0 \quad \text{and} \quad d_2 = 0,$$

which have an infinite number of nontrivial solutions,

$$\mathbf{d} = \begin{bmatrix} c \\ 0 \\ c \end{bmatrix}.$$

We can normalize this result to say the projection direction is

$$\mathbf{d} = \begin{bmatrix} 1/\sqrt{2} \\ 0 \\ 1/\sqrt{2} \end{bmatrix}.$$

In computer graphics, we often speak of *eye coordinates* which assumes that the projection plane is $x_3 = 0$, which has normal

$$\mathbf{n} = \begin{bmatrix} 0 \\ 0 \\ 1 \end{bmatrix}.$$

The angle that the projection direction forms with this plane is found by calculating

$$\cos(\theta) = \mathbf{d} \cdot \mathbf{n} = 1/\sqrt{2}.$$

This corresponds to a 45° angle, so this is an oblique parallel projection.

12. The kernel is all vectors of the form

$$\begin{bmatrix} c \\ 0 \\ c \end{bmatrix};$$

these are all mapped to the zero vector.

13. The inverse matrices are as follows.

$$\text{rotation:} \quad \begin{bmatrix} 1/\sqrt{2} & 0 & -1/\sqrt{2} \\ 0 & 1 & 0 \\ 1/\sqrt{2} & 0 & 1/\sqrt{2} \end{bmatrix}$$

$$\text{scale:} \quad \begin{bmatrix} 2 & 0 & 0 \\ 0 & 4 & 0 \\ 0 & 0 & 1/2 \end{bmatrix}$$

$$\text{projection:} \quad \text{No inverse exists}$$

Chapter 13

1. As always, draw a sketch when possible! Construct the parallel projection map defined in (13.8):

$$\mathbf{x}' = \begin{bmatrix} 8/14 & 0 & -8/14 \\ 0 & 1 & 0 \\ -6/14 & 0 & 6/14 \end{bmatrix} \mathbf{x} + \begin{bmatrix} 6/14 \\ 0 \\ 6/14 \end{bmatrix}.$$

The point

$$\mathbf{x}_1 = \begin{bmatrix} 1 \\ 0 \\ 0 \end{bmatrix}$$

is mapped to $\mathbf{x}_1' = \mathbf{x}_1$. By inspection of your sketch and the plane equation, this is clear since $\mathbf{x}_1$ is already in the plane.
The point

$$\mathbf{x}_2 = \begin{bmatrix} 0 \\ 1 \\ 0 \end{bmatrix}$$

is mapped to the point

$$\mathbf{x}_2' = \begin{bmatrix} 3/7 \\ 1 \\ 3/7 \end{bmatrix}.$$

Check that $\mathbf{x}_2'$ lies in the projection plane. This should seem reasonable, too. The projection direction $\mathbf{v}$ is parallel to the $\mathbf{e}_2$-axis, thus this coordinate is unchanged. Additionally, $\mathbf{v}$ projects equally into the $\mathbf{e}_1$- and $\mathbf{e}_3$-axes.

Try the other points yourself, and rationalize each resulting point as we have done here.

3. Construct the perspective projection vector equation defined in (13.10). This will be different for each of the $\mathbf{x}_i$:

$$\mathbf{x}_1' = \frac{3/5}{3/5}\mathbf{x}_1 = \begin{bmatrix} 1 \\ 0 \\ 0 \end{bmatrix}$$

$$\mathbf{x}_2' = \frac{3/5}{0}\mathbf{x}_2$$

$$\mathbf{x}_3' = \frac{3/5}{-4/5}\mathbf{x}_3 = \begin{bmatrix} 0 \\ 0 \\ 3/4 \end{bmatrix}$$

$$\mathbf{x}_4' = \frac{3/5}{4/5}\mathbf{x}_4 = \begin{bmatrix} 0 \\ 0 \\ 3/4 \end{bmatrix}.$$

There is no solution for $\mathbf{x}_2'$ because $\mathbf{x}_2$ is projected parallel to the plane.

6. Construct the matrices in (13.6). For this problem, use the notation $A = YX^{-1}$, where

$$Y = \begin{bmatrix} \mathbf{y}_2 - \mathbf{y}_1 & \mathbf{y}_3 - \mathbf{y}_1 & \mathbf{y}_4 - \mathbf{y}_1 \end{bmatrix}$$
$$= \begin{bmatrix} 1 & 1 & 1 \\ -1 & 0 & 0 \\ 0 & -1 & 1 \end{bmatrix}$$

and

$$X = \begin{bmatrix} \mathbf{x}_2 - \mathbf{x}_1 & \mathbf{x}_3 - \mathbf{x}_1 & \mathbf{x}_4 - \mathbf{x}_1 \end{bmatrix}$$
$$= \begin{bmatrix} -1 & -1 & -1 \\ 1 & 0 & 0 \\ 0 & -1 & 1 \end{bmatrix}.$$

The first task is to find X^{-1}:

$$X^{-1} = \begin{bmatrix} 0 & 1 & 0 \\ -1/2 & -1/2 & -1/2 \\ -1/2 & -1/2 & 1/2 \end{bmatrix}.$$

Always check that $XX^{-1} = I$. Now A takes the form:

$$A = \begin{bmatrix} -1 & 0 & 0 \\ 0 & -1 & 0 \\ 0 & 0 & 1 \end{bmatrix},$$

which isn't surprising at all if you sketched the tetrahedra formed by the $\mathbf{x}_i$ and $\mathbf{y}_i$.

7. The point $\begin{bmatrix} 1 \\ 1 \\ 1 \end{bmatrix}$ is mapped to $\begin{bmatrix} -1 \\ -1 \\ 1 \end{bmatrix}$.

10. First, translate so that a point $\mathbf{p}$ on the line is at the origin. Second, rotate $90°$ about the $\mathbf{e}_1$-axis with the matrix R. Third, remove the initial translation. Let's define

$$\mathbf{p} = \begin{bmatrix} 1 \\ 1 \\ 0 \end{bmatrix}, \qquad R = \begin{bmatrix} 1 & 0 & 0 \\ 0 & \cos(90°) & -\sin(90°) \\ 0 & \sin(90°) & \cos(90°) \end{bmatrix},$$

and then the affine map is defined as

$$\mathbf{q}' = R(\mathbf{q} - \mathbf{p}) + \mathbf{p} = \begin{bmatrix} 1 \\ 1 \\ -1 \end{bmatrix}.$$

Chapter 14

1. The solution vector $\mathbf{v} = \begin{bmatrix} 1 \\ -4 \\ 0 \\ -1 \end{bmatrix}$.

2. The solution vector $\mathbf{v} = \begin{bmatrix} 0 \\ 0 \\ -1 \end{bmatrix}$.

4. Use the explicit form of the line, $x_2 = ax_1 + b$. The overdetermined system is

$$\begin{bmatrix} 1 & 1 \\ 0 & 1 \\ -3 & 1 \\ 3 & 1 \\ 0 & 1 \end{bmatrix} \begin{bmatrix} a \\ b \end{bmatrix} = \begin{bmatrix} -1 \\ 0 \\ 0 \\ -1 \\ 1 \end{bmatrix}.$$

We want to form a square matrix, and the solution should be the least squares solution. Following (14.7), the linear system becomes

$$\begin{bmatrix} 19 & 1 \\ 1 & 5 \end{bmatrix} \begin{bmatrix} a \\ b \end{bmatrix} = \begin{bmatrix} -4 \\ -1 \end{bmatrix}.$$

Thus, the least squares line is $x_2 = -2.2x_1 + 0.24$. Sketch the data and the line to convince yourself.

6. The determinant is equal to 5.

8. The rank is 3 since the determinant is nonzero.

9. The forward substitution algorithm for solving the lower triangular linear system $L\mathbf{y} = \mathbf{b}$ is as follows.

> **Forward substitution:**
> $y_1 = b_1$
> For $j = 2, \ldots, n$
> $\qquad y_j = b_j - y_1 l_{j,1} - \ldots - y_{j-1} l_{n,j-1}$

10. The solution is $\begin{bmatrix} 2 \\ 2 \\ 2 \end{bmatrix}$.

1. Yes. We have to check the four defining properties from Section 15.3. Each is easily verified.

Chapter 15

2. A simple inner product is the following: if A and B are two matrices in the space, then set $AB = a_{11}b_{11} + a_{12}b_{12} \ldots + a_{33}b_{33}$. This inner product satisfies all properties 1-4. Note that it is *not* the product of the two matrices!

4. One such basis is given by the four matrices

$$\begin{bmatrix} 1 & 0 \\ 0 & 0 \end{bmatrix}, \begin{bmatrix} 0 & 1 \\ 0 & 0 \end{bmatrix}, \begin{bmatrix} 0 & 0 \\ 1 & 0 \end{bmatrix}, \begin{bmatrix} 0 & 0 \\ 0 & 1 \end{bmatrix}.$$

6. No. If we multiply an element of that set by -1, we produce a vector which is not in the set, a violation of the linearity condition.

7. Gauss elimination produces the matrix

$$\begin{bmatrix} 1 & 2 & 0 \\ 0 & 0 & 1 \\ 0 & 0 & 0 \\ 0 & 0 & 0 \end{bmatrix}$$

from which we conclude the rank is 2.

9. No. The linearity conditions are violated. For example $\Phi(-\mathbf{u}) = \Phi(\mathbf{u})$, contradicting the linearity condition which would demand $\Phi(\alpha\mathbf{u}) = \alpha\Phi(\mathbf{u})$ with $\alpha = -1$.

11. They are $-2, -1, 1, 2$. An easy way to see this is to exchange rows of the matrix (keeping track of determinant sign changes) until it is of diagonal form.

Chapter 16

1. We have

$$D = \begin{bmatrix} 4 & 0 & 0 \\ 0 & -8 & 0 \\ 0 & 0 & 2 \end{bmatrix} \quad \text{and} \quad R = \begin{bmatrix} 0 & 0 & 1 \\ 2 & 0 & 2 \\ 1 & 0 & 0 \end{bmatrix}.$$

Hence,

$$
\begin{aligned}
\mathbf{u}^{(2)} &= \begin{bmatrix} 0.25 & 0 & 0 \\ 0 & -0.125 & 0 \\ 0 & 0 & 0.5 \end{bmatrix} \left(\begin{bmatrix} 0 \\ -2 \\ 0 \end{bmatrix} - \begin{bmatrix} 0 & 0 & -1 \\ 2 & 0 & 2 \\ 1 & 0 & 0 \end{bmatrix} \begin{bmatrix} 0 \\ 0 \\ 0 \end{bmatrix} \right) \\
&= \begin{bmatrix} 0 \\ -0.25 \\ 0 \end{bmatrix}.
\end{aligned}
$$

Next,

$$
\begin{aligned}
\mathbf{u}^{(3)} &= \begin{bmatrix} 0.25 & 0 & 0 \\ 0 & 0.125 & 0 \\ 0 & 0 & 0.5 \end{bmatrix} \left(\begin{bmatrix} 2 \\ -2 \\ 0 \end{bmatrix} - \begin{bmatrix} 0 & 0 & -1 \\ 2 & 0 & 2 \\ 1 & 0 & 0 \end{bmatrix} \begin{bmatrix} 0.5 \\ -0.25 \\ 0 \end{bmatrix} \right) \\
&= \begin{bmatrix} 0.5 \\ -0.275 \\ -0.25 \end{bmatrix}.
\end{aligned}
$$

Similarly, we find

$$\mathbf{u}^{(4)} = \begin{bmatrix} 0.4375 \\ -0.3125 \\ -0.25 \end{bmatrix}.$$

The true solution is

$$\mathbf{u} = \begin{bmatrix} 0.444 \\ -0.306 \\ -0.222 \end{bmatrix}.$$

3. We have

$$\mathbf{r}^{(2)} = \begin{bmatrix} 5 \\ 3 \\ 5 \end{bmatrix}, \quad \mathbf{r}^{(3)} = \begin{bmatrix} 25 \\ 13 \\ 21 \end{bmatrix}, \quad \mathbf{r}^{(4)} = \begin{bmatrix} 121 \\ 55 \\ 93 \end{bmatrix}.$$

The last two give the ratios 4.84, 4.23, 4.43. The true dominant eigenvalue is 4.646.

4. We have

$$\mathbf{r}^{(2)} = \begin{bmatrix} 0 \\ -1 \\ 6 \end{bmatrix}, \quad \mathbf{r}^{(3)} = \begin{bmatrix} 48 \\ -13 \\ 2 \end{bmatrix}, \quad \mathbf{r}^{(4)} = \begin{bmatrix} -368 \\ -17 \\ 410 \end{bmatrix}.$$

The last two give the ratios -7.66, 1.31, 205. This is not close yet to revealing the true dominant eigenvalue -13.02.

5. The first three iterations yield the vectors

$$
\begin{bmatrix} -0.5 \\ -0.25 \\ -0.25 \\ -0.5 \end{bmatrix}, \quad
\begin{bmatrix} 0.187 \\ 0.437 \\ 0.4375 \\ 0.187 \end{bmatrix}, \quad
\begin{bmatrix} -0.156 \\ 0.093 \\ 0.0937 \\ -0.156 \end{bmatrix}.
$$

The actual solution is

$$
\begin{bmatrix} -0.041 \\ 0.208 \\ 0.208 \\ -0.041 \end{bmatrix}.
$$

6. For any i,

$$
\mathbf{u}^i = \begin{bmatrix} 1/i \\ 1 \\ -1/i \end{bmatrix}.
$$

Hence there is a limit, namely $[0, 1, 0]^{\mathrm{T}}$.

9. The outline of the graph of all 2D unit vectors in the Manhattan norm is a diamond. Notice that the following vectors are unit length in this norm:

$$
\begin{bmatrix} 1 \\ 0 \end{bmatrix} \quad
\begin{bmatrix} 0 \\ 1 \end{bmatrix} \quad
\begin{bmatrix} 0.5 \\ 0.5 \end{bmatrix} \quad
\begin{bmatrix} -1 \\ 0 \end{bmatrix} \quad
\begin{bmatrix} -0.5 \\ -0.5 \end{bmatrix},
$$

and make a sketch for yourself.

10. The outline of the graph of all 2D unit vectors in the L_∞ norm is a square. Notice that the following vectors are unit length in this norm:

$$
\begin{bmatrix} 1 \\ 0 \end{bmatrix} \quad
\begin{bmatrix} 0 \\ 1 \end{bmatrix} \quad
\begin{bmatrix} 1.0 \\ 1.0 \end{bmatrix} \quad
\begin{bmatrix} -1 \\ 0 \end{bmatrix} \quad
\begin{bmatrix} -1.0 \\ -1.0 \end{bmatrix},
$$

and make a sketch for yourself.

12. The linear system after running through the algorithm for $j = 1$ is

$$
\begin{bmatrix} -1.41 & -1.41 & -1.41 \\ 0 & -0.14 & -0.14 \\ 0 & -0.1 & 0.2 \end{bmatrix} \mathbf{u} = \begin{bmatrix} -1.41 \\ 0 \\ 0.3 \end{bmatrix}.
$$

The linear system for $j = 2$ is

$$
\begin{bmatrix} -1.41 & -1.41 & -1.41 \\ 0 & 0.17 & -0.02 \\ 0 & 0 & 0.24 \end{bmatrix} \mathbf{u} = \begin{bmatrix} -1.41 \\ -0.19 \\ 0.24 \end{bmatrix},
$$

and from this one we can use back substitution to find the solution

$$
\mathbf{u} = \begin{bmatrix} 1 \\ -1 \\ 1 \end{bmatrix}.
$$

Chapter 17

3. A rhombus is equilateral but not equiangular.

5. The winding number is 0.

6. The area is 3.

8. The estimate normal to the polygon, using the methods from Section 17.7, is

$$\mathbf{n} = \frac{1}{\sqrt{6}} \begin{bmatrix} 1 \\ 1 \\ 2 \end{bmatrix}.$$

9. The outlier is $\mathbf{p}_4$; it should be

$$\begin{bmatrix} 0 \\ 3 \\ -3/2 \end{bmatrix}.$$

Three ideas for planarity tests were given in Section 17.8. There is not just one way to solve this problem, but one is not suitable for finding the outlier. If we use the "average normal test," which calculates the centroid of all points, then it will lie outside the plane since it is calculated using the outlier. Either the "volume test" or the "plane test" could be used.

For so few points, we can practically determine the true plane and the outlier. In "real world" applications, where we would have thousands of points, the question above is ill-posed. We would be inclined to calculate an average plane and then check if any points deviate significantly from this plane. Can you think of a good method to construct such an average plane?

Chapter 18

2. The evaluation point at $t = 1/4$ is calculated as follows:

$$\mathbf{b}_0 = \begin{bmatrix} 0 \\ 0 \\ 0 \end{bmatrix}$$

$$\mathbf{b}_1 = \begin{bmatrix} 4 \\ 0 \\ 0 \end{bmatrix} \quad \begin{bmatrix} 1 \\ 0 \\ 0 \end{bmatrix}$$

$$\mathbf{b}_2 = \begin{bmatrix} 4 \\ 4 \\ 0 \end{bmatrix} \quad \begin{bmatrix} 4 \\ 1 \\ 0 \end{bmatrix} \quad \begin{bmatrix} 7/4 \\ 1/4 \\ 0 \end{bmatrix}$$

$$\mathbf{b}_3 = \begin{bmatrix} 4 \\ 4 \\ 4 \end{bmatrix} \quad \begin{bmatrix} 4 \\ 4 \\ 1 \end{bmatrix} \quad \begin{bmatrix} 4 \\ 7/4 \\ 1/4 \end{bmatrix} \quad \begin{bmatrix} 37/16 \\ 10/16 \\ 1/16 \end{bmatrix}.$$

3. The first derivative is

$$\dot{\mathbf{b}}(1/4) = 3 \left(\begin{bmatrix} 4 \\ 7/4 \\ 1/4 \end{bmatrix} - \begin{bmatrix} 7/4 \\ 1/4 \\ 0 \end{bmatrix} \right) = 3 \begin{bmatrix} 9/4 \\ 6/4 \\ 1/4 \end{bmatrix}.$$

The second derivative is

$$\ddot{\mathbf{b}}(1/4) = 3 \times 2 \left(\begin{bmatrix} 4 \\ 4 \\ 1 \end{bmatrix} - 2 \begin{bmatrix} 4 \\ 1 \\ 0 \end{bmatrix} + \begin{bmatrix} 1 \\ 0 \\ 0 \end{bmatrix} \right) = 6 \begin{bmatrix} -3 \\ 2 \\ 1 \end{bmatrix}.$$

5. The monomial coefficients $\mathbf{a}_i$ are

$$\mathbf{a}_0 = \begin{bmatrix} 0 \\ 0 \\ 0 \end{bmatrix} \quad \mathbf{a}_1 = \begin{bmatrix} 12 \\ 0 \\ 0 \end{bmatrix} \quad \mathbf{a}_2 = \begin{bmatrix} -12 \\ 12 \\ 0 \end{bmatrix} \quad \mathbf{a}_3 = \begin{bmatrix} 4 \\ -8 \\ 4 \end{bmatrix}.$$

For additional insight, compare these vectors to the point and derivatives of the Bézier curve at $t = 0$.

8. The Frenet frame at $t = 1$ is computed by first finding the first and second derivatives of the curve there:

$$\dot{\mathbf{b}}(1) = 3 \begin{bmatrix} 0 \\ 0 \\ 4 \end{bmatrix} \quad \ddot{\mathbf{b}}(1) = 6 \begin{bmatrix} 0 \\ -4 \\ 4 \end{bmatrix}.$$

Because of the homogeneous property of the cross product, see Section 10.2, the 3 and 6 factors from the derivatives simply cancel in the calculation of the Frenet frame. The Frenet frame:

$$\mathbf{f}_1 = \begin{bmatrix} 0 \\ 0 \\ 1 \end{bmatrix} \quad \mathbf{f}_2 = \begin{bmatrix} 0 \\ 1 \\ 0 \end{bmatrix} \quad \mathbf{f}_3 = \begin{bmatrix} 1 \\ 0 \\ 0 \end{bmatrix}.$$

9. We will evaluate the curve at $t = 2$ using the de Casteljau algorithm. Let's use the triangular schematic to guide the evaluation.

$$\mathbf{b}_0 = \begin{bmatrix} 0 \\ 0 \\ 0 \end{bmatrix}$$

$$\mathbf{b}_1 = \begin{bmatrix} 4 \\ 0 \\ 0 \end{bmatrix} \quad \begin{bmatrix} 8 \\ 0 \\ 0 \end{bmatrix}$$

$$\mathbf{b}_2 = \begin{bmatrix} 4 \\ 4 \\ 0 \end{bmatrix} \quad \begin{bmatrix} 4 \\ 8 \\ 0 \end{bmatrix} \quad \begin{bmatrix} 0 \\ 16 \\ 0 \end{bmatrix}$$

$$\mathbf{b}_3 = \begin{bmatrix} 4 \\ 4 \\ 4 \end{bmatrix} \quad \begin{bmatrix} 4 \\ 4 \\ 8 \end{bmatrix} \quad \begin{bmatrix} 4 \\ 0 \\ 16 \end{bmatrix} \quad \begin{bmatrix} 8 \\ -16 \\ 32 \end{bmatrix}.$$

We have moved fairly far from the polygon!

10. To achieve tangent continuity, the curve $\mathbf{c}(t)$ must have

$$\mathbf{c}_3 = \mathbf{b}_0,$$

and $\mathbf{c}_2$ must be on the line formed by $\mathbf{b}_0$ and $\mathbf{b}_1$:

$$\mathbf{c}_2 = \mathbf{c}_0 + c[\mathbf{b}_0 - \mathbf{b}_1],$$

for positive c. Let's choose $c = 1$, then

$$\mathbf{c}_2 = \begin{bmatrix} -4 \\ 0 \\ 0 \end{bmatrix}.$$

We are free to choose $\mathbf{c}_1$ and $\mathbf{c}_0$ anywhere!

Glossary

In this Glossary, we give brief definitions of the major concepts in the book. We try to avoid equations here, so that we give a slightly different perspective compared to what you find in the text.

affine map: A map which leaves geometric (i.e., linear) relationships between points unchanged. For instance, midpoints are mapped to midpoints. In a given coordinate system, an affine map is described by a transformation matrix and a translation vector.

affine space: A set of points with the property that any barycentric combination of two points is again in the space.

barycentric combination: A weighted average of points where the sum of the weights equals one.

barycentric coordinates: When a point is expressed as a barycentric combination of the three vertices of a triangle, the coefficients in that combination are called barycentric coordinates.

basis: For a linear space of dimension n, any set of n linearly independent vectors is a basis, meaning that every vector in the space may be uniquely expressed as a linear combination of these n basis vectors.

basis transformation: The linear map taking one set of n basis vectors to another set of n basis vectors.

centroid: The center of mass, or the average of a set of points, with all weights being equal (and summing to one).

collinear: A set of points is collinear if they all lie on the same straight line.

condition number: A function measuring how sensitive a map is to changes in its input. If a small changes in the input cause large changes in the output, then the condition number is large.

conic section: The intersection curve of a double cone with a plane. A nondegenerate conic section is either an ellipse, a parabola, or a hyperbola.

convex: A point set is convex if the straight line segment through any of its points is completely contained inside the set. Example: all points on and inside of a sphere form a convex set; all points on and inside of an hourglass do not.

coordinates: A vector in an n-dimensional linear space may be uniquely written as a linear combination of a set of basis vectors. The coefficients in that combination are the vector's coordinates with respect to that basis.

coplanar: A set of points is coplanar if all points lie on the same plane.

Cramer's Rule: A method for solving a linear system explicitly using ratios of determinants.

cross product: The cross product of two 3D vectors results in a third vector which is perpendicular to them.

curve: The locus of a moving point.

determinant: A linear map takes a geometric object to another geometric object. The ratio of their volumes is the map's determinant.

dimension: The number of linearly independent vectors needed to span a linear space.

dot product: A value of two vectors which, in the case of unit vectors, is equal to the cosine of the angle between the vectors.

dual space: Consider all linear maps from a linear space into the 1D linear space of scalars. All these maps form a linear space themselves, the dual space of the original space.

eigenvalue: If a linear map happens to take some vector to itself, multiplied by some constant, then that constant is an eigenvalue of the map.

eigenvector: A vector whose direction is unchanged by a linear map.

ellipse: A bounded conic section. When written in implicit form, its 2×2 matrix has two positive eigenvalues.

Gauss elimination: The process of transforming a linear system into an equivalent linear system whose coefficient matrix is of upper triangular form.

Gauss-Seidel iteration: Solving a linear system by successively improving an initial guess for the solution. Similar to Gauss-Jacobi iteration.

homogeneous coordinates: Points in 2D affine space may be viewed as projections of points in 3D affine space, all being multiples of each other. The coordinates of any of these points are the homogeneous coordinates of the given point.

homogeneous linear system: A linear system whose right-hand side consists of zeroes only.

hyperbola: An unbounded conic section with two branches. When written in implicit form, its 2×2 matrix has a positive and a negative eigenvalue.

idempotent: A map is idempotent if repeated applications of the map yield the same result as only one application: for example, projections.

identity matrix: A square matrix with entries 1 on the diagonal and entries 0 elsewhere. This matrix maps every object to itself.

image: The result of a map.

incenter: There is exactly one circle inside a triangle which has its three edges as tangents. The center of this circle is the triangle's incenter.

inner product: Given two elements of a linear space, their inner product is a scalar. The inner product of two orthogonal vectors is zero.

inverse matrix: A matrix maps an object to another object. The inverse matrix undoes this map.

kernel: The set of vectors being mapped to the zero vector by a linear map. Also called the null space.

least squares: A method for finding the best approximate solution to an overdetermined problem.

line: Given two points in affine space, the set of all barycentric combinations is a line.

line segment: As above, but with all coefficients of the barycentric combinations being nonnegative.

linear combination: A weighted sum of vectors.

linear independence: A set of vectors is linearly independent if none of its elements may be written as a linear combination of the remaining ones.

linear interpolation: A weighted average of two points, where the weights sum to one and are linear functions of a parameter.

linear map: A map of a linear space to another linear space such that linear relationships between points are not changed by the map. In a given coordinate system, a linear map is described by a matrix.

linear space: A set (whose elements are called vectors) with the property that any linear combination of any two vectors is also in the set.

linear system: If we attempt to write a given vector as a linear combination (with unknown coefficients) of a set of vectors, then the resulting set of equations is called a linear system.

length: The distance between two points forming a line segment.

local coordinates: A specific coordinate system used to define a geometric object. This object may then be placed in a global coordinate system.

map: The process of changing objects. Example: rotating and scaling an object. The object being mapped is called the preimage, the result of the map is called the image.

matrix: The coordinates of a linear map, written in a rectangular array of scalars.

nonlinear map: A map which does not preserve linear relationships. Example: a perspective map.

norm: A function which assigns a length to a vector.

null space: The set of vectors mapped to the zero vector by a linear map is called that map's null space. Also called the kernel.

orthogonality: Two vectors are orthogonal if their dot product vanishes.

parabola: An unbounded conic section with one branch. When written in implicit form, its 2×2 matrix has one zero eigenvalue.

parallel: Two lines or two planes are parallel if they have no point in common. Two vectors are parallel if they are multiples of each other.

plane: Given three points in affine space, the set of all barycentric combinations is a plane.

point: A location, i.e., an element of an affine space.

point cloud: A set of 3D points without any additional structure.

polygon: The set of edges formed by connecting a set of points.

projection: A linear map which reduces the dimension of an object.

range: The set of possible images resulting from a given map.

rank: For a set of vectors, the maximal number of linearly independent vectors. For a matrix, the maximal number of linearly independent rows or columns.

ratio: A measure of how three collinear points are distributed. If one is the midpoint of the other two, the ratio is 1.

rigid body motion: An affine map which leaves distances and angles unchanged.

singular matrix: A matrix describing a linear map which is a projection, meaning that it reduces the dimension of an object.

span: For a given set of vectors, its span is the set (space) of all vectors that can be obtained as linear combinations of these vectors.

subspace: A set of linearly independent vectors defines a linear space. Any subset of these vectors defines a subspace of that linear space.

triangulation: A set of 2D or 3D points which is faceted into nonoverlapping triangles. Also called triangle mesh.

unit vector: A vector whose length is 1.

vector: An element of a linear space. Also the difference of two points in an affine space.

zero vector: A vector of zero length. Every linear space contains a zero vector.

Bibliography

[1] H. Anton and Chris Rorres. *Elementary Linear Algebra, Applications Version*. John Wiley & Sons, 1999. Eighth edition.

[2] W. Boehm and H. Prautzsch. *Geometric Concepts for Geometric Design*. A K Peters Ltd., 1992.

[3] H. M. S. Coxeter. *Introduction to Geometry, 2nd Edition*. Wiley Text Books, 1989.

[4] M. de Berg, M. van Kreveld, M. Overmars, and O. Schwarzkopf. *Computational Geometry Algorithms and Applications*. Berlin: Springer–Verlag, 1997.

[5] M. Escher and J. Locher. *The Infinite World of M.C. Escher*. New York: Abradale Press/Harry N. Abrams, Inc., 1971.

[6] G. Farin. *NURBS: From Projective Geometry to Practical Use*. A K Peters, Ltd., 1999. Second edition.

[7] G. Farin and D. Hansford. *The Essentials of CAGD*. A K Peters, Ltd., 2000.

[8] R. Goldman. Triangles. In A. Glassner, editor, *Graphics Gems, Volume 1*, pages 20–23. Academic Press, 1990.

[9] G. Golub and C. Van Loan. *Matrix Computations*. Baltimore, MD: The Johns Hopkins University Press, 1986.

[10] M. Goossens, F. Mittelbach, and A. Samarin. *The LaTeX Companion*. Reading, MA: Addison-Wesley Publishing Company, Inc., 1994.

[11] D. Hearn and M. Baker. *Computer Graphics with OpenGL, 3/E*. Prentice-Hall, 2003.

[12] Adobe Systems Inc. *PostScript Language Reference Manual*. Addison-Wesley Publishing Company, Inc., 1985.

[13] Adobe Systems Inc. *PostScript Language Tutorial and Cookbook*. Addison-Wesley Publishing Company, Inc., 1985.

[14] L. Johnson and R. Riess. *Numerical Analysis*. Addison-Wesley Publishing Company, Inc., 1982. Second edition.

[15] E. Kästner. *Erich Kästner erzaehlt Die Schildbürger*. Cecilie Dressler Verlag, 1995.

[16] L. Lamport. *LaTeX User's Guide and Reference Manual*. Addison-Wesley Publishing Company, Inc., 1994.

[17] D. Lay. *Linear Algebra and its Applications, 3/E*. Addison-Wesley Publishing Company, Inc., 2003.

[18] S. Leon. *Linear Algebra with Applications, 6/E*. Prentice Hall, 2003.

[19] P. Shirley. *Fundamentals of Computer Graphics*. A K Peters, Ltd., 2002.

[20] D. Shreiner, M. Woo, J. Neider, and T. Davis. *OpenGL Programming Guide*. Addison-Wesley Publishing Company, Inc., 2004. fourth edition.

[21] G. Thomas and R. Finney. *Calculus and Analytic Geometry*. Addison-Wesley Publishing Company, Inc., 1980.

Index